The Physics of Particle Accelerators

The Physics of Particle Accelerators

An Introduction

KLAUS WILLE
Professor Physics Department
University of Dortmund

Translated by
JASON McFALL
Physics Department
University of Bristol

This book has been printed digitally and produced in a standard specification in order to ensure its continuing availability

OXFORD
UNIVERSITY PRESS

Great Clarendon Street, Oxford OX2 6DP

Oxford University Press is a department of the University of Oxford.
It furthers the University's objective of excellence in research, scholarship, and education by publishing worldwide in

Oxford New York

Auckland Cape Town Dar es Salaam Hong Kong Karachi
Kuala Lumpur Madrid Melbourne Mexico City Nairobi
New Delhi Shanghai Taipei Toronto
With offices in
Argentina Austria Brazil Chile Czech Republic France Greece
Guatemala Hungary Italy Japan South Korea Poland Portugal
Singapore Switzerland Thailand Turkey Ukraine Vietnam

Oxford is a registered trade mark of Oxford University Press
in the UK and in certain other countries

Published in the United States
by Oxford University Press Inc., New York

Oxford is a registered trade mark of Oxford University Press
in the UK and in certain other countries

Published in the United States
by Oxford University Press Inc., New York

Reprinted 2005

ISBN 0-19-850549-3

Preface

Since the 1920s, particle accelerators have played an important role in research into the structure of matter, delivering particle beams with well defined properties to be used in experiments with atomic nuclei or elementary particles. The machines developed for this purpose have become ever larger over time, driven by the need for very high particle energies, and have grown to sizes of 10 km and above. In circular electron-beam accelerators an intense form of electromagnetic radiation, known as synchrotron radiation, is emitted at energies above a few tens of MeV. This radiation has some very useful properties, and for the past three decades has primarily been used in fixed target experiments. The importance of this synchrotron radiation has grown so much around the world that today many machines are built exclusively for this purpose.

This book sets out to explain systematically the key physical principles behind particle accelerators and the basics of high-energy particle physics as well as the production of synchrotron radiation. There are so many different kinds of accelerator, with such a variety of different uses, that it is not possible to consider all aspects of current accelerator technology here. Instead, after introducing the fundamental principles common to all types of accelerators, we concentrate upon the electron storage ring. This type of accelerator has proven to be extraordinarily successful, both for elementary particle physics and for the production of synchrotron radiation. The criteria used to optimize these machines for these two different uses are discussed extensively. Throughout the text the aim is to present all derivations clearly and to carefully justify any approximations that are needed. Wherever possible, the description is illustrated with figures and diagrams.

The first ideas for this book were developed over many years of building and operating accelerators at the German Electron-Synchrotron Laboratory DESY in Hamburg. It often proved difficult, in the short time available, to bring young physicists and engineers up to speed with the physics of modern accelerators, since hardly any suitable literature was available. Furthermore, there were always misunderstandings with the experimental groups using the accelerators, due to a lack of understanding of how accelerators work. This was addressed first by seminars and then by courses of lectures at autumn schools. From these beginnings arose a special lecture course on the physics of particle accelerators, which has been given since 1987 at the University of Dortmund. The experiences gained from this, along with the comments of the students, have shaped the treatment presented in this text.

This book is aimed at all students wishing either to become directly involved in the development and construction of accelerators, or to use them to experiment in nuclear and particle physics or use synchrotron radiation in fixed target experiments. In addition, it is recommended to all practising physicists and engineers who wish to become familiar with the operation and physical limitations of the particle accelerators at which they perform their experiments.

In putting together the various topics in this book, I benefited from numerous suggestions and discussions with my colleagues, with whom I have enjoyed many years of very fruitful collaboration. Particular thanks are due to the former director of the accelerator division at the German Electron Synchrotron DESY, Professor G. A. Voss, under whose leadership I gained much valuable experience.

Dortmund
January 1996 K. W.

Translator's note

I am grateful to Dr Roman Walczak, who recommended the original edition of this book to his students at Oxford University. I am even more grateful to the students for rebelling because it was in German and demanding a translation!

J. McF.

Contents

Symbols

A	acceptance
$\boldsymbol{A}$	vector potential
$a_x,\ a_z,\ a_s$	damping constants
$\boldsymbol{B},\ B$	magnetic flux density
$\tilde{B}$	wiggler field along the beam axis
B	brilliance
$\mathbf{B}$	beta matrix
C	capacitance
$c = 2.99793 \times 10^8\ \mathrm{m\,s^{-1}}$	speed of light in a vacuum
$D(s)$	dispersive trajectory
$e = 1.60203 \times 10^{-19}\ \mathrm{C}$	elementary charge
E	energy
E_γ	photon energy
E_p	proton energy
E_e	electron energy
$\boldsymbol{E}$	electric field strength
$E(s)$	beam envelope
$\boldsymbol{F},\ F$	force
F	photon flux
f_rev	revolution frequency
$G,\ G_N$	FEL gain
$G(x)$	horizontal field distribution function
$g = \partial B_x/\partial z = \partial B_z/\partial x$	field gradient
$h = 6.6252 \times 10^{-34}\ \mathrm{J\,s}$	Planck's constant
$\boldsymbol{H},\ H$	magnetic field strength
$\mathcal{H}(s)$	optical function
I_beam	beam current
$I(\omega)$	intensity distribution
$J_\mathrm{i}(\xi)$	Bessel function
$J_x,\ J_z,\ J_s$	damping numbers
K	wiggler parameter
K_L	FEL field parameter
k	quadrupole strength
$k,\ k_x,\ k_y,\ k_z$	wavenumbers in a waveguide
k_c	cut-off wavenumber
$\mathcal{L}$	luminosity
L	circumference of circular accelerator
L	inductance
$\mathbf{M}$	transfer matrix
m	sextupole strength
m	particle mass
m_0	particle rest mass

$m_e = 9.1081 \times 10^{-31}$ kg	rest mass of electron
$m_p = 1.67236 \times 10^{-27}$ kg	rest mass of proton
N	number of particles
$\boldsymbol{p}$, p	particle momentum
P_{RF}	RF power
P_L	FEL power
P_μ	four-momentum
P_0, P_s	radiated power
$\Delta p/p$	relative momentum deviation
$q = \nu_{RF}/f_{rev}$	harmonic number
Q, Q_x, Q_z	tune or working point
ΔQ	tune shift
Q	quality factor of a resonator
$\boldsymbol{r}$	position vector
R	bending radius
R_s	shunt impedance
S	brightness
$S_s(\xi)$	spectral function
U	voltage
$\boldsymbol{u}$, u	particle velocity
$\boldsymbol{v}$, v	particle velocity
$W(\boldsymbol{r}, t)$	wavefunction
W, W_0	energy loss during acceleration
s	coordinate in beam direction
T	temperature
T, T_0	revolution time
t	time
$\boldsymbol{X}$	trajectory vector
x	horizontal beam coordinate
z	horizontal beam coordinate
α	momentum compaction factor
$\alpha(s) = -\beta(s)/2$	gradient of beta function
$\beta = v/c$	relative velocity
$\beta(s)$	beta function
$\gamma = E/m_0c^2$	relative energy
γ_t	transition energy
$\gamma(s) = (1 + \alpha(s))/\beta(s)$	optical function of the beam
$\varepsilon_0 = 8.85419 \times 10^{-12}$ As/Vm	permittivity of free space
ε, ε_x, ε_z	beam emittance
ϵ_c	critical energy
η	efficiency
$\eta(s)$	Floquet variable
Θ	radiation opening angle
κ	bending angle (of steering coils)
λ	wavelength

λ_{c}	cut-off wavelength
λ_{RF}	RF wavelength
λ_{u}	undulator period
$\mu_0 = 4\pi 10^{-7}$ Vs/Am	permeability of free space
μ_{r}	relative permeability
ν	frequency
ν_{RF}	RF frequency
$\xi,\ \xi_x,\ \xi_z$	chromaticity
$\rho(x, z, s)$	density distribution function
σ	interaction cross-section
$\sigma_x,\ \sigma_z,\ \sigma_s$	beam sizes
τ	pulse length
τ_{RF}	RF period
$\Phi(x, z, s)$	scalar potential
$\phi(s)$	Floquet variable
$\varphi(x, z, s)$	scalar potential
$\Psi(s)$	betatron phase
Ψ_0	nominal phase
ω	frequency
ω_{c}	critical frequency
ω_{char}	characteristic frequency
$\omega_z = eB/m$	cyclotron frequency

1 Introduction

1.1 The importance of high energy particles in fundamental research

The study of the basic building blocks of matter and the forces which act between them is a fundamental area of physics. The structures under scrutiny are extraordinarily small, sometimes well below 10^{-15} m, and in order to perform experiments at this scale probes with correspondingly high spatial resolution are needed. Visible light, with a wavelength $\lambda \approx 500$ nm, is wholly inadequate. Instead, high energy photon or particle beams have proven to be excellent tools, and the results of elementary particle physics would be unimaginable without them. High energy particle beams are thus essential to this field of experimental physics.

In general a microstructure may only be resolved by a probe, for example electromagnetic radiation, if the wavelength is small compared to the size of the structure. Thus wavelengths below $\lambda < 10^{-15}$ m are required in elementary particle physics. The photon energy of this radiation is

$$E_\gamma = h\nu = \frac{hc}{\lambda} = 2 \times 10^{-10} \text{ J}. \tag{1.1}$$

If these photons are produced via bremsstrahlung from energetic electron beams, then particle energies of

$$E_\mathrm{e} = eU \qquad \text{with} \qquad E_\mathrm{e} > E_\gamma \tag{1.2}$$

are required. To achieve such energies, the electrons in the beam must cross a total electrical potential $U > E_\mathrm{e}/e = 1.2 \times 10^9$ V. Similar considerations apply to particle beams, for which the de Broglie wavelength must again be small compared to the size of the structure. This wavelength is given by the relation

$$\lambda_\mathrm{B} = \frac{h}{p} = \frac{hc}{E}, \tag{1.3}$$

where p and E are the momentum and energy of the particle, respectively. Comparing this with relation (1.1) shows that similarly high particle energies are also necessary here.

In physics, energy is usually measured in the unit of the Joule (J). However, this unit is not very convenient when describing particle beams, and in general

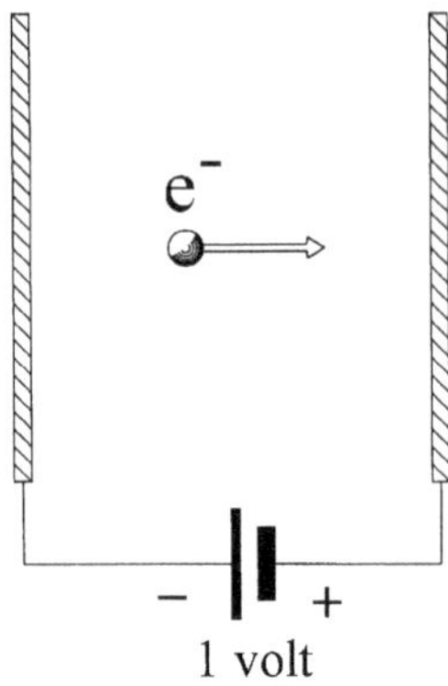

Fig. 1.1 Definition of the eV. (1 keV = 10^3 eV, 1 MeV = 10^6 eV, 1 GeV = 10^9 eV, 1 TeV = 10^{12} eV).

the unit of the eV (**electron volt**) is preferred. This is the kinetic energy gained by a particle of elementary charge $e = 1.602 \times 10^{-19}$ C as it crosses a potential difference of $\Delta U = 1$ V. The conversion factor is thus 1 eV $= 1.602 \times 10^{-19}$ J. To describe higher particle energies the convenient units keV, MeV, GeV, and TeV are also used (see Fig. 1.1). It should be noted at this point that the SI (MKSA) system of units will generally be used in this book. Where different units are used in particular cases, this will be clearly stated.

Another important aspect of elementary particle physics, as well as the resolution of the finest structures of matter, is the production of new, mostly very short-lived, particles. The amount of energy needed to produce a particle follows directly from the fundamental relation

$$E = mc^2. \tag{1.4}$$

Note that most particles can only be produced along with their antiparticles in pairs; for example electrons and positrons are produced together from high-energy γ rays (Fig. 1.2).

As a result of the conservation of momentum, a reaction of this kind can only take place in the vicinity of a heavy nucleus. The nucleus itself gains some

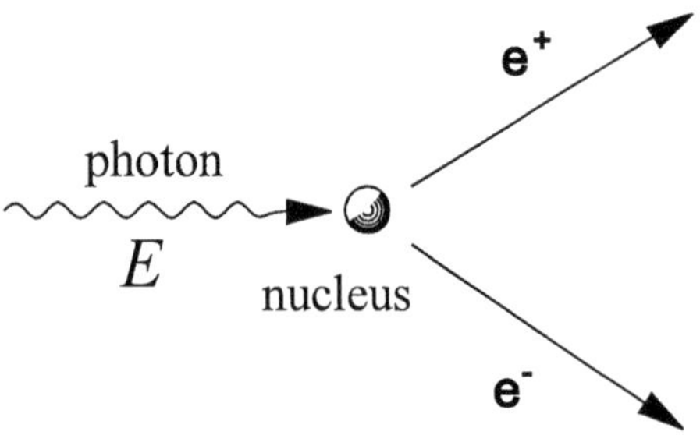

Fig. 1.2 Production of an e^+-e^- pair in the collision of a high-energy photon with a heavy nucleus.

momentum, and hence also energy, which is then not available for particle production. The γ-ray energy needed for particle production is therefore always higher than that given by relation (1.4), namely

$$E_\gamma > 2m_e c^2 = 1.637 \times 10^{-13}\ \mathrm{J} = 1.02\ \mathrm{MeV}. \tag{1.5}$$

This is the threshold energy for the production of electron–positron pairs, where each particle has a mass $m_e = 9.108 \times 10^{-31}$ kg, corresponding to a rest energy of $E_0 = 511$ keV. The rest energies of elementary particles investigated nowadays are considerably higher. A few examples are:

proton	p	:	E_0	=	938 MeV
b quark	b	:	E_0	=	4735 MeV
vector boson	Z_0	:	E_0	=	91 190 MeV
t quark	t	:	E_0	=	174 000 MeV

In order to produce these particles, correspondingly high energies must be available.

1.2 Forces used in particle acceleration

Since the velocity v of elementary particles studied in collisions is generally close to the velocity of light ($c = 2.997925 \times 10^8$ m/s), the energy must be written in the relativistically invariant form

$$E = \sqrt{m_0^2 c^4 + p^2 c^2} \qquad (m_0 = \text{rest mass}). \tag{1.6}$$

Here the only free parameter is the momentum p of the particle. In the usual notation $\beta = v/c$ and $\gamma = (1 - \beta^2)^{-1/2}$, the relativistic particle momentum is given by the relation

$$p = mv = \gamma m_0 v \tag{1.7}$$

and the energy-dependent particle mass $m = \gamma m_0$. The increase in energy E in (1.6) is the same as the increase in the particle momentum p. The momentum can only be changed by the action of a force $\boldsymbol{F}$ on the particle, as described by Newton's second law of motion

$$\dot{\boldsymbol{p}} = \boldsymbol{F}. \tag{1.8}$$

In order to reach high kinetic energies, a sufficiently strong force must be exerted on the particle for a sufficient period of time. Nature offers us four different forces, listed along with their most important properties in Table 1.1. It is clear that forces with a range below 10^{-15} m are of no practical use for particle acceleration. The **strong force**, which might have been useful because of its relative strength, is therefore ruled out, as of course is the **weak force**. **Gravity** is many orders of magnitude too weak. The only possible choice left is the **electromagnetic force**. When a particle of velocity $\boldsymbol{v}$ passes through a volume containing a magnetic field $\boldsymbol{B}$ and an electric field $\boldsymbol{E}$ it is acted upon by the **Lorentz force**

$$\boxed{\boldsymbol{F} = e(\boldsymbol{v} \times \boldsymbol{B} + \boldsymbol{E}).} \tag{1.9}$$

As the particle moves from point $\boldsymbol{r}_1$ to $\boldsymbol{r}_2$, its energy changes by the amount

Table 1.1 The four forces of nature

force	relative strength	range [m]	particles affected
gravity	6×10^{-39}	∞	all particles
electromagnetism	$1/137$	∞	charged particles
strong force	≈ 1	$10^{-15} - 10^{-16}$	hadrons
weak force	10^{-5}	$\ll 10^{-16}$	hadrons & leptons

$$\Delta E = \int_{\boldsymbol{r}_1}^{\boldsymbol{r}_2} \boldsymbol{F} \cdot d\boldsymbol{r} = e \int_{\boldsymbol{r}_1}^{\boldsymbol{r}_2} (\boldsymbol{v} \times \boldsymbol{B} + \boldsymbol{E}) \cdot d\boldsymbol{r}. \tag{1.10}$$

During the motion the path element $d\boldsymbol{r}$ is always parallel to the velocity vector $\boldsymbol{v}$. The vector $\boldsymbol{v} \times \boldsymbol{B}$ is thus perpendicular to $d\boldsymbol{r}$, i.e. $(\boldsymbol{v} \times \boldsymbol{B}) \cdot d\boldsymbol{r} \equiv 0$. Hence the magnetic field $\boldsymbol{B}$ does not change the energy of the particle. Acceleration involving an increase in energy can thus only be achieved by the use of **electric fields**. The gain in energy follows directly from (1.10) and is

$$\Delta E = e \int_{\boldsymbol{r}_1}^{\boldsymbol{r}_2} \boldsymbol{E} \cdot d\boldsymbol{r} = eU, \tag{1.11}$$

where U is the voltage crossed by the particle.

Although **magnetic fields** do not contribute to the energy of the particle, they play a very important role when forces are required which act perpendicular to the particle's direction of motion. Such forces are used to steer, bend and focus particle beams.

Accelerator physics is concerned with these two problems: the acceleration and steering of particle beams. Both processes rely on the electromagnetic force and hence on the foundations of classical electrodynamics. Here **Maxwell's equations** are of fundamental importance. In addition, the most important results of the **special theory of relativity** are required. A knowledge of these fundamentals will be assumed in what follows; a summary of the most important relations can be found in the appendices.

1.3 Overview of the development of accelerators

Since the 1920s various machines have been developed to accelerate particle beams for experimental physics, always with the principal objective of reaching ever higher energies. An overview of the most significant advances will be given in this chapter, and the various types of accelerator which have played an important part in physics so far will be introduced. A detailed description of the early developments in accelerator physics can be found in Livingston and Blewett [1].

1.3.1 The direct-voltage accelerator

The simplest particle accelerators use a constant electric field between two electrodes, produced by a high voltage generator. This principle is illustrated in Fig. 1.3. One of the electrodes contains the particle source. In the case of electron beams this is a thermionic cathode, widely used in vacuum tube technology. Protons, as well as light and heavy ions, are extracted from the gas phase by using a further DC or high frequency voltage to ionize a very rarified gas and so produce a plasma inside the particle source. Charged particles are then continuously emitted from the plasma, and are accelerated by the electric field. In the accelerating region there is a relatively good vacuum, in order to avoid particle collisions with residual gas molecules. The particles are thus continuously accelerated without any loss of energy until they reach the second electrode. There they exit the accelerator and usually traverse a further field-free drift region, through which they travel at constant energy until they reach a target. This principle is very widely employed in research and technology, and all modern VDU screens and oscilloscopes are based upon it. The particle energies which can be reached in this way are, however, very limited by modern standards.

In electrostatic accelerators the maximum achievable energy is directly proportional to the maximum voltage which can be developed, and this gives the energy limit. This may be seen from the curves in Fig. 1.4, which show the dependence of current on voltage in electrostatic accelerators. The current may be divided into essentially three components. Since the conductivity of an insulator is never quite zero, there is always an **ohmic** component, which increases in proportion to the voltage. This element can be made very small by a judicious choice of materials and careful construction. The second component arises due to the **ions** which are always present in the residual gas. It very quickly reaches a constant saturation level if the applied voltage is so high that space charge effects become negligible and all the ions are sucked out. The accelerated particle beam is also part of this component of the current. The third

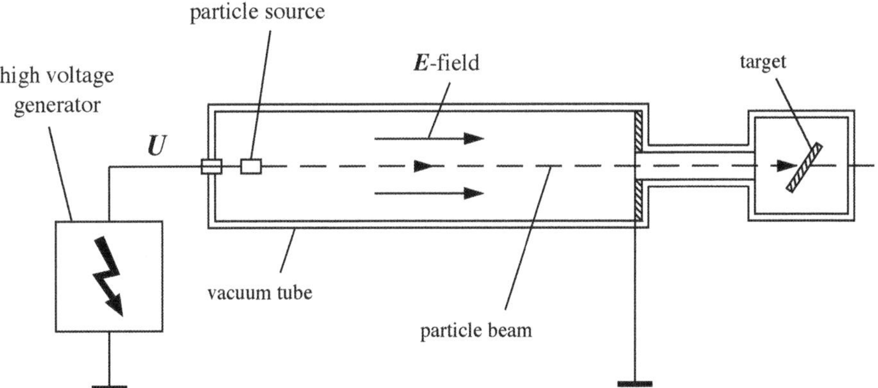

Fig. 1.3 General principle of the electrostatic accelerator.

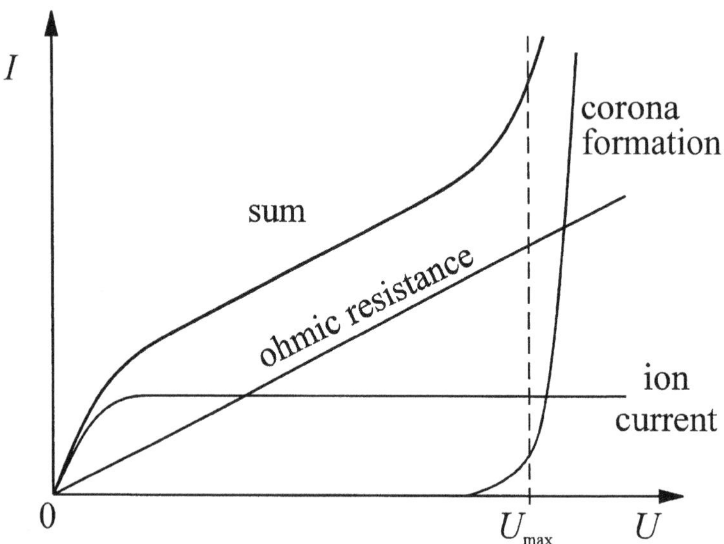

Fig. 1.4 Dependence of the current on the applied high voltage in electrostatic accelerators.

component, **corona formation**, leads to the actual energy limit. At low voltages it is completely unmeasurable, but at higher voltages the field strength close to the electrodes grows so much that ions and electrons produced in this region are accelerated to considerable energies. They collide with gas molecules and so produce many more ions, which themselves undergo the same process. The result is an avalanche of charge carriers causing spark discharge and the breakdown of the high voltage. The current grows exponentially in the region of the corona formation. It is this effect which limits the maximum achievable energy. Voltages of a few MV are technically possible, and hence particles of unit charge can reach a maximum energy of a few MeV. This is the general limit for electrostatic accelerators: substantially higher energies cannot be achieved with this technology.

1.3.2 The Cockroft–Walton cascade generator

A significant problem for electrostatic accelerators is the production of sufficiently high accelerating voltages. At the beginning of the 1930s, Cockroft and Walton [2] developed a high voltage generator based on a system of multiple rectifiers. At their first attempt they achieved a voltage of around 400 000 V. The operation of this generator [3, 4], also known as the **Greinacker circuit**, is explained in Fig. 1.5. At point A a transformer produces a sinusoidally varying voltage $U(t) = U \sin \omega t$ of frequency ω. The first rectifying diode ensures that at point B the voltage never goes negative. Thus the capacitor C_1 charges up to a potential U. At point B the voltage now oscillates between the values 0 and $2U$. The capacitor C_2 is then charged up via the second rectifier to a potential

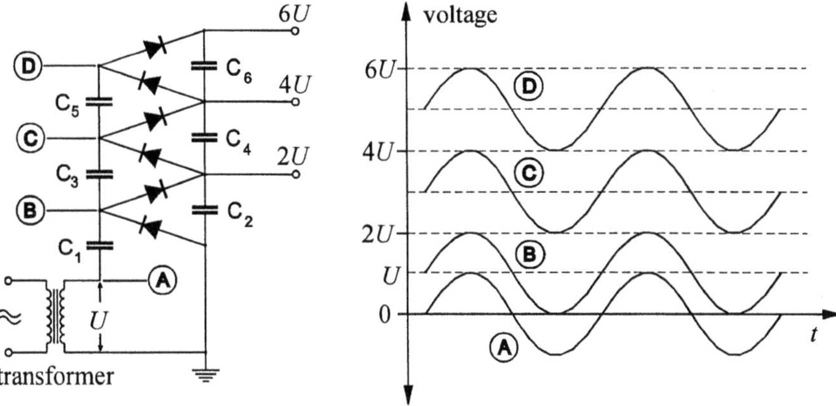

Fig. 1.5 Operation of the Cockroft–Walton cascade generator.

of $2U$. In the same way as before, the third diode ensures that the potential at point C does not fall below $2U$. Here it varies between $2U$ and $4U$, and so with the help of the fourth rectifier a voltage of $4U$ is generated. The pattern is repeated, with many such rectifier stages arranged one after another. Without loading, the maximum achievable voltage is then $2nU$, where n is the number of rectifier stages.

It must be noted here that a current I must always be drawn from the generator. This always discharges the capacitors slightly when the diodes are in reverse bias, leading to a somewhat lower generated voltage than that expected simply from the number of rectifier stages. The current-dependent voltage generated by the cascade circuit is given more exactly by the relation

$$U_{\text{tot}} = 2Un - \frac{2\pi I}{\omega C}\left(\frac{2}{3}n^3 + \frac{1}{4}n^2 + \frac{1}{12}n\right). \tag{1.12}$$

It is immediately apparent that a high capacitance C and a high operating frequency ω strongly reduce the dependence on the current.

In **Cockroft–Walton** accelerators, voltages up to about 4 MV can be reached. By using pulsed particle beams with pulse lengths of a few μs, beam currents of several hundred mA have been achieved.

1.3.3 The Marx generator

An alternative way to produce high voltages for particle acceleration is that used by the **Marx generator**. It too has a cascade design, but only delivers short voltage pulses. However, this allows very high particle currents. As Fig. 1.6 shows, the Marx generator consists of a network of resistors and capacitors. The capacitors C_1 to C_4, which are connected quasi in parallel, are charged across the resistors R up to a voltage U by a high voltage supply. As the applied voltage increases it approaches the firing voltage of the spark gaps. When this firing voltage is reached, spark discharge occurs and the spark gaps act as very low-resistance switches. The value of the load resistors R is, however, very large. As

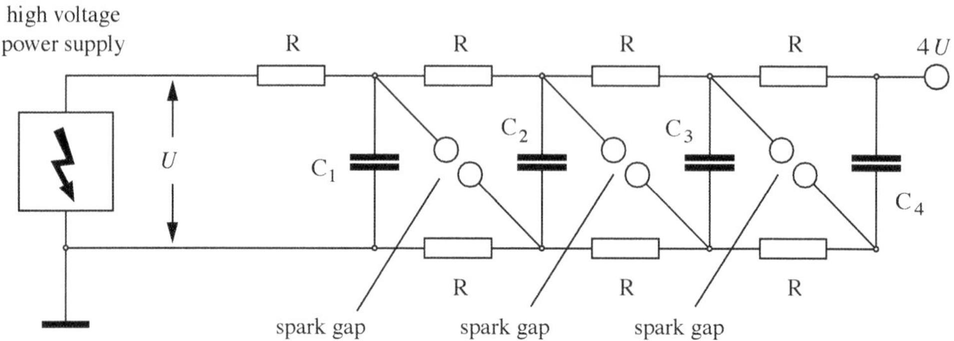

Fig. 1.6 The Marx generator.

the spark gap fires the capacitors become connected in series and the combined voltage then reaches the value

$$U_{\mathrm{tot}} = nU, \tag{1.13}$$

where n is the number of capacitors used. Since this process takes place very rapidly, lasting much less than 1 μs, discharging of the capacitors is not significant and may be neglected. Using 100 capacitors with capacitances of 2μF and an applied voltage of $U = 20$ kV, pulse lengths of 40 ns and beam currents of up to $I = 500$ kA can be achieved. In 1932 peak voltages of around 6 MV were reached using a Marx generator built by General Electric [5].

1.3.4 The Van de Graaff accelerator

In 1930 R.J. Van de Graaff [6] began development work on the high voltage generator that was later named after him. The key element is a belt made of isolating material looped around two rollers (Fig. 1.7), which is driven like a conveyor belt by an electric motor. Charge produced in corona formation around a sharp electrode is transferred onto the belt. The belt carries this charge up to an isolated conducting dome, where it is then discharged through a second electrode. The dome thus charges up continuously until the critical voltage limit is reached. The dome is connected to the upper electrode of the particle accelerator proper, which also contains the particle source. The accelerator consists of a large number of circular electrodes arranged in a line and connected to one another by high-value resistors. This design reduces the risk of spark discharge, since the voltage between the individual electrodes is comparatively low, resulting in a relatively even field distribution. In addition, the electrodes act as electrostatic lenses which focus the beam to a certain extent.

Under normal circumstances **Van de Graaff** generators can produce voltages of up to 2 MV. Considerably higher values, up to 10 MV, become possible if the generator and accelerator section are placed in a tank filled with an insulating gas, such as SF_6, at a pressure of around 1 MPa.

By stripping charge from the ions during acceleration it is possible to make use of the potential twice and hence produce particle beams with twice as much

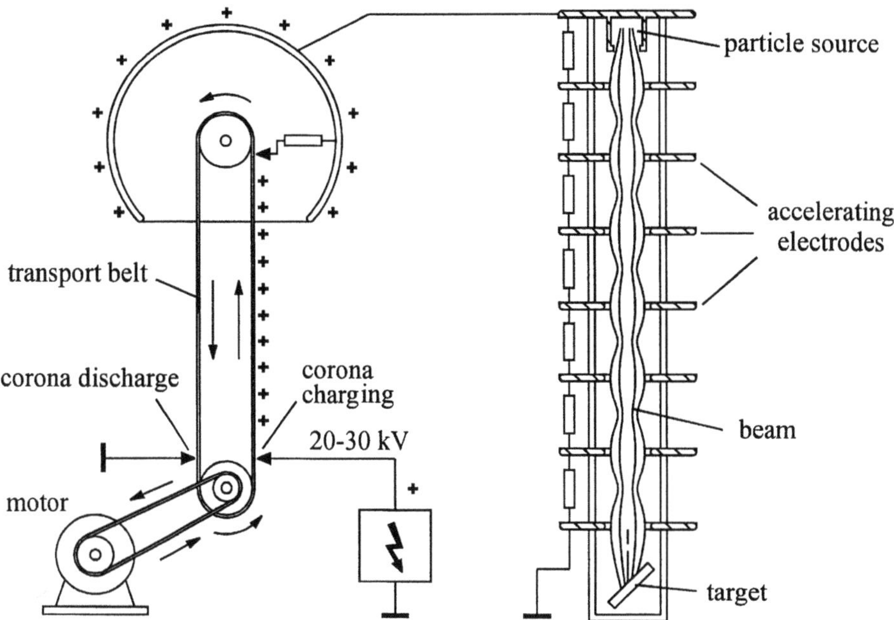

Fig. 1.7 The Van de Graaff accelerator.

energy. In 1936 Van de Graaff and collaborators built the first accelerator based on this principle, also known as a **tandem accelerator**. Figure 1.8 shows the basic layout. The ions, which are initially positive, come from an ion source and pass into a charge exchange region where they gain electrons until they have an excess and become **negative** ions. A homogeneous magnetic field selects ions with a particular e/m ratio which then enter the accelerator. The particles cross the potential difference once and then collide with gas molecules and lose their added electrons, so that when they leave this so-called 'gas stripper' they are once again **positive** ions. Because their charge has been reversed they gain energy a second time in the accelerating section as they cross back through the potential difference down to earth. As the particles exit the accelerator a further bending magnet selects out those with the requisite charge and energy. This principle allows energies of up to 1000 MeV to be reached with multiply ionized ions.

1.3.5 The linear accelerator

All electrostatic accelerators which produce high voltages are limited by corona formation and discharge, as described above. In 1925 the Swede Ising suggested using rapidly changing high frequency voltages instead of direct voltages, in order to avoid this problem [7]. Three years later Wideröe performed the first successful test of a linear accelerator based on this principle [8]. Illustrated in Fig. 1.9, it consists of a series of metal drift tubes arranged along the beam axis and connected, with alternating polarity, to a radiofrequency (RF) supply. The supply delivers a high-frequency alternating voltage of the form $U(t) = U_{\max} \sin \omega t$.

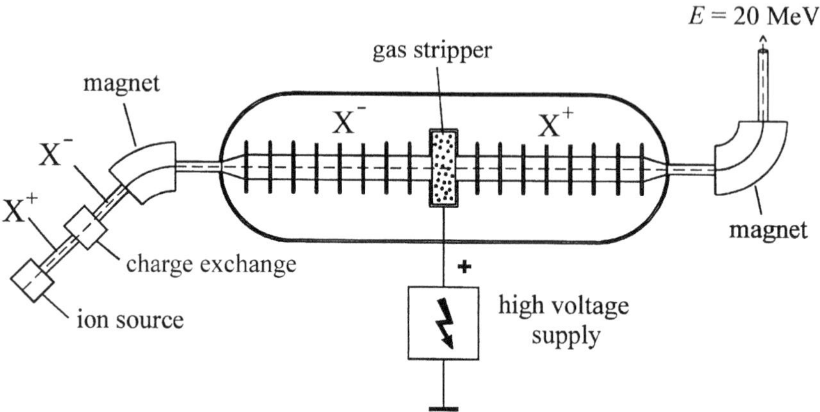

Fig. 1.8 The tandem accelerator.

During a half period the voltage applied to the first drift tube acts to accelerate the particles leaving the ion source. The particles reach the first drift tube with a velocity v_1. They then pass through this drift tube, which acts as a Faraday cage and shields them from external fields. Meanwhile the direction of the RF field is reversed without the particles feeling any effect. When they reach the gap between the first and second drift tubes, they again undergo an acceleration. This process is repeated for each of the drift tubes. After the i-th drift tube the particles of charge q have reached an energy

$$E_i = iqU_{\mathrm{max}} \sin \Psi_0, \tag{1.14}$$

where Ψ_0 is the average phase of the RF voltage that the particles see as they cross the gaps. It is immediately evident that the energy is again proportional to the number of stages i traversed by the particles. The important point, however, is that the largest voltage in the entire system is never greater than U_{max}. It is therefore possible in principle to produce arbitrarily high particle energies without encountering the problem of voltage discharge. This is the decisive advantage RF accelerators have over electrostatic systems. For this reason almost all particle accelerators nowadays use high-frequency alternating voltages, produced by powerful RF supplies.

During the acceleration the velocity increases monotonically, but the frequency of the alternating voltage must remain constant, in order to keep the costs of the already very expensive RF power supply to within reasonable limits. This means that the size of the gaps between the drift tubes must increase. In the ith drift tube the velocity v_i is reached, which for a particle of mass m corresponds to an energy

$$E_i = \frac{1}{2} m v_i^2, \tag{1.15}$$

assuming non-relativistic velocities (i.e. $v \ll c$). In addition, the RF voltage moves through exactly half a period $\tau_{\mathrm{RF}}/2$ as the particle travels through one

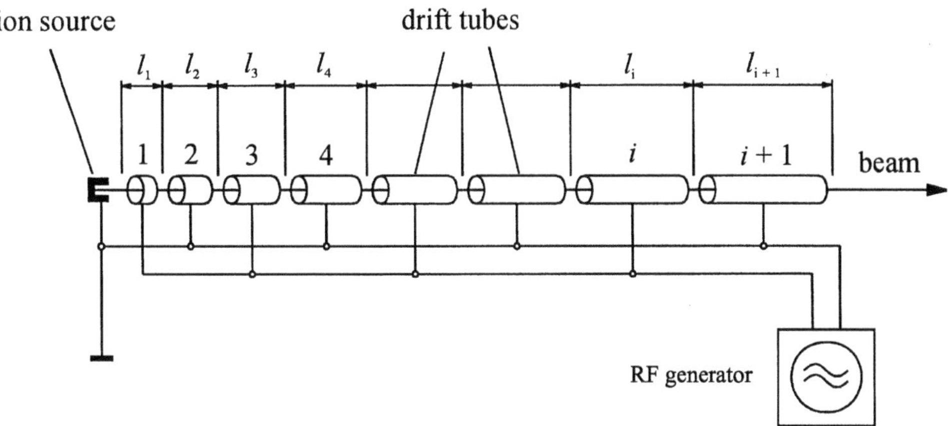

Fig. 1.9 Wideröe linear accelerator.

drift section. This immediately fixes the required separation between the ith and $(i+1)$th gaps to be

$$l_i = \frac{v_i \tau_{\mathrm{RF}}}{2} = \frac{v_i}{2\nu_{\mathrm{RF}}} = \frac{v_i \lambda_{\mathrm{RF}}}{2c} = \beta_i \frac{\lambda_{\mathrm{RF}}}{2}. \tag{1.16}$$

Here we assume that the frequency $\nu_{\mathrm{RF}} = c/\lambda_{\mathrm{RF}}$ is strictly constant. β_i is the relative velocity v_i/c. From (1.14) to (1.16) it then immediately follows that

$$l_i = \frac{1}{\nu_{\mathrm{RF}}} \sqrt{\frac{iqU_{\max} \sin \Psi_0}{2m}}. \tag{1.17}$$

The spacing of the accelerating gaps between the drift tubes must thus increase in proportion with $\sqrt{i}$.

In this high-frequency linear accelerator there is still one other problem to solve, as can easily be seen from the relation (1.14). The energy transferred to the particles depends critically on the voltage $U_{\max}$ and the nominal phase Ψ_0. When a very large number of stages are used, a small deviation from the nominal voltage $U_{\max}$ means that the particle velocity no longer matches the design velocity fixed by the length of the drift sections, so that the particles undergo a phase shift relative to the RF voltage. The synchronization of the particle motion and the RF field is then lost. A mechanism is thus required to automatically bring the particles back to the nominal phase in the event of any deviation.

Fortunately there is a very simple principle which satisfies this requirement, which we will describe with the help of Fig. 1.10. The key principle is not to use the phase $\Psi_0 = \pi/2$ and hence the peak voltage $U_{\max}$ to accelerate the particles, but instead to use a value $\Psi_0 < \pi/2$. The effective accelerating voltage is then $U_{\mathrm{eff}} < U_{\max}$. Let us assume that a particle has gained too much energy in the preceding stage and so is travelling faster than an ideal particle and hence

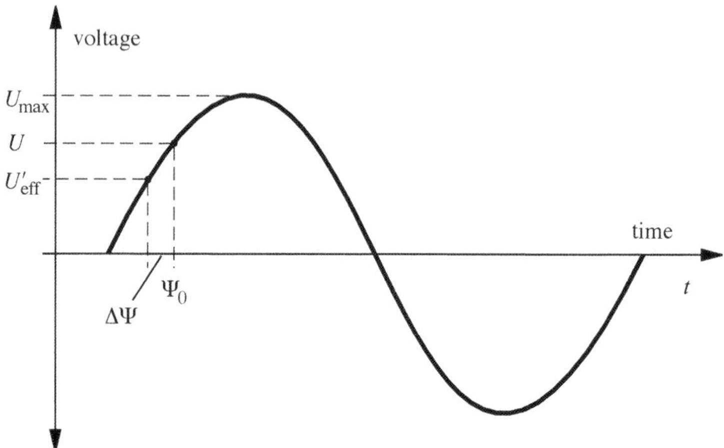

Fig. 1.10 Phase focusing in linear accelerators, based on the time-dependence of the RF voltage between two drift sections.

arrives earlier. It sees an average RF phase $\Psi = \Psi_0 - \Delta\Psi$ and is accelerated by a voltage

$$U'_{\text{eff}} = U_{\max} \sin(\Psi_0 - \Delta\Psi) < U_{\max} \sin \Psi_0, \tag{1.18}$$

which is below the ideal voltage. The particle therefore gains less energy and slows down again until it returns to the nominal velocity. The opposite happens for particles whose energy is too low. In practice all particles oscillate about the nominal phase Ψ_0. This principle of **phase focusing** is of crucial importance in the design of all accelerators using RF voltage.

Nowadays simple drift tubes are no longer used and have generally been replaced by cavity structures. The first studies of the use of cavities in linear accelerators were performed as early as 1933 and 1934 by J.W. Beams and co-workers at the University of Virginia [9] and at around the same time by W.W. Hansen at Stanford University [10]. Hansen used klystron tubes as drivers for the first time, a technique which has been very successfully employed up to the present day. These first cavity linacs (linac = linear accelerator) accelerated electrons and exploited the fact that by energies of a few MeV the particles have already reached velocities very close to the speed of light ($v \approx c$). As they are accelerated the electrons increase in mass but their velocity remains almost constant. This allows cavity structures of the same size to be used along the whole length of the linac, leading to relatively a simple design. Today the largest electron linac is situated at the Stanford Linear Accelerator Center SLAC in California [11]. It is over 3 km long and reaches final energies of about 50 GeV.

In the case of protons and heavy ions, however, the particles still have non-relativistic velocities in the first few stages of a linac and so a Wideröe-type design is required in which the length of the drift tubes increases in proportion to $\sqrt{i}$. These drift tubes are nowadays arranged in a tank, made of a good conductor, in which a cavity wave is induced. The drift tubes, which have no field inside them, also contain the magnets required to focus the beam. The development of this structure, nowadays termed the **Alvarez structure**, for proton and heavy-ion linacs was begun by L. Alvarez and W.K.H. Panofsky [12] after the Second World War. Today such systems are used to accelerate particles ranging from protons up to the heaviest ions, reaching energies of up to 20 MeV per nucleon.

1.3.6 The cyclotron

Although linear accelerators can in principle reach arbitrarily high particle energies, the length and hence the cost of the machine grow with the energy. It is therefore desirable to drive the particles around a circular path and so use the same accelerating structure many times. The first circular accelerator to be developed according to this principle was the **cyclotron**, proposed by E.O. Lawrence at the University of California in 1930 [13]. A year later Livingston succeeded in demonstrating the operation of such a machine experimentally. In 1932 they together built the first cyclotron suitable for experiments, with a peak energy of 1.2 MeV.

To make the particles follow a circular path the cyclotron uses an iron magnet which produces a homogeneous field with a strength of $B \approx 2$ T between its two round poles. The particles circulate in a plane between the poles. In order to derive the equation of motion of the particles in a homogeneous field we will choose, without loss of generality, a coordinate system in which the x- and y-axes lie in the plane of the path. The magnetic field then has only one component, perpendicular to these axes (Fig. 1.11) and may be written in the form

$$\boldsymbol{B} = \begin{pmatrix} 0 \\ 0 \\ B_z \end{pmatrix} . \tag{1.19}$$

We obtain the equation of motion from the expression for the Lorentz force (1.9), setting the electric field to zero, $\boldsymbol{E} = 0$. It then follows that

$$\boldsymbol{F} = \dot{\boldsymbol{p}} = \frac{d}{dt}(m\boldsymbol{v}) = e\boldsymbol{v} \times \boldsymbol{B}. \tag{1.20}$$

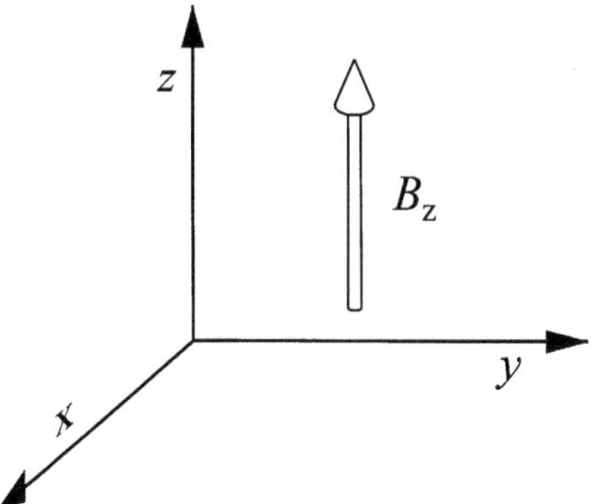

Fig. 1.11 Coordinate system in the B-field.

We assume that the motion is confined to the x-y plane. The particle momentum then has the form

$$\boldsymbol{p} = \begin{pmatrix} p_x \\ p_y \\ 0 \end{pmatrix} = m \begin{pmatrix} v_x \\ v_y \\ 0 \end{pmatrix}. \tag{1.21}$$

Using these expressions for $\boldsymbol{B}$ and $\boldsymbol{p}$ (or $\boldsymbol{v}$) we may calculate the cross-product in (1.20) and obtain

$$\dot{\boldsymbol{p}} = e \begin{pmatrix} v_y B_z \\ -v_x B_z \\ 0 \end{pmatrix}, \tag{1.22}$$

or, written in terms of components,

$$\begin{aligned} \dot{p}_x &= m\dot{v}_x = e v_y B_z \\ \dot{p}_y &= m\dot{v}_y = -e v_x B_z. \end{aligned} \tag{1.23}$$

Differentiating these equations again with respect to time and combining the resulting expressions with (1.23), we finally obtain the required equation of motion

$$\begin{aligned} \ddot{v}_x + \frac{e^2}{m^2} B_z^2 v_x &= 0 \\ \ddot{v}_y + \frac{e^2}{m^2} B_z^2 v_y &= 0 \end{aligned} \tag{1.24}$$

with solutions

$$\begin{aligned} v_x(t) &= v_0 \cos \omega_z t \\ v_y(t) &= v_0 \sin \omega_z t. \end{aligned} \tag{1.25}$$

The particles thus follow a circular orbit between the poles with a revolution frequency

$$\omega_z = \frac{e}{m} B_z, \tag{1.26}$$

which is also known as the **cyclotron frequency**. Notice that ω_z does not depend on the particle velocity at all. This is because as the energy increases, the orbit radius and hence the circumference along which the particles travel increase in proportion. A higher velocity is exactly compensated for by a larger radius, provided that the mass m remains constant. This condition is of course only satisfied for non-relativistic particles.

The cyclotron, shown in Fig. 1.12, relies on the fact that the cyclotron frequency remains constant. It consists of a large H-shaped magnet with constant current flowing through the coils. Between the poles of this magnet there is the vacuum chamber, which also contains the D-shaped electrodes, known as DEEs, which are needed for particle acceleration. They are shaped rather like a flat tin can, cut through the middle. The RF voltage from a generator is applied between these two halves. The particles are emitted from an ion source in the

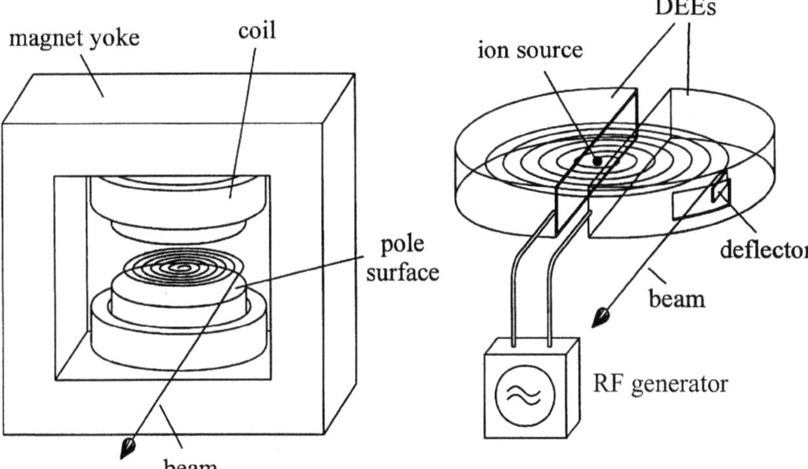

Fig. 1.12 The cyclotron.

centre, between the two poles. As their circular trajectory passes through the gap between the two DEEs they are accelerated and continue along a new path with a larger radius, until they arrive back at the gap half an orbit later. The high frequency of the supply is chosen to be exactly equal to the cyclotron frequency, i.e. $\omega_{\mathrm{RF}} = \omega_z$. This value is usually around 10 MHz for an RF power of around 100 kW. This ensures that the particles always encounter an accelerating field in the gap. As the particles gain energy, they spiral outwards until they reach the edge of the magnet. There they are deflected by means of a small electrode or small deflector magnet and are steered to the experiment.

Classical cyclotrons can accelerate protons, deuterons, and alpha particles up to about 22 MeV per electron charge. At these energies the motion is still sufficiently non-relativistic ($v \approx 0.15c$) for the revolution frequency to remain approximately constant during acceleration and hence to fulfil the cyclotron condition.

At higher energies the cyclotron frequency decreases in inverse proportion to the increasing particle mass $m(E)$. If the frequency of the RF supply is decreased accordingly, much higher energies can be reached. This principle is employed in the **synchrocyclotron**. It turns out that the high frequency is only ever optimal for particles in a limited energy range, and so the beam may only be accelerated in short pulses or bunches. Hence the beam intensity is lower.

A more effective method is adopted in the **isocyclotron**, in which the radial magnetic field is increased in such a way that the cyclotron frequency remains constant, namely

$$\omega_z = \frac{q\,B_z(r(E))}{m(E)} = \text{const}, \tag{1.27}$$

where $r(E)$ is the orbit radius, E is the energy of a particle and q is its charge. However, this approach has the problem that the beam becomes defocused by the changing magnetic field. To resolve this, isocyclotrons use magnets with rather complex pole shapes, which compensate for this loss of focusing with so-called 'edge-focusing'. Isocyclosynchrotrons can reach energies of over 600 MeV.

1.3.7 The microtron

The principle of the cyclotron, which relies on the revolution frequency ω_z being constant, cannot be directly applied to electrons because their very small rest mass ($m_e c^2 = 511$ keV) means that they very quickly reach relativistic velocities, i.e. $v \approx c$. As a result, their mass increases almost proportionally to their energy and the cyclotron frequency ω_z decreases in inverse proportion. This effect cannot be counterbalanced by varying the high frequency or shaping the magnetic field.

Strictly speaking it is not actually necessary for the revolution frequency to be constant: it is sufficient merely for the particles to see the same phase of the RF voltage on each revolution. This can be achieved by choosing a relatively high accelerating frequency and correspondingly short wavelength — frequencies of around $\nu_{\mathrm{RF}} = 3$ GHz are typical — and tuning the energy gain per revolution so that the total circumference of the particle orbit always increases by an exactly integer number of RF wavelengths. This principle is employed in the **microtron**, a type of cyclotron for electrons [14]. We will use the example of the **racetrack microtron** to explain its operation (Fig. 1.13). The name 'racetrack' refers to the shape of the machine, which consists of two semicircles connected by two straight sections. The electrons are emitted by a cathode and an injector magnet then directs them into the accelerating region for the first time. At the end of the accelerating section they encounter a bending magnet which turns them through 180°. They travel in a straight line until they arrive at a second bending magnet which turns them around and back into the same accelerating section. This process is repeated a number of times with ever-increasing bending radii in the magnets until the beam finally passes through an ejector magnet and is deflected towards the experiment. Special focusing magnets are installed on either side of the accelerating section to keep the transverse beam size small during acceleration. If the two bending magnets have a separation l and the bending radius on the ith revolution through these magnets is R_i, then electrons travelling at a velocity v_i complete the orbit in a time

$$t_i = \frac{2(\pi R_i + l)}{v_i}. \tag{1.28}$$

Since the centripetal force is equal to the Lorentz force, the bending radius is given by

$$R_i = \frac{v_i m_i c^2}{ec^2 B} = \frac{v_i}{ec^2 B} E_i, \tag{1.29}$$

where m_i is the relativistic particle mass, E_i is the energy and B is the strength of the magnetic field. Inserting this result into (1.28) and calculating the difference in the periods of the ith and $(i+1)$th revolutions gives

$$\Delta t = t_{i+1} - t_i = \frac{2\pi}{ec^2 B}(E_{i+1} - E_i) = \frac{2\pi}{ec^2 B}\Delta E. \tag{1.30}$$

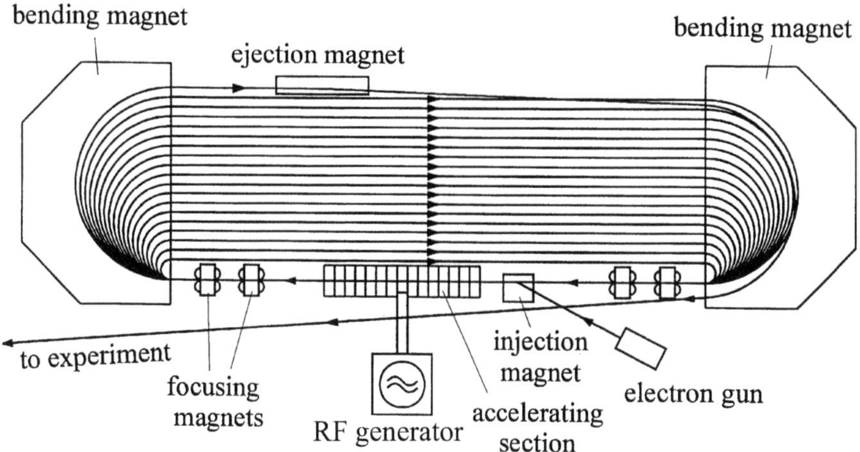

Fig. 1.13 Electron acceleration in the racetrack microtron. The name comes from the shape of the path taken by the particles.

This difference must be an exact integer multiple k of the period of the RF voltage, i.e. $\Delta t = k/\nu_{\mathrm{RF}}$. Imposing this condition we can calculate the amount of energy the electron has to gain per revolution

$$\Delta E = k\frac{ec^2 B}{2\pi\nu_{\mathrm{RF}}}. \tag{1.31}$$

For a magnetic field of strength $B = 1$ T, a high frequency of $\nu_{\mathrm{RF}} = 3$ GHz and with $k = 1$ we obtain $\Delta E = 4.78$ MeV. This relatively high value is nowadays obtained with a short linear accelerator section in which gradients of $dE/ds > 10\ \mathrm{MeV\,m^{-1}}$ are possible.

The microtron — a compact design with only one bending magnet and circular paths with no straight sections — is nowadays used for energies up to 20 MeV or so, mostly in medical applications. To reach higher energies of up to 100 MeV or so, race-track microtrons are used. The largest machine of this type, called **'MAMI'**, is located at the University of Mainz [15] and is designed to reach energies up to $E = 820$ MeV.

1.3.8 The betatron

In all the accelerators we have considered so far, the magnetic field remains constant and the radius of the particle trajectory increases with energy. In the **betatron**, on the other hand, the magnetic field is increased as the particles accelerate so that the circular path remains the same size. The accelerating electric field arises naturally from the magnetic field, which is changing very rapidly with time, in accordance with the law of induction. It is therefore not necessary to construct a special accelerating section. The betatron, illustrated in Fig. 1.14, is in essence an AC transformer with the secondary winding replaced by a circulating electron beam.

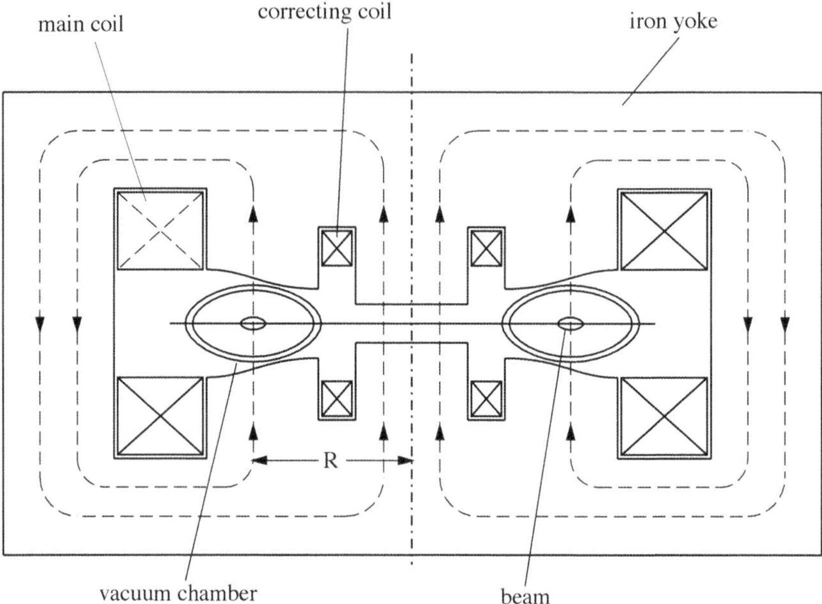

Fig. 1.14 Cross-section through a betatron. The accelerator is rotationally symmetric about the perpendicular axis. The electron beam travels in a circular vacuum chamber between the poles of the AC magnet.

The first working accelerator based on this principle was built in 1940 by D.W. Kerst at the University of Illinois [16]. It was able to accelerate electrons up to an energy of 2.3 MeV. Just two years later Kerst and his co-workers succeeded in building a 20 MeV betatron. Nowadays a series of betatrons operate in this energy range, used mostly for medical applications.

It is crucial to the operation of the betatron that the particle trajectory should on average remain constant during acceleration. A stability condition must therefore be satisfied, which we will now derive. As mentioned above, the betatron is based on the application of the law of induction

$$\oint \boldsymbol{E} \cdot d\boldsymbol{r} = -\iint_A \dot{\boldsymbol{B}} \cdot d\boldsymbol{s}. \tag{1.32}$$

The magnetic field $\boldsymbol{B}(t)$ is produced by an alternating current through the primary winding and so has a sinusoidal time-dependence $\boldsymbol{B}(t) = \boldsymbol{B}_0 \sin \omega t$. Betatrons have a simple rotationally symmetric geometry, and to a good approximation we may assume a circular trajectory of constant radius R, as shown in Fig. 1.15. The surface enclosed by this path is A. Owing to the rotational symmetry, the azimuthal distribution of the magnetic field is constant, i.e. $d\boldsymbol{B}/d\Theta = 0$. The field does, however, vary with radial distance, $B(r) \neq$ const. The average magnetic field crossing perpendicular to the surface A is given by

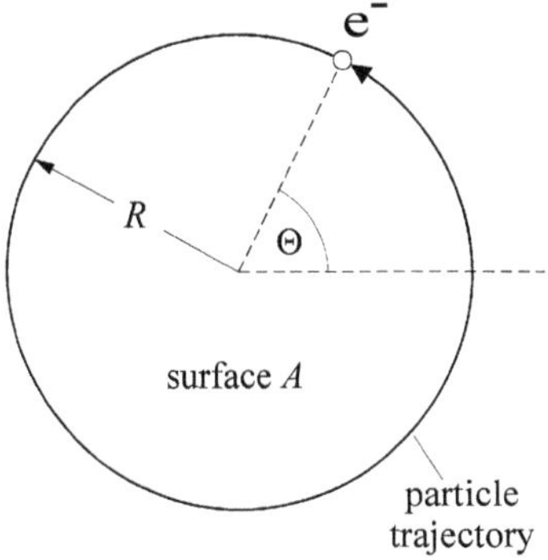

Fig. 1.15 Particle trajectory in a betatron.

$$\langle|\boldsymbol{B}|\rangle = \frac{1}{\pi R^2}\iint_A \boldsymbol{B}(r)\cdot d\boldsymbol{s}. \tag{1.33}$$

According to the law of induction (1.32) this field induces an electric field $\boldsymbol{E}(r)$ along the particle trajectory. The rotational symmetry again ensures that $d\boldsymbol{E}/d\Theta = 0$. With this symmetry (1.32) simplifies to

$$2\pi R|\boldsymbol{E}| = -\pi R^2 \frac{d}{dt}\langle|\boldsymbol{B}|\rangle \qquad \Longrightarrow \qquad |\boldsymbol{E}| = -\frac{R}{2}\langle|\dot{\boldsymbol{B}}|\rangle. \tag{1.34}$$

The electrons thus experience an accelerating force

$$|\boldsymbol{F}| = |\dot{\boldsymbol{p}}| = -e|\boldsymbol{E}| = \frac{eR}{2}\langle|\dot{\boldsymbol{B}}|\rangle. \tag{1.35}$$

Using $p = E/c$ and $v = c$ it immediately follows from (1.29) that $1/R = e|\boldsymbol{B}|/|\boldsymbol{p}|$ and hence that $|\boldsymbol{p}| = eR|\boldsymbol{B}|$. Comparing this expression with (1.35) yields the simple relation $|\dot{\boldsymbol{B}}| = \langle|\dot{\boldsymbol{B}}|\rangle/2$, and integrating with respect to time leads to the important condition known as **Wideröe's betatron condition** [17]

$$\boxed{|\boldsymbol{B}(t)| = \frac{1}{2}\langle|\boldsymbol{B}(t)|\rangle + |\boldsymbol{B}_0|.} \tag{1.36}$$

Wideröe's condition describes the circumstances under which stable particle motion may be achieved during acceleration, and the shape of the magnet pole must thus be designed in accordance with it. The constant field $|\boldsymbol{B}_0|$, which may be fine-tuned by correcting coils, allows the particle trajectory to be adjusted. The particles execute transverse oscillations, known as **betatron oscillations**, about the nominal trajectory. Nowadays this name is used as a general term for all transverse particle oscillations in accelerators.

1.3.9 The synchrotron

The advancement of elementary particle physics requires ever higher beam energies, which cannot be achieved with the relatively compact machines described so far. For relativistic particles ($v = c$) the orbit radius increases with energy, according to equation (1.29), as

$$R = \frac{E}{ecB}. \tag{1.37}$$

There is a technical limit to the magnetic field that can be produced in practice — currently around $B = 1.5$ T in conventional magnets and $B = 5$ T in superconducting magnets. This means that at energies of $E > 1$ GeV the orbit radius increases to several metres, and it is hardly feasible to produce a magnet of such a size. Hence the concept was developed of a fixed particle orbit of arbitrary size but with constant radius R, passing through individual narrow bending magnets which only have a field in the region of the beam. Since the radius R is constant it follows immediately from (1.37) that the ratio E/B must be constant. In other words the magnetic field B be must increase **synchronously** with the energy E. This type of accelerator is hence also called the **synchrotron**. The principle of the synchrotron was developed almost simultaneously in 1945 by E.M. McMillan at the University of California [18] and by V. Veksler in the Soviet Union [19]. In that same year construction of the first 320 MeV electron synchrotron began at the University of California. A year later, using a very small machine with a maximum energy of 8 MeV, F.G. Gouard and D.E. Barnes [20] in England succeeded in confirming the theoretical predictions of the function of a synchrotron. The first design studies for a proton synchrotron were produced in 1947 by M.L. Oliphant *et al.* [21]. Then at the beginning of the fifties the cosmotron, a 3 GeV proton synchrotron [22], was built in Brookhaven. Following these early successes a whole series of synchrotrons were designed and built at the end of the fifties to accelerate protons as well as electrons. Figure 1.16 shows the structure and important elements of a synchrotron. Individual narrow bending magnets with pole separations of around 0.2 m and homogeneous fields are positioned around the almost circular orbit. Since the particles circulate many thousands of times, their unavoidable divergence makes it necessary to focus them using separate special magnets. The focusing was at one time integrated into the bending magnets, using so-called **combined function** magnets, which resulted in relatively simple magnet structures. This has the disadvantage, however, that the focusing is fixed from the outset and cannot be adapted to special circumstances. As a result, separate elements (**separated function** magnets) are nowadays chosen for beam bending and focusing, especially in electron synchrotrons. The synchrotron only needs a few accelerating regions — one is enough as a minimum — which can be supplied by one or more RF generators. If the circumference L of the machine is an exact multiple q of the RF wavelength λ_{RF}, i.e. $L = q\lambda_{\mathrm{RF}}$, the particles always reach the accelerating structure with the phase Ψ_0, and the energy gain per turn is

$$\Delta E_{\mathrm{beam}} = eU_{\mathrm{max}} \sin \Psi_0 - \Delta E_{\mathrm{loss}}. \tag{1.38}$$

In proton accelerators there are no losses ($\Delta E_{\mathrm{loss}} = 0$) and so very high energies over 1000 GeV can be reached. Here superconducting magnets are used in order

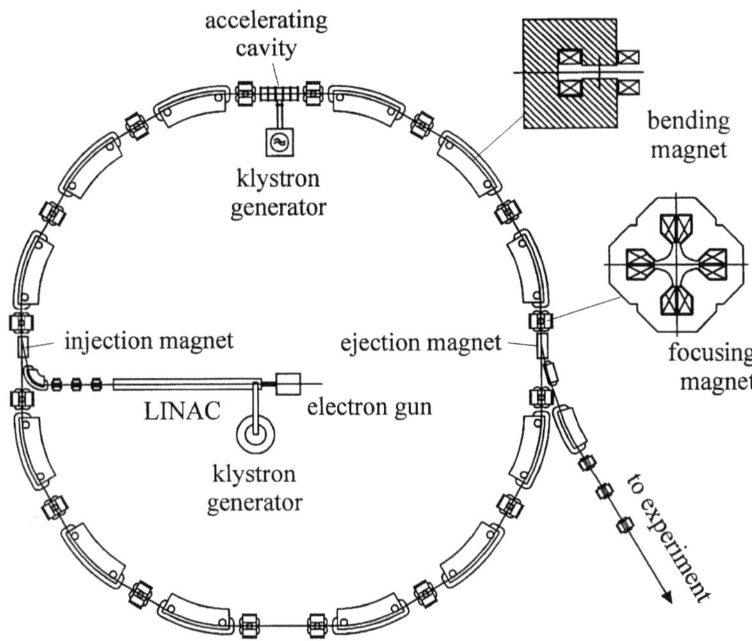

Fig. 1.16 Basic layout of a modern synchrotron. The particle trajectory is controlled by bending magnets with homogeneous fields, while the beam focusing is performed by specially designed magnets. The acceleration takes place in one or more short RF structures. The particles are supplied by a pre-accelerator (linac or microtron).

to keep the radius of the accelerator to a reasonable size. In the case of electrons there is the disadvantage of the emission of electromagnetic radiation, which increases strongly with the beam energy. The energy lost due to this so-called **synchrotron radiation**, described in Chapter 2, grows according to $\Delta E_{\text{loss}} \propto E^4$ and for energies of order $E \approx 10$ GeV becomes so dominant that this must be considered the approximate upper limit for electron synchrotrons. Such an energy loss could only be compensated by providing considerably more RF power, which would lead to disproportionately high construction and running costs.

Synchrotrons cannot accelerate particles from a starting energy $E = 0$. This is because it is not possible, simply by using low currents in the magnets, to produce a magnetic field that starts at exactly $B = 0$ and then increases linearly and precisely enough. One source of problems is the coercive fields of the ferromagnets, which are strongly excitation-dependent. These cause significant and uncontrollable deviations of the particle position from the ideal trajectory and leads to beam losses. The magnetic field of the earth also causes disturbances. As a result of these problems, synchrotrons start accelerating from a minimum energy of not less than 20 MeV. As a general rule, the problems of acceleration in synchrotrons lessen as the injection energy increases. In a 5–10 GeV electron synchrotron an optimum solution, taking cost into account, is to inject at an

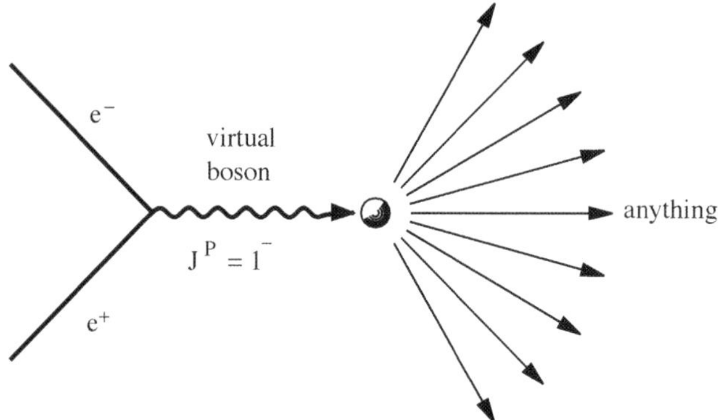

Fig. 1.17 Inelastic electron–proton collision.

energy of around 200 MeV. This is why synchrotrons are always filled from a linac or a microtron.

The particle beams delivered from a pre-accelerator must be deflected onto the orbit around the synchrotron by a magnet. By the end of one revolution this magnetic field must be removed or the particles will immediately be bent away from the orbit again and into the walls of the vacuum chamber. To achieve this, a very rapidly pulsed magnet called a **kicker** is employed, with a pulse duration of around 1 μs. For the same reasons, this type of fast kicker magnet is also used to deflect the beam out of the machine at the end of the accelerating cycle.

1.4 Particle production by colliding beams

1.4.1 The physics of particle collisions

A very basic way to produce heavy particles is by deep-inelastic collision of highly energetic electron–positron pairs, as shown in Fig. 1.17. In this particle–antiparticle reaction both are completely annihilated and a virtual photon is produced which contains the entire energy of the process. This consists of the rest energy of the two particles, $2m_ec^2$, plus their kinetic energy before the collision, which in highly relativistic collisions is the dominant component. Because the dynamics of this process are very clear, it is very well suited to the study of complex particle structures. Among other topics, very significant research into quark–antiquark systems has been undertaken using these e^+–e^- reactions, especially the charm and B-quark systems. One of the challenges of accelerator physics is to develop machines which are optimized to produce particle interactions of this kind at high event rates.

In principle, e^+–e^- collisions could be achieved very simply, as Fig. 1.18 shows. A particle accelerator is used to bring positrons up to the required energy E_1. The positrons may be produced by β-decay or preferably by pair production. If they are then fired against a fixed target of solid material they will collide with the orbiting electrons in the atoms, giving the required e^+–e^- collision. The disadvantage of such a system is that, especially at extremely relativistic

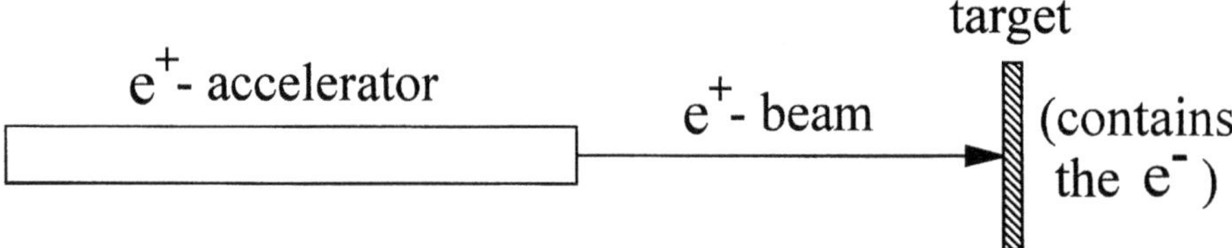

Fig. 1.18 e^+–e^- collision in a fixed target.

energies $\gamma = E/m_e c^2 \gg 1$, almost all the energy is lost through recoil. We will now show this quantitatively.

We begin by choosing a frame of reference K, at rest in the laboratory, in which the positron moving with velocity u has momentum $p = mu$. Let us list the velocities, momenta and energies of both particles before the collision:

Laboratory frame K :

$$e^+ \quad \oplus \quad \overset{p}{\Longrightarrow} \qquad\qquad \bigcirc \quad e^- \text{ (at rest)}$$

velocity	:	u_1	$=$	u	$u_2 =$	0
momentum	:	p_1	$=$	p	$p_2 =$	0
energy	:	E_1	$=$	pc	$E_2 =$	0
		γ	$=$	$\dfrac{E_1}{m_e c^2}$		

We now move to the centre of mass frame K', in which the sum of the particle momenta is by definition zero, i.e. $\sum p_i = 0$. In this reference frame both particles move with the same velocity v, but in opposing directions, towards one another. In this frame our list before collision becomes:

Centre of mass frame K' :

$$e^+ \quad \oplus \quad \overset{p'}{\Longrightarrow} \qquad \overset{-p'}{\Longleftarrow} \quad \bigcirc \quad e^-$$

velocity	:	u'_1	$=$	v	$u'_2 =$	$-v$
momentum	:	p'_1	$=$	p'	$p'_2 =$	$-p'$
energy	:	E'_1	$=$	$p'c$	$E'_2 =$	$p'c$
		γ'	$=$	$\dfrac{E'_1}{m_e c^2}$		

In this choice of reference frames the centre of mass system K' moves with a velocity v relative to the laboratory frame. In the frame K' both particles come to a complete rest at the point of collision and the energy transferred in the collision

$$E^* = E'_1 + E'_2 = 2p'c = 2\gamma' m_e c^2 \tag{1.39}$$

is available for particle reactions. Part of the total energy, the kinetic energy of the centre of mass system in the laboratory frame, is however wasted. In order to calculate this portion, or alternatively the energy E^* available to the reaction, we need to determine γ'. We calculate this by applying a Lorentz transformation to obtain the relativistically invariant momentum of the centre of mass in the laboratory frame. Since the centre of mass moves in a straight line, we only need to consider one spatial coordinate x. Using $\beta' = v'/c$ we have

$$\begin{pmatrix} E_1/c \\ p_1 \end{pmatrix} = \begin{pmatrix} \gamma' & \beta'\gamma' \\ \beta'\gamma' & \gamma' \end{pmatrix} \cdot \begin{pmatrix} E_1'/c \\ p_1' \end{pmatrix}. \tag{1.40}$$

With $p_1' = E_1'/c$ we immediately obtain

$$p_1 = \gamma'(\beta' + 1)p_1'. \tag{1.41}$$

For extremely relativistic particles $\beta' = 1$ to very good approximation, and so the momentum takes the form

$$p_1 = 2\gamma' p_1' = 2\gamma'(\gamma' m_e c) = 2\gamma'^2 m_e c. \tag{1.42}$$

Using the general expression $p_1 = \gamma m_e c$ we finally obtain the relative energy of the particles in the centre of mass system

$$\gamma' = \sqrt{\frac{\gamma}{2}}. \tag{1.43}$$

The fraction of useful energy E^* available to the reaction out of the total supplied particle energy E_1 is then

$$\eta = \frac{E^*}{E_1} = \frac{2\gamma' m_e c^2}{\gamma m_e c^2} = \frac{2\gamma'}{\gamma} = \sqrt{\frac{2}{\gamma}}. \tag{1.44}$$

At relativistic energies with $\gamma > 1000$ the relative effectiveness of the procedure becomes very small. To put it another way, how large must the original particle energy E_1 be in order to achieve a particular reaction energy E^*? Rearranging equation (1.44) gives us the answer:

$$E_1 = \frac{E^{*2}}{2m_e c^2}. \tag{1.45}$$

To produce B mesons a minimum energy of $E^* = 9.47$ GeV is required. With $m_e c^2 = 5.11 \times 10^{-4}$ GeV we immediately see that $E_1 = 87\,750$ GeV (!). An electron accelerator this powerful will most certainly not be built in the foreseeable future! The solution to this problem is in principle very simple: two beams with the same energy E are used, and fired at one another. The laboratory and centre of mass frames are then identical, and the total energy of **both** beams is available for particle production, namely

$$\begin{array}{ccccccc} & & E & & E & & \\ e^+ & \oplus & \Longrightarrow & \times & \Longleftarrow & \bigcirc & e^- \\ & & & & & & \\ & & E^* & = & 2E. & & \end{array}$$

In this case B mesons can be produced by beams with an energy of around $E \approx 5$ GeV, easily achievable in modern accelerators.

1.4.2 The storage ring

The idea of firing two beams of the same energy against one another is simple in principle, but is not so easy to achieve in practice. The particle density in highly energetic beams is very low compared to solid matter, and so there is only a very small interaction probability in colliding beams. The head-on collisions needed for particle reactions are extremely rare. Consequently, strenuous efforts must be made to produce a sufficiently high event rate for experiments. The number of events per second is given by the relation

$$\dot{N}_{\mathrm{p}} = \sigma_{\mathrm{p}}\mathcal{L}. \tag{1.46}$$

The cross-section σ of the particle reaction is given by the laws of nature, while the factor $\mathcal{L}$ is a measure of the interaction probability in the colliding beams. This factor, called the **luminosity**, determines the performance of the accelerator. It is calculated as follows:

$$\mathcal{L} = \frac{1}{4\pi}\frac{f_{\mathrm{rev}}N_1N_2}{\sigma_x\sigma_z}. \tag{1.47}$$

It may be seen immediately that the luminosity $\mathcal{L}$ increases in proportion to the number of particles N_i per particle beam, and hence to the product N_1N_2 for two beams. Reducing the horizontal and vertical size of the beam σ_x and σ_z in the collision region also increases the luminosity. Furthermore, $\mathcal{L}$ is proportional to the frequency with which the beams are fired at one another. Particle accelerators must therefore be designed both to achieve high energy beams and to collide them with high luminosity. For this reason the storage ring has proven to be the most successful accelerator design to date.

The storage ring consists of a circular accelerator, outwardly very similar to the synchrotron, but with several key differences. The first is that two beams, electrons and positrons, circulate in the accelerator at the same time. This relies on the fact that electrons circulating clockwise through a magnetic field experience the same force as positrons circulating in the opposite direction, i.e.

$$\boldsymbol{F} = e(\boldsymbol{v} \times \boldsymbol{B}) = -e(-\boldsymbol{v} \times \boldsymbol{B}). \tag{1.48}$$

As a result, the same magnet structure and vacuum chamber can be used for both beams. As the particles in a given experiment are required to collide at a fixed energy, the beam energies must not vary during data-taking. This means, for example, that the current levels in all the magnets must be kept constant to within very narrow limits. Strictly speaking, the storage ring is therefore not really an accelerator at all.

As mentioned above, only a few particles actually undergo inelastic collisions in each beam crossing, and so the beam intensity is not significantly reduced by these interactions. The same beams can circulate continuously in opposing directions for long periods and lifetimes of several hours are now common. Once injected into the storage ring the particles remain there at constant energy, without refilling, for the duration of the data-taking run. They are **stored**, which explains how this type of accelerator gets its name. In storage rings the low interaction probability between the colliding beams is partly compensated for by the high frequency f_{rev} of beam crossings, since the beams travel at close to the speed of light. In a storage ring 300 m in circumference, for example, the frequency has a value $f_{\text{rev}} = 1$ MHz.

To achieve the very long beam lifetimes obtained in a storage ring compared to those in a synchrotron, additional measures must be taken. Firstly, the vacuum pressure must be more than three orders of magnitude lower than in other types of accelerator, in order to minimize the number of unwanted collisions with residual gas molecules. Extreme ultra-high vacuum at a pressure below 10^{-7} Pa is required. Secondly, the beam steering and focusing must be very carefully controlled, which places stringent requirements on the tolerance of the magnets and their power supplies. The long lifetime of the beams means that even relatively small uncertainties can have significant detrimental effects. Thirdly, in electron storage rings it is still necessary to construct high frequency accelerating sections even though the beams are not really being accelerated at all. This is because the circulating electrons radiate electromagnetic energy which must be replaced. These accelerating sections are designed for considerably higher beam currents than are common in synchrotrons, in order to achieve the highest possible luminosities.

Since storage rings operate at fixed energies, the beams must be brought up to the correct energy by a pre-accelerator, usually a synchrotron or linac. They are then injected into the storage ring by means of a pulsed magnet, similar to that found in a synchrotron. If new particles can be injected without the particles already circulating inside the storage ring being lost, then the injection process can in principle be repeated as often as desired, allowing very high beam currents to be reached. This principle of **accumulation** is in fact a considerable advantage of storage rings and allows the high particle numbers N_i necessary from (1.47) to be achieved.

The typical layout of an electron-positron storage ring is shown in Fig. 1.19. The two semicircles are separated by two rather longer straight sections, each with a collision region in the middle. Each collision region, also called an **interaction region**, is surrounded by a particle detector, which consists of a multitude of different specialized subdetectors. These cover almost the entire solid angle, to allow a complete analysis of the particle interactions. Powerful focusing magnets are arranged on both sides of the detectors to produce very small beam cross-sections σ_x and σ_z. In particular these can achieve extremely small vertical beam sizes of less than $\sigma_z = 25\ \mu$m. Depending on the beam energies, luminosities in the range $\mathcal{L} = 10^{30}$ to 10^{33} cm^{-2} s^{-1} are nowadays achieved in storage rings.

The idea of allowing particle beams to circulate at constant energy in a ring accelerator without significant loss of intensity, thereby storing them, was proposed as early as 1943 by Kollath, Touschek, and Wideröe. However, 13 years passed before Kerst in 1955 and soon afterwards also O'Neill [23] worked out a detailed design for such a machine. In 1958 construction of storage rings began in Stanford and Moscow. The first successful attempt to store an electron beam for a long period was made in 1961 at the small storage ring A.d.A. [24] in Italy

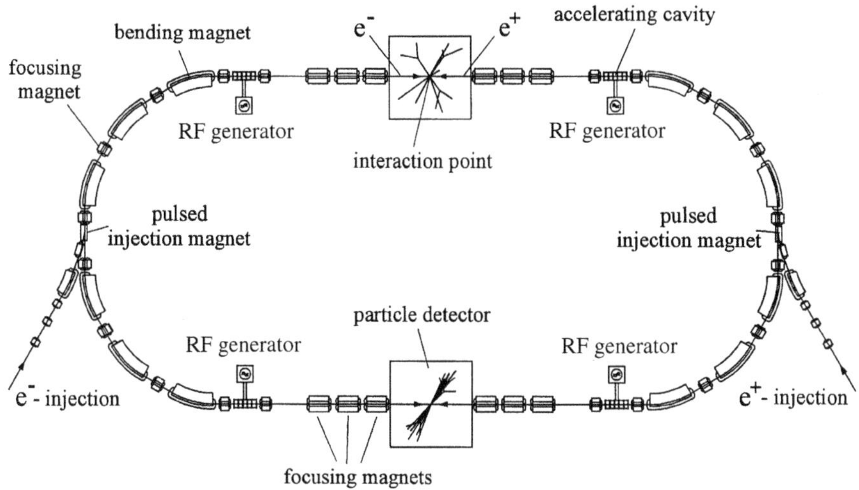

Fig. 1.19 Layout of an electron-positron storage ring with two interaction regions surrounded by particle detectors. The very small beam sizes required in these interaction regions are achieved using special focusing magnets on either side of the detectors. The beams (e^+ and e^-), which circulate in opposite directions, are supplied by a pre-accelerator of appropriate energy, such as a synchrotron. Accelerating regions replace the energy lost due to radiation of electromagnetic energy.

and at around the same time in the Soviet Union [25]. Many storage rings have since been constructed around the world, most for electron–positron collisions. In a few other machines protons are collided with protons or antiprotons, and at the German Electron Synchrotron Laboratory DESY in Hamburg the double ring HERA is the first machine in the world to collide electrons with high energy protons [26]. In this machine, however, the two types of particles circulate in separate rings because of their very different masses, and are only steered onto the same path close to the interaction regions. The largest storage ring is currently LEP at the European research centre CERN in Geneva [27]. Here electrons and positrons are collided at energies of over 50 GeV, with an upper limit of around 100 GeV. This machine, situated underground, has a circumference of 27 km.

1.4.3 The linear collider

The rate at which electrons circulating in storage rings lose energy through electromagnetic radiation increases very steeply with particle energy, namely $\propto E^4$. We will examine the characteristics of this radiation more closely in Chapter 2. A correspondingly large amount of RF power is required to make up this energy loss, and the technical and financial limit is reached at beam energies of a few tens of GeV. If we wish to study $e^+ - e^-$ reactions at energies above 100 GeV then the particles must be accelerated in a straight line in linacs, since the amount of electromagnetic radiation induced by linear acceleration is negligible even for very high energies. In the last few years ideas have therefore been developed to collide particles using linacs. These machines are therefore known as **linear colliders**.

The luminosity in linear colliders is given by the same relation (1.47) as for storage rings. However, linacs have the disadvantage that the collision frequency f_0 is generally very low, usually not more than a few hundred hertz. Furthermore, the number of particles N_i cannot be made very large because it is extremely difficult to stably and reliably bring high-intensity beams all the way from an energy $E = 0$ up to the final beam energy.

To overcome this disadvantage linear colliders have compared to storage rings, it is of prime importance to drastically reduce the cross-section of the beam. To achieve this, cathodes have recently been developed which produce electron beams with extremely small dimensions and angular divergence. This process cannot be used for positrons, however, which are created by pair production. The transverse size and angular divergence of the positron beam are strongly increased by collisions with nuclei inside the target after pair production. To remedy this, the partly accelerated positrons are injected into a small storage ring, where they circulate for a few ms. The transverse betatron oscillations of the beam are damped by the emission of electromagnetic radiation, considerably reducing the beam cross-section. Finally the particles are returned to the linac and are accelerated up to the final beam energy. This use of a storage ring, termed a **damping ring**, is necessary in order for a linear collider to deliver sufficient luminosity.

The need for very small beam sizes, $\sigma_{x,z} < 1\mu$m, places very strong demands on the final focusing of the beam at the interaction point. All the imperfections of the focusing system must be understood and compensated for, as must the effect of energy loss in the beam. The reduction of all non-linear effects in the beam optics up to higher orders requires very costly control systems. The final focusing system in a linear collider is therefore considerably more expensive than in a storage ring.

The very strong beam compression at the interaction point causes a strong transverse force to be exerted on the beam particles as they cross, due to the space charge effect. As a result, they are strongly deflected and so emit bremsstrahlung radiation. This broadens the energy distribution within the beam and of course also increases the level of background in the detectors. On the other hand, computer simulations have shown that the space charge effect causes the beams to be squeezed together during collision, very much like the **pinch effect** observed, for example, in electric arc discharge. This effect, which should increase the luminosity by a considerable factor, has so far only been investigated theoretically.

The first linear collider began operation in Stanford at the end of the 1980s and was called the Stanford Linear Collider SLC [28]. It used the existing 3 km long linac, which has a maximum energy of 50 GeV. Electron and positron beams were accelerated in the linac, one closely after the other, and separated by a magnet at the end. Each beam then travelled though one of two semicircular magnet structures, moving apart and then back together again in the shape of a pair of pincers, arriving finally at the special focusing system. The SLC demonstrated the principle of the linear collider, with all the elements working successfully together for the first time. For future projects it will, however, be necessary to increase the gradient of accelerating cavities from the present $dE/ds \approx 15$ $\mathrm{MeV\,m^{-1}}$ to over 100 $\mathrm{MeV\,m^{-1}}$, in order to keep the length of the linac, and hence the costs, to within reasonable limits.

2
Synchrotron radiation

According to the fundamental laws of classical electrodynamics, a charge undergoing acceleration will radiate energy in the form of electromagnetic waves. This is of course also true of any charged particle moving in an accelerator. As we shall see in what follows, this radiation plays an important role for electrons of sufficiently high energy, while for protons and all other heavy particles it can in general be neglected. Our aim in this chapter is to understand the key physical principles of this **synchrotron radiation**. A full theoretical treatment of the physics of this radiation and its properties may be found in Jackson [29] and Hofmann [30].

Consider first an accelerated particle of charge e, moving with momentum $\boldsymbol{p} = m_0\boldsymbol{v}$. We assume the particle's velocity is non-relativistic, i.e. $v \ll c$. The total radiated power in this case was calculated by Larmor at the end of the nineteenth century. He found the important relation

$$P_\mathrm{s} = \frac{e^2}{6\pi\varepsilon_0 m_0^2 c^3}\left(\frac{d\boldsymbol{p}}{dt}\right)^2, \tag{2.1}$$

where $\varepsilon_0 = 8.85419 \times 10^{-12}\ \mathrm{A\,s\,V^{-1}\,m^{-1}}$ is the permittivity of free space. One immediately sees that electromagnetic energy is only emitted when the particle's momentum changes as a result of some applied force, i.e. $d\boldsymbol{p}/dt \neq 0$. Here we are not concerned with the details of this force and the resulting acceleration. The azimuthal angular distribution of the radiation is identical to that of the **Hertz dipole**, namely

$$\frac{dP_\mathrm{s}}{d\Omega} = \frac{e^2}{16\pi^2\varepsilon_0 m_0^2 c^3}\left(\frac{d\boldsymbol{p}}{dt}\right)^2 \sin^2\Psi. \tag{2.2}$$

The radiation emitted by non-relativistic charged particles is completely described by classical electrodynamics, and in any case is so weak that it may be neglected.

2.1 Radiation from relativistic particles

Let us now consider this case where the particles reach velocities comparable to that of light. Here the situation is fundamentally different. We require the

Lorentz-invariant form of equation (2.1). This is most easily found by transforming time according to

$$dt \quad \longrightarrow \quad d\tau = \frac{1}{\gamma} dt \qquad \text{with} \qquad \gamma = \frac{E}{m_0 c^2} = \frac{1}{\sqrt{1-\beta^2}} \tag{2.3}$$

and replacing the classical momentum $\boldsymbol{p}$ with the four-momentum P_μ of relativistic electrodynamics:

$$\left(\frac{dP_\mu}{d\tau}\right)^2 \quad \longrightarrow \quad \left(\frac{d\boldsymbol{p}}{d\tau}\right)^2 - \frac{1}{c^2}\left(\frac{dE}{d\tau}\right)^2 . \tag{2.4}$$

Using this transformation we may rewrite (2.1) and obtain the relativistically invariant form of the radiation equation

$$\boxed{P_\mathrm{s} = \frac{e^2 c}{6\pi\varepsilon_0} \frac{1}{(m_0 c^2)^2} \left[\left(\frac{d\boldsymbol{p}}{d\tau}\right)^2 - \frac{1}{c^2}\left(\frac{dE}{d\tau}\right)^2\right].} \tag{2.5}$$

The radiated power depends principally on the angle between the direction of motion of the particle $\boldsymbol{v}$ and the direction of the acceleration $d\boldsymbol{v}/d\tau$. We will consider the following limiting cases:

a. linear acceleration : $\dfrac{d\boldsymbol{v}}{d\tau} \parallel \boldsymbol{v}$

b. circular acceleration : $\dfrac{d\boldsymbol{v}}{d\tau} \perp \boldsymbol{v}$.

2.1.1 Linear acceleration

The relativistically invariant expression for the particle energy is

$$E^2 = (m_0 c^2)^2 + p^2 c^2 \tag{2.6}$$

which, when differentiated with respect to τ, immediately gives the relation

$$E\frac{dE}{d\tau} = c^2 p \frac{dp}{d\tau}. \tag{2.7}$$

Using $E = \gamma m_0 c^2$ and $p = \gamma m_0 v$, (2.7) simplifies to

$$\frac{dE}{d\tau} = v\frac{dp}{d\tau}. \tag{2.8}$$

Inserting this into the radiation equation (2.5), one obtains

$$\begin{aligned} P_{\mathrm{s}} &= \frac{e^2 c}{6\pi\varepsilon_0} \frac{1}{(m_0 c^2)^2} \left[\left(\frac{dp}{d\tau}\right)^2 - \left(\frac{v}{c}\right)^2 \left(\frac{dp}{d\tau}\right)^2 \right] \\ &= \frac{e^2 c}{6\pi\varepsilon_0} \frac{1}{(m_0 c^2)^2} (1-\beta^2) \left(\frac{dp}{d\tau}\right)^2 \end{aligned} \tag{2.9}$$

and using $1 - \beta^2 = 1/\gamma^2$,

$$P_{\mathrm{s}} = \frac{e^2 c}{6\pi\varepsilon_0 (m_0 c^2)^2} \left(\frac{dp}{\gamma d\tau}\right)^2 = \frac{e^2 c}{6\pi\varepsilon_0 (m_0 c^2)^2} \left(\frac{dp}{dt}\right)^2 \tag{2.10}$$

immediately follows. In the case of linear acceleration, the energy gained per unit distance is usually known. Using $dp/dt = dE/dx$ we obtain the following expression for the power radiated from linearly accelerated particles:

$$P_{\mathrm{s}} = \frac{e^2 c}{6\pi\varepsilon_0 (m_0 c^2)^2} \left(\frac{dE}{dx}\right)^2 . \tag{2.11}$$

In accelerators built today the energy gained per metre is approximately $dE/dx \approx 15 \mathrm{MeV\,m^{-1}} = 2.4 \times 10^{-12}\ \mathrm{J\,m^{-1}}$, giving rise to a radiated power $P_{\mathrm{s}} \approx 4 \times 10^{-17}$ watts. Comparing the radiated power with the power delivered by the accelerator gives

$$\eta = \frac{P_{\mathrm{s}}}{dE/dt} = \frac{P_{\mathrm{s}}}{v\,dE/dx} = \frac{e^2}{6\pi\varepsilon_0 (m_0 c^2)^2} \frac{1}{\beta} \frac{dE}{dx} . \tag{2.12}$$

Using the value of dE/dx given above we obtain a relative loss of $\eta = 5.5 \times 10^{-14}$ for extremely relativistic particles ($v \approx c$). The radiation of electromagnetic energy during longitudinal acceleration is thus completely negligible.

2.1.2 Circular acceleration

The situation is quite different when particles are bent **perpendicular** to their direction of motion and so travel in a circular path. In this case the particle energy remains constant and the general radiation formula (2.5) reduces to

$$P_{\mathrm{s}} = \frac{e^2 c}{6\pi\varepsilon_0} \frac{1}{(m_0 c^2)^2} \left(\frac{dp}{d\tau}\right)^2 = \frac{e^2 c \gamma^2}{6\pi\varepsilon_0} \frac{1}{(m_0 c^2)^2} \left(\frac{dp}{dt}\right)^2 . \tag{2.13}$$

During circular motion through an angle $d\alpha$ the momentum of the particle changes by an amount $dp = p\,d\alpha$. It follows immediately that

$$\frac{dp}{dt} = p\omega = p\frac{v}{R} . \tag{2.14}$$

Here R is the bending radius of the particle orbit. We now insert this relation into (2.13). The radiation of electromagnetic waves increases very strongly as a function of the particle energy and so in this case it is sufficient to consider only extremely relativistic velocities, i.e. $v = c$. Under these circumstances $E = pc$. Replacing γ by E/m_0c^2 one obtains the following expression for the radiated power during transverse acceleration:

$$\boxed{P_s = \frac{e^2 c}{6\pi\varepsilon_0} \frac{1}{(m_0c^2)^4} \frac{E^4}{R^2}.} \tag{2.15}$$

This relation was discovered back at the end of the nineteenth century by **Liénard** [31]. For particles of a given energy with elementary charge e, the radiated power varies inversely with the fourth power of the rest mass m_0. Comparing the power radiated from an electron ($m_ec^2 = 0.511$ MeV) with that from a proton of the same energy ($m_pc^2 = 938.19$ MeV) gives

$$\frac{P_{s,e}}{P_{s,p}} = \left(\frac{m_pc^2}{m_ec^2}\right)^4 = 1.13 \times 10^{13} \quad (!) \tag{2.16}$$

It is evident that in practice this radiation is only important in the case of electrons. With protons it can only be observed at energies of at least several hundred GeV. In circular accelerators it is often important to know the energy loss ΔE that a particle undergoes during one complete revolution. Here it is assumed that on average the radiation is emitted quasi-continuously at a constant rate in the bending magnets, in which the bending radius R is the same everywhere. Under these assumptions, which are generally very reliable, one obtains the rather simple relation

$$\Delta E = \oint P_s\, dt = P_s\, t_b = P_s \frac{2\pi R}{c}. \tag{2.17}$$

Here t_b is the time per revolution that a particle spends in the bending magnets. This time is less than the total period of a revolution. Here one must remember that radiation is only emitted during transverse acceleration, i.e. during bending. Inserting (2.15) into (2.17), one obtains

$$\boxed{\Delta E = \frac{e^2}{3\varepsilon_0 (m_0c^2)^4} \frac{E^4}{R}.} \tag{2.18}$$

Using the values for the electron and choosing convenient units gives the following, easy to remember, formula

$$\boxed{\Delta E\ [\mathrm{keV}] = 88.5\, \frac{E^4\ [\mathrm{GeV}^4]}{R\ [\mathrm{m}]}.} \tag{2.19}$$

One sees that the radiation increases with the fourth power of the beam energy. At low, virtually non-relativistic, particle energies this radiation is negligible and only becomes significant when the electrons have energies of at least a few tens of MeV. This is why it was not observed experimentally until circular electron accelerators of sufficiently high energy were developed, namely the synchrotrons.

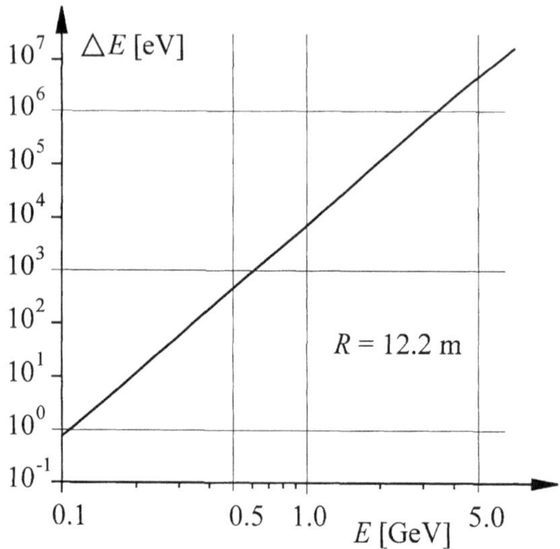

Fig. 2.1 Electron energy loss per revolution as a function of particle energy. We assume a circular accelerator in which the path through the bending magnets has a bending radius $R = 12.2$ m.

Hence this form of electromagnetic radiation was given the name **synchrotron radiation**. The first experimental observation came at the end of the 1940s at General Electric's 70-MeV synchrotron in the USA [32].

The strong increase in power of the synchrotron radiation with the particle energy, evident in equations (2.18) and (2.19), is illustrated in Fig. 2.1, in which the energy loss per turn is plotted against the electron energy. Here the example is taken of a circular accelerator with bending magnets of radius $R = 12.2$ m. Varying the energy from 0.1 GeV to 5.0 GeV changes the energy loss by almost seven orders of magnitude.

This strongly energy-dependent emission of synchrotron radiation has an important consequence for the construction of circular electron accelerators, as can be seen in Table 2.1. A few ring-shaped electron accelerators are listed, with maximum energies between 0.8 and 70 GeV. It turns out that the maximum field of the bending magnets used decreases with energy. Thus the storage ring LEP in Geneva has the highest maximum energy, $E_{\max} = 70$ GeV, but by far the weakest magnetic field, $B = 0.078$ T.

The power lost through synchrotron radiation must be replaced by the high-frequency cavities, otherwise the beam will soon be lost. High-frequency power is very expensive, so every effort must be made to minimize the energy loss due to radiation. Since this loss is proportional to E^4/R according to (2.18), the bending radius must be increased faster than the energy, which means that weaker bending magnets suffice. Hence the circumference of accelerators also increases faster than the energy. As a general rule, at energies of around 100 GeV the power loss due to synchrotron radiation becomes so high that it is no longer

Table 2.1 A few important circular electron accelerators. L is the circumference of the machine, E the maximum beam energy, R the bending radius, B the field in the bending magnets and ΔE the energy loss per revolution.

accelerator	L [m]	E [GeV]	R [m]	B [T]	ΔE [keV]
BESSY I (Berlin)	62.4	0.80	1.78	1.50	20.3
DELTA (Dortmund)	115	1.50	3.34	1.50	134.1
DORIS II (Hamburg)	288	5.00	12.2	1.37	4.53×10^3
ESRF (Grenoble)	844	6.00	23.4	0.855	4.90×10^3
PETRA (Hamburg)	2304	23.50	195	0.40	1.38×10^5
LEP (Geneva)	27×10^3	70.00	3000	0.078	7.08×10^5

practical to replace it. Consequently, accelerators at energies above 100 GeV must use other particles, such as protons. If electrons are absolutely required then a linear accelerator must be employed.

2.2 Angular distribution of synchrotron radiation

The angular distribution of synchrotron radiation from relativistic electrons is more complicated than the $\sin^2 \Phi$ dependence (2.2) familiar from the Hertz dipole. To calculate it we must first transform into a frame of reference K' moving alongside the electron. In this frame the electron is only accelerated along the x-axis and the radiation has the characteristics of a normal Hertz dipole (Fig. 2.2). Of all possible directions in which synchrotron radiation may be emitted, we will choose the particular case in which the radiation is perpendicular to both the direction of acceleration and the direction of motion, in order to simplify the calculation as far as possible. We thus observe a photon emitted

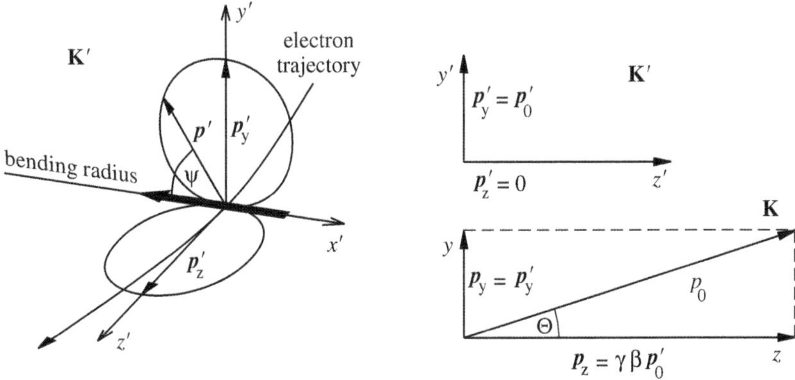

Fig. 2.2 Transformation of the spatial intensity distribution of synchrotron radiation from a co-moving frame of reference K' into the laboratory frame K.

exactly parallel to the y-axis. This photon has the momentum

$$\boldsymbol{p}_y{}' = \boldsymbol{p}_0{}' = \frac{E_s'}{c}\boldsymbol{n}, \tag{2.20}$$

where E_s' is the energy of the photon and $\boldsymbol{n}$ is the direction of motion. The four-momentum of this photon has the form

$$P_\mu' = \begin{pmatrix} p_t \\ p_x \\ p_y \\ p_z \end{pmatrix} = \begin{pmatrix} E_s'/c \\ 0 \\ p_0' \\ 0 \end{pmatrix}. \tag{2.21}$$

The photon's direction of motion in the laboratory system is obtained by a Lorentz transformation of this four-momentum:

$$P_\mu = \begin{pmatrix} \gamma & 0 & 0 & \beta\gamma \\ 0 & 1 & 0 & 0 \\ 0 & 0 & 1 & 0 \\ \beta\gamma & 0 & 0 & \gamma \end{pmatrix} \begin{pmatrix} E_s'/c \\ 0 \\ p_0' \\ 0 \end{pmatrix} = \begin{pmatrix} \gamma E_s'/c \\ 0 \\ p_0' \\ \gamma\beta E_s'/c \end{pmatrix}. \tag{2.22}$$

Using the relationship (2.20) between the energy and momentum of the photon we find that the angle Θ of the emitted photon relative to the electron beam direction follows the relation

$$\tan\Theta = \frac{p_y}{p_z} = \frac{p_0'}{\beta\gamma p_0'} \approx \frac{1}{\gamma}. \tag{2.23}$$

Since Θ is very small $\tan\Theta \approx \Theta$ to a good approximation. The axially-symmetric radiation distribution in the moving frame K' transforms into a sharply forward-peaked distribution in the laboratory frame, with a half opening-angle $\Theta = 1/\gamma$.

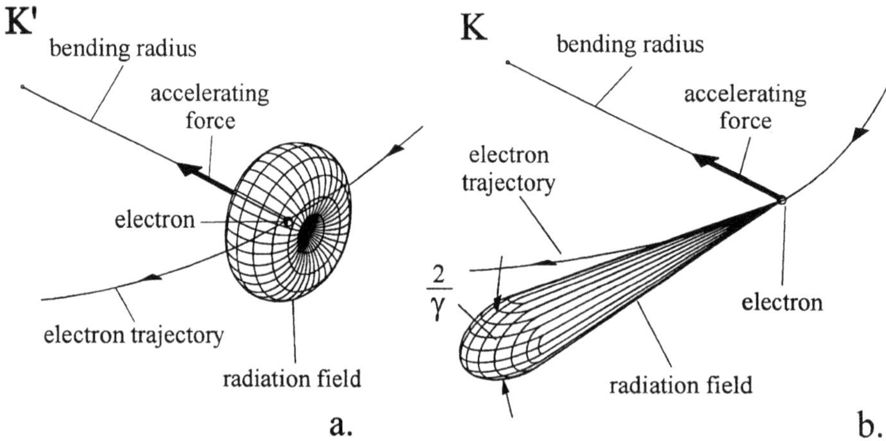

Fig. 2.3 Transformation of the axially symmetric radiation distribution in the centre of mass frame (a) into the sharply forward peaked distribution in the laboratory frame K (b).

This transformation is illustrated in Fig. 2.3. Electrons with energy $E = 1\,\text{GeV}$, i.e. with $\gamma = 1957$, thus emit synchrotron radiation in a cone of half opening-angle $\Theta = 0.5$ mrad $= 0.03°$. This sharp peaking of the radiation is one of its most useful features.

2.3 Time dependence and frequency spectrum of the radiation

We now turn to the frequency spectrum of synchrotron radiation. A complete calculation is beyond the scope of this book, and the reader is referred to the excellent treatments by Jackson [29] and Hofmann [30]. Here we confine ourselves to an approximate calculation and collect together the most important relations concerning electron accelerators.

The idea behind the calculation of the frequency spectrum is relatively simple. During each revolution the electron emits an electromagnetic pulse of duration Δt as it passes the observer. The pulses are periodic, with a frequency given by the revolution frequency f_{rev}. The spectrum thus consists of harmonics of the revolution frequency, the intensities of which may be determined by Fourier analysis of the pulse. The width of the spectrum depends primarily on the length of the pulse Δt, and this can be determined fairly easily with the help of the sketch in Fig. 2.4. In this region the electron passes through a magnetic field, assumed to be homogeneous, and so follows a curved path of bending radius R. Because the radiation is strongly forward-peaked the observer first sees it when the edge

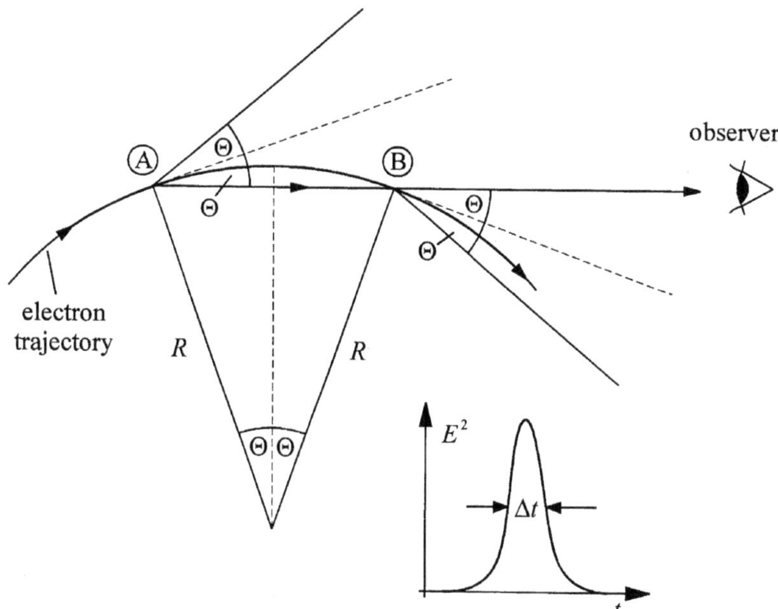

Fig. 2.4 Estimate of the length of the electromagnetic pulse produced by a relativistic electron as it flies past an observer.

of the radiation cone reaches him. This edge has an angle $\Theta = -1/\gamma$ to the electron direction. At this moment the electron is at point A along its trajectory. As it continues forward the emitted radiation passes across the observer until the opposite edge of the cone is reached, at an angle $\Theta = +1/\gamma$ (point B on the electron's trajectory). After that the observer does not see any more radiation until the next revolution. The first photon to reach the observer is emitted at point A, and the last at point B. The time difference between the arrival of the two photons gives the length of the electromagnetic pulse, and this is the same as the difference in the time taken by the photon and the electron to travel from point A to point B:

$$\begin{aligned}\Delta t &= t_{\mathrm{e}} - t_{\gamma} = \frac{2R\Theta}{c\beta} - \frac{2R\sin\Theta}{c} \\ &= \frac{2R}{c}\left(\frac{\Theta}{\beta} - \Theta + \frac{\Theta^3}{3!} - \frac{\Theta^5}{5!} + \dots\right).\end{aligned} \tag{2.24}$$

Since $\Theta \approx 1/\gamma$ and $\gamma\beta \approx \gamma - 1/2\gamma$ one obtains, to a good approximation

$$\Delta t \approx \frac{2R}{c}\left(\frac{1}{\beta\gamma} - \frac{1}{\gamma} + \frac{1}{6\gamma^3}\right) \approx \frac{4R}{3c\gamma^3}. \tag{2.25}$$

This short electromagnetic pulse results in a broad spectrum with a characteristic frequency

$$\omega_{\mathrm{char}} = \frac{2\pi}{\Delta t} = \frac{3\pi c\gamma^3}{2R}. \tag{2.26}$$

In general the **critical frequency** ω_{c} is used rather than the characteristic frequency to describe the spectral region of the synchrotron radiation. This is defined as

$$\omega_{\mathrm{c}} = \frac{\omega_{\mathrm{char}}}{\pi} = \frac{3c\gamma^3}{2R}. \tag{2.27}$$

The exact formula describing the electromagnetic spectrum emitted from relativistic electrons following a curved path was first calculated by Schwinger [33]. He obtained the following expression for the spectral photon density:

$$\frac{d\dot{N}}{d\epsilon/\epsilon} = \frac{P_0}{\omega_{\mathrm{c}}\hbar} S_{\mathrm{s}}\left(\frac{\omega}{\omega_{\mathrm{c}}}\right). \tag{2.28}$$

Here P_0 is the total power radiated from N electrons per revolution. This can be obtained immediately with the help of equation (2.15):

$$P_0 = \frac{e^2 c\gamma^4}{6\pi\varepsilon_0 R^2} N = \frac{e\gamma^4}{3\varepsilon_0 R} I_{\mathrm{beam}}. \tag{2.29}$$

I_{beam} is the electron beam current, which is usually more convenient to use in calculations than the number of electrons. The spectral function S_{s} in (2.28) has the form

$$S_{\mathrm{s}}(\xi) = \frac{9\sqrt{3}}{8\pi}\xi \int_{\xi}^{\infty} K_{5/3}(\xi)d\xi. \tag{2.30}$$

$K_{5/3}$ is the modified Bessel function. The spectral function naturally satisfies the normalization condition

$$\int_0^{\infty} S_{\mathrm{s}}(\xi)d\xi = 1, \tag{2.31}$$

as may be easily verified. Integrating up to the upper limit $\xi = 1$, i.e. $\omega = \omega_{\mathrm{c}}$, gives

$$\int_0^{1} S_{\mathrm{s}}(\xi)d\xi = \frac{1}{2}. \tag{2.32}$$

This result means that the critical frequency ω_{c} divides the synchrotron radiation into two regions of equal radiated power. A typical spectrum is shown in Fig. 2.5, in which the photon density is plotted as a function of the photon energy E_{γ}. The **critical energy** $\epsilon_{\mathrm{c}} = \hbar\omega_{\mathrm{c}}$ corresponding to the critical frequency is indicated. It is easy to appreciate how the strong forward peaking, extremely high intensity and very broad spectrum of synchrotron radiation have made it an exceedingly powerful tool for fundamental research and, increasingly, for industrial applications.

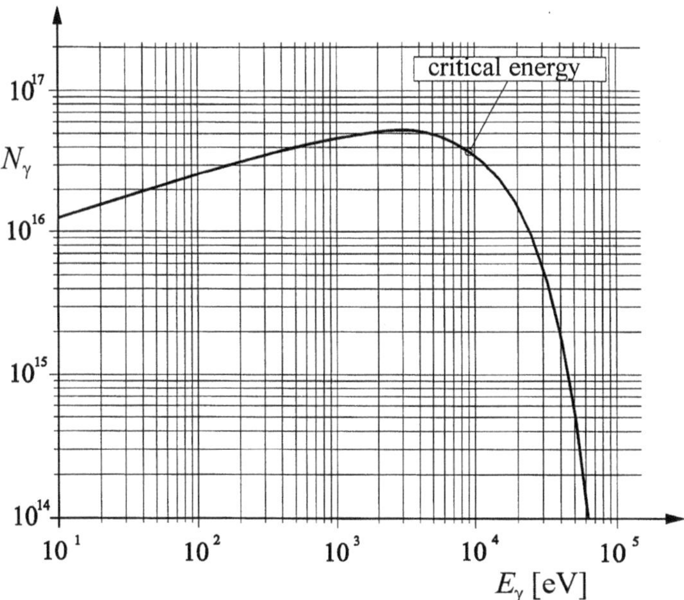

Fig. 2.5 Spectral photon density of synchrotron radiation as a function of photon energy. In this example the electron beam passes through a bending magnet with a bending radius $R = 12.2$ m at an energy $E = 5$ GeV.

2.4 Storage rings for synchrotron radiation

As already mentioned, the first experiments with synchrotron radiation were carried out at synchrotrons. However, these accelerators could not keep pace with the requirements of the increasingly refined experiments. During each accelerating cycle the electron beam is injected at very low energy, and emits almost no radiation at this stage. Only when the electron energy grows does the radiated power increase according to the E^4 law. As a result, the full intensity of synchrotron radiation is only available for a very short time, when the beam is at maximum energy. The average yield of photons is thus not very large. In addition the energy spectrum changes constantly during the acceleration, since the critical energy is proportional to γ^3. A further disadvantage is the limited beam focusing in synchrotrons, which cannot achieve the very narrow beam cross-section desirable for experiments. What is more, because the field strength in the individual magnets does not increase exactly linearly, the electron beam oscillates during certain stages of acceleration. Because of the long lever arm this can lead to large variations in the position of the synchrotron beam in the experiment.

As a result, storage rings have taken the place of synchrotrons as sources of synchrotron radiation. They operate at a fixed energy and the beams circulate for several hours without interruption. In addition, very high beam currents can be achieved by accumulation. Operating at a constant energy allows a very stable well-behaved beam to be produced and also allows the electron beam to be very strongly focused; such a beam could not be achieved in a synchrotron with very rapidly changing magnetic fields. Specially optimized storage rings have a beam spot two orders of magnitude smaller than in a synchrotron, coming very close to the ideal of a point-like radiation source. For these reasons storage rings are nowadays the standard choice as powerful sources of synchrotron radiation.

The first storage ring to be constructed exclusively for this purpose was TANTALUS, designed and built by E. Rowe and co-workers in 1968 [34]. There-

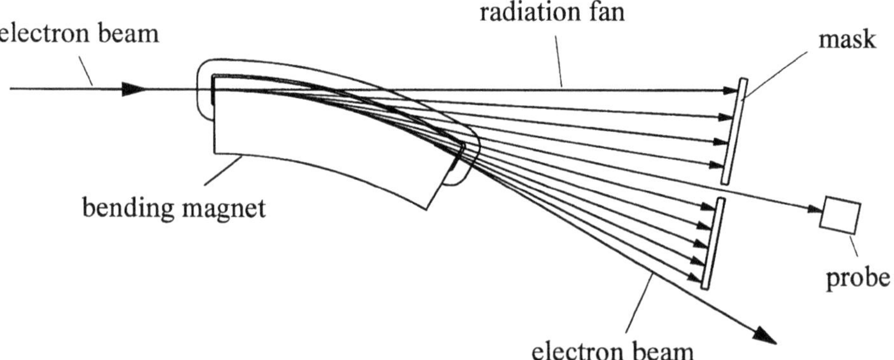

Fig. 2.6 Horizontal fan of synchrotron radiation emitted from an electron beam as it passes through a bending magnet.

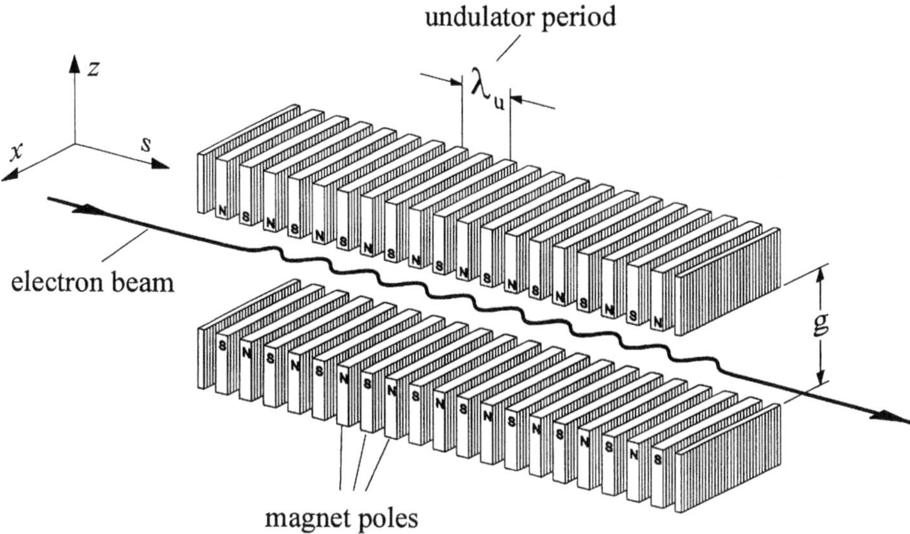

Fig. 2.7 The wiggler or undulator, consisting of a periodic arrangement of short bending magnets of alternating polarity. The bending of the electron beam in the magnets is exaggerated in the diagram.

after storage rings built for high energy physics were mostly used parasitically, such as SPEAR at the Stanford Synchrotron Radiation Laboratory [35] and DORIS at the German Electron Synchrotron Laboratory DESY in Hamburg [36]. The accelerators developed for high energy physics are, however, not optimized for the use of synchrotron radiation. In general the beam dimensions are far too large and do not allow the precise position resolution necessary for many experiments. As a result, proposals were soon developed for storage rings with extremely small beam sizes, to be used exclusively for the production of synchrotron radiation. These 'second generation' synchrotron radiation sources were developed during the mid-1970s, among them Aladdin in Madison (USA) [37], Super-ACO in Orsay (France) [38], BESSY in Berlin [39], the National Synchrotron Light Source in Brookhaven (USA) [40], and the Photon Factory in Tsukuba (Japan) [41]. In these storage rings synchrotron radiation is produced as the beam trajectory curves inside the bending magnets. A broad horizontal fan of radiation results, as sketched in Fig. 2.6. Since the sample under illumination is usually small, much of the radiation hits the screen and is lost, especially at large distances from the source. Furthermore the synchrotron beam does not have sharp horizontal edges. Thus the extremely small opening angle $\Theta = 1/\gamma$ can only be exploited in the vertical plane. Because of this problem, attention soon turned to how to achieve both high photon density and strong horizontal bunching of the radiation. The solution nowadays is to use a periodic arrangement of special magnets with poles of alternating polarity. The principle behind these magnets, called **wigglers** or **undulators**, is illustrated in Fig. 2.7.

Undulators and wigglers use the same basic principle and differ only in the strength of their bending field, with the undulator generally being much weaker than the wiggler. A distinction is only made between the two versions because of the different characteristics of the radiation they produce. This question will be discussed in detail in Chapter 8. Here it is sufficient to remark that the strong wiggler produces a spectrum similar to that from a bending magnet, while in the undulator, because of the weaker bending, constructive coherent radiation is produced with very high intensity. It is precisely this property that makes the undulator so useful.

The intensity of radiation produced by the undulators and wigglers depends essentially on their length and number of periods. Nowadays magnets of this kind may be several metres in length. As they do not contribute to the bending of the beam in the storage ring, long straight sections must be provided within the accelerator to accommodate them. In existing high energy machines such straight sections are short or non-existent. As a result, special 'third generation' storage rings were developed, with long sections kept free for wigglers and undulators. Here the only purpose of the bending magnets is to steer the electron beam along a closed path from one wiggler to the next (Fig. 2.8). Several storage rings

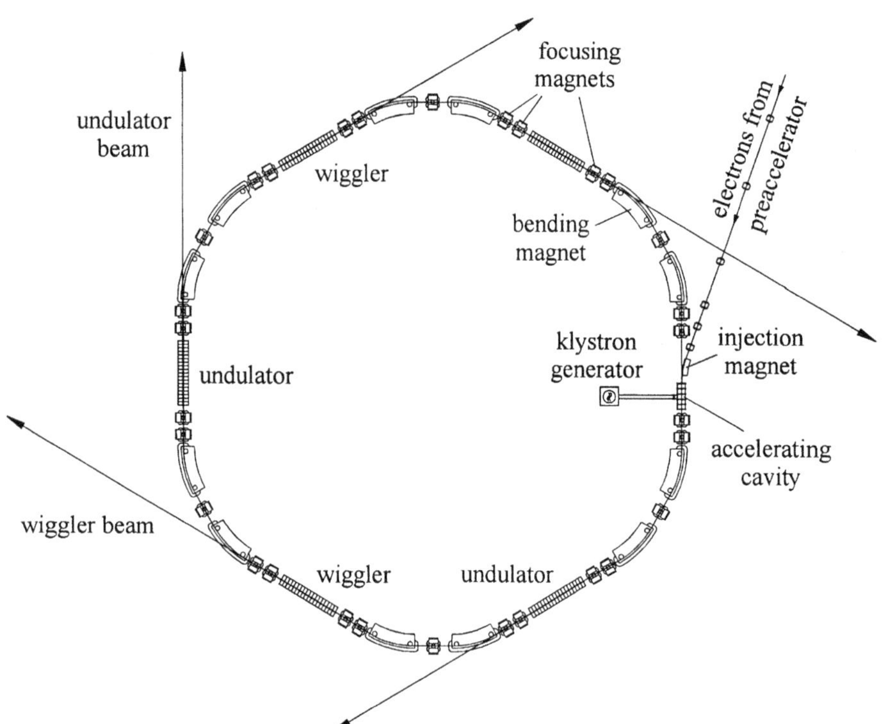

Fig. 2.8 Layout of a 'third generation' storage ring with long straight sections to accommodate wigglers and undulators.

have been constructed which use wigglers and undulators. Some are smaller machines with beam energies between 1.5 and 2 GeV, such as the Advanced Light Source in Berkeley (USA) [42], the Trieste storage ring ELETTRA [43], and the Dortmund storage ring DELTA [44]. Then there are large machines with maximum energies of 6 to 8 GeV, which deliver radiation mainly in the X-ray region. Some are being built through international collaboration, among them the European Synchrotron Radiation Facility in Grenoble (France) [45], the Advanced Photon Source at the Argonne National Laboratory (USA) [46] and the SPring-8 project in Japan [47].

3

Linear beam optics

When an accelerator is constructed, the nominal trajectory of the particle beam is fixed. This trajectory may simply be a straight line, as is the case in linear accelerators. In circular machines such as storage rings, however, it may have a very complicated shape consisting of numerous curves connected by straight sections of various lengths. The particles follow the resulting closed path over and over again and so cover an extremely large total path length within the accelerator. On the other hand, the trajectories of individual particles within a beam always have a certain angular divergence, and without further measures the particles would eventually hit the wall of the vacuum chamber and be lost.

It is therefore necessary first of all to fix the particle trajectory, in general an arbitrary curve, and then to repeatedly steer the diverging particles back onto the ideal trajectory. In most general terms, this is done by means of electromagnetic fields ($\boldsymbol{E}$ and $\boldsymbol{B}$), in which particles of charge e and velocity $\boldsymbol{v}$ experience the **Lorentz force**

$$\boldsymbol{F} = e(\boldsymbol{E} + \boldsymbol{v}\times\boldsymbol{B}) = \dot{\boldsymbol{p}}. \tag{3.1}$$

At relativistic velocities electric fields $\boldsymbol{E}$ and magnetic fields $\boldsymbol{B}$ have the same effect if $\boldsymbol{E} = c\boldsymbol{B}$. That is to say, a magnetic field of strength $B = 1$ T is equivalent to an electric field of strength $E = 3\times 10^8\ \mathrm{V\,m^{-1}}$. Nowadays it is relatively easy to produce magnetic fields over 1 T by conventional means, whereas an electric field of $3\times 10^8\ \mathrm{V\,m^{-1}}$ is far beyond technical limits. As a result, magnets are almost always used to steer the beams in modern accelerators. Electric fields are employed only at very low energies, or to separate particles according to their charge.

The physical fundamentals of beam steering and focusing, which by analogy with light are called **beam optics**, were developed by Courant and Snyder [48]. Further treatments and reviews may be found in Persico, Ferrari and Segre [49], Steffen [50], Sands [51], Kolomenski and Lebedev [52], Wilson [53] and Bruck [54]. The reader is also referred to the Proceedings of the CERN Accelerator School (CAS) [55].

3.1 Charged particle motion in a magnetic field

To describe the motion of a particle in the vicinity of the nominal trajectory we introduce a Cartesian coordinate system $K = (x, z, s)$ whose origin moves

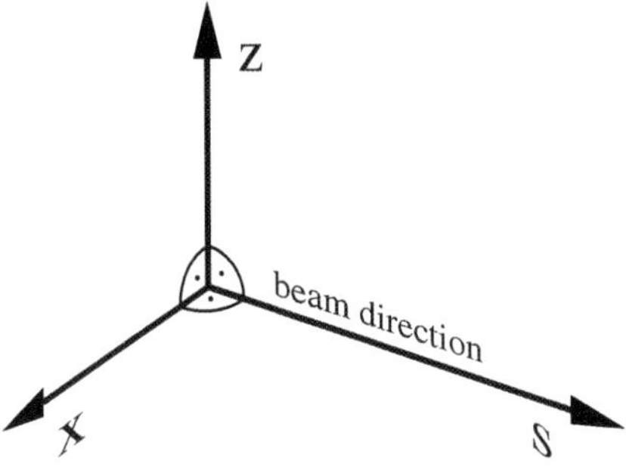

Fig. 3.1 Coordinate system to describe the motion of particles in the vicinity of the nominal trajectory.

along the trajectory of the beam (Fig. 3.1). The axis along the beam direction is s, while the horizontal and vertical axes are labelled x and z respectively. For simplicity we will assume that the particles move essentially parallel to the s-direction, i.e. $\boldsymbol{v} = (0, 0, v_s)$, and that the magnetic field only has transverse components and so has the form $\boldsymbol{B} = (B_x, B_z, 0)$. For a particle moving in the horizontal plane through the magnetic field there is then a balance between the Lorentz force $F_x = -ev_sB_z$ and the centrifugal force $F_r = mv_s^2/R$. Here m is the particle mass and R is the radius of curvature of the trajectory. Using $p = mv_s$, this balance of forces leads directly to the relation

$$\frac{1}{R(x,z,s)} = \frac{e}{p} B_z(x,z,s). \tag{3.2}$$

There is a corresponding expression for the vertical deflection. Since the transverse dimensions of the beam are small compared to the radius of curvature of the particle trajectory, we may expand the magnetic field in the vicinity of the nominal trajectory:

$$B_z(x) = B_{z0} + \frac{dB_z}{dx}x + \frac{1}{2!}\frac{d^2B_z}{dx^2}x^2 + \frac{1}{3!}\frac{d^3B_z}{dx^3}x^3 + \dots . \tag{3.3}$$

Multiplying by e/p

$$\begin{array}{ccccccccccc} \frac{e}{p}B_z(x) & = & \frac{e}{p}B_{z0} & + & \frac{e}{p}\frac{dB_z}{dx}x & + & \frac{1}{2!}\frac{e}{p}\frac{d^2B_z}{dx^2}x^2 & + & \frac{1}{3!}\frac{e}{p}\frac{d^3B_z}{dx^3}x^3 & + \dots \\ & = & \frac{1}{R} & + & kx & + & \frac{1}{2!}mx^2 & + & \frac{1}{3!}ox^3 & + \dots . \\ & & \text{dipole} & & \text{quadrupole} & & \text{sextupole} & & \text{octupole} & \end{array} \tag{3.4}$$

The magnetic field around the beam may therefore be regarded as a sum of multipoles, each of which has a different effect on the path of the particle. The most important multipoles and their effects are listed in Table 3.1. If only the two lowest multipoles are used for beam steering in an accelerator then one speaks of **linear beam optics**, since the only bending forces present are either constant

Table 3.1 The most important multipoles in beam steering and their principal effects on the motion of the beam.

multipole	definition	effect
dipole	$\frac{1}{R} = \frac{e}{p} B_{z0}$	beam steering
quadrupole	$k = \frac{e}{p} \frac{dB_z}{dx}$	beam focusing
sextupole	$m = \frac{e}{p} \frac{d^2 B_z}{dx^2}$	chromaticity compensation
octupole	$o = \frac{e}{p} \frac{d^3 B_z}{dx^3}$	field errors or field compensation
etc.		

(dipole field, entering through the bending radius R) or increase linearly with the transverse displacement from the ideal trajectory (quadrupole field, described by the quadrupole strength k). Higher multipoles (sextupole, octupole etc.) are either unwanted field errors or are deliberately introduced for compensation or field correction. We will begin by considering the physics of linear beam optics, since this is the foundation of all beam steering.

3.2 Equation of motion in a co-moving coordinate system

The transverse beam size is in general very small compared to the dimensions of an accelerator. As a result it is useful to consider the motion of individual

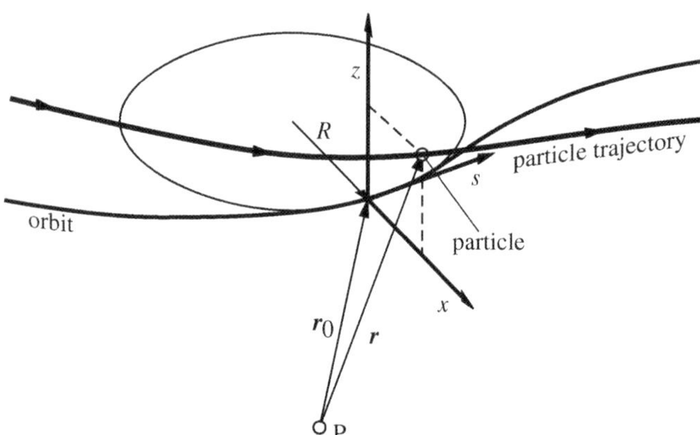

Fig. 3.2 Rotated co-moving coordinate system to describe the beam motion relative to the orbit.

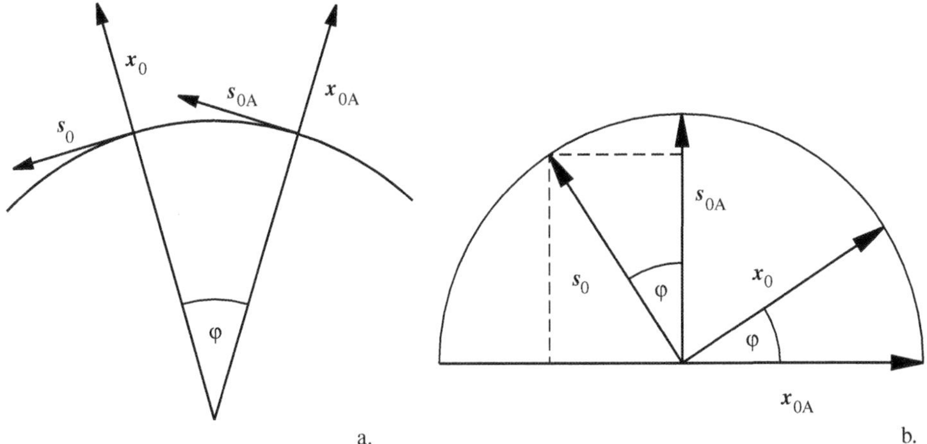

Fig. 3.3 Rotation of a Cartesian coordinate system (x, z, s) around the z-axis.

particles only in the immediate vicinity of the ideal trajectory. This ideal trajectory, which is fixed by the construction of the accelerator, is termed the **orbit**. The trajectory of an individual particle is described using a coordinate system $K = (x, z, s)$ whose origin moves along the orbit and follows the longitudinal motion of the particle (Fig. 3.2). In this system x is the horizontal and z the vertical position of the particle relative to the orbit.

In regions where the beam is bent by a magnetic field, the coordinate system must be rotated accordingly. Our first important task is to develop, in this rotated coordinate system, a general equation of particle motion due to the Lorentz force (3.1). Let us begin by collecting a few important properties of the rotated coordinate system, which may easily be derived with the help of Fig. 3.3.

We start by noting that the beam steering only acts in the horizontal plane, i.e. the coordinate system only rotates around the z-axis. Initially the system is fixed in the x-s plane with unit vectors $\boldsymbol{x}_{0\mathrm{A}}$ and $\boldsymbol{s}_{0\mathrm{A}}$ which, following a rotation through the angle φ, are transformed into the vectors $\boldsymbol{x}_0$ and $\boldsymbol{s}_0$. This transformation can be directly obtained by inspecting Fig. 3.3:

$$
\begin{aligned}
\boldsymbol{x}_0 &= \boldsymbol{x}_{0\mathrm{A}} \cos\varphi + \boldsymbol{s}_{0\mathrm{A}} \sin\varphi \\
\boldsymbol{s}_0 &= -\boldsymbol{x}_{0\mathrm{A}} \sin\varphi + \boldsymbol{s}_{0\mathrm{A}}. \cos\varphi.
\end{aligned} \tag{3.5}
$$

Differentiating these vectors with respect to φ yields

$$
\frac{d\boldsymbol{s}_0}{d\varphi} = -\boldsymbol{x}_0 \qquad\qquad \frac{d\boldsymbol{x}_0}{d\varphi} = \boldsymbol{s}_0. \tag{3.6}
$$

The path element $ds = R d\varphi$ of a curved trajectory immediately gives

$$
\frac{d\varphi}{dt} = \frac{1}{R}\frac{ds}{dt}. \tag{3.7}
$$

With the help of these relations we may calculate the time derivatives of the unit vectors:

$$\begin{aligned} \dot{\boldsymbol{x}}_0 &= \frac{d\boldsymbol{x}_0}{d\varphi}\frac{d\varphi}{dt} = \frac{1}{R}\dot{s}\boldsymbol{s}_0 \\ \dot{\boldsymbol{s}}_0 &= \frac{d\boldsymbol{s}_0}{d\varphi}\frac{d\varphi}{dt} = -\frac{1}{R}\dot{s}\boldsymbol{x}_0 \\ \dot{\boldsymbol{z}}_0 &= 0. \end{aligned} \tag{3.8}$$

The origin of coordinates moves along the orbit by $d\boldsymbol{r}_0 = \boldsymbol{s}_0\, ds$ and it follows that $\dot{\boldsymbol{r}}_0 = \dot{s}\, \boldsymbol{s}_0$. The position vector $\boldsymbol{r}$ of the particle can be written in the general form

$$\boldsymbol{r} = \boldsymbol{r}_0 + x\, \boldsymbol{x}_0 + z\, \boldsymbol{z}_0. \tag{3.9}$$

To formulate the equation of motion we also need the first and second time derivatives, which we may easily obtain using equations (3.8):

$$\begin{aligned} \dot{\boldsymbol{r}} &= \dot{x}\, \boldsymbol{x}_0 + \dot{z}\, \boldsymbol{z}_0 + \left(1 + \frac{x}{R}\right) \dot{s}\, \boldsymbol{s}_0 \\ \ddot{\boldsymbol{r}} &= \left[\ddot{x} - \left(1 + \frac{x}{R}\right)\frac{\dot{s}^2}{R}\right] \boldsymbol{x}_0 + \ddot{z}\, \boldsymbol{z}_0 + \left[\frac{2}{R}\dot{x}\dot{s} + \left(1 + \frac{x}{R}\right)\ddot{s}\right] \boldsymbol{s}_0. \end{aligned} \tag{3.10}$$

As the particle travels through the magnet structure, its position s is uniquely defined for any time t. This allows us to replace the time derivative by a derivative with respect to the spatial coordinate s as follows:

$$\begin{aligned} \dot{x} &= \frac{dx}{ds}\frac{ds}{dt} = x'\, \dot{s} \\ \ddot{x} &= \dot{x}'\, \dot{s} + x'\, \ddot{s} = x''\, \dot{s}^2 + x'\, \ddot{s}. \end{aligned} \tag{3.11}$$

Using these relations it follows from (3.10) that

$$\begin{aligned} \dot{\boldsymbol{r}} &= x'\dot{s}\boldsymbol{x}_0 + z'\dot{s}\boldsymbol{z}_0 + \left(1 + \frac{x}{R}\right)\dot{s}\boldsymbol{s}_0 \\ \ddot{\boldsymbol{r}} &= \left[x''\dot{s}^2 + x'\ddot{s} - \left(1 + \frac{x}{R}\right)\frac{\dot{s}^2}{R}\right]\boldsymbol{x}_0 + (z''\dot{s}^2 + z'\ddot{s})\boldsymbol{z}_0 \\ &\quad + \left[\frac{2}{R}x'\dot{s}^2 + \left(1 + \frac{x}{R}\right)\ddot{s}\right]\boldsymbol{s}_0. \end{aligned} \tag{3.12}$$

This is the general expression for the trajectory vector $\boldsymbol{r}$ in a co-moving rotated coordinate system. Substituting $\dot{\boldsymbol{p}} = m\, \ddot{\boldsymbol{r}}$ and $\boldsymbol{v} = \dot{\boldsymbol{r}}$ into (3.1), we obtain the equation of motion of a charged particle in a pure magnetic field in the form

$$\ddot{\boldsymbol{r}} = \frac{e}{m}\left(\dot{\boldsymbol{r}} \times \boldsymbol{B}\right). \tag{3.13}$$

We now assume that only the transverse components of the magnetic field are non-zero, i.e. $\boldsymbol{B} = (B_x, B_z, 0)$. This assumption is generally very closely satisfied in particle accelerators. From (3.13) it then follows that

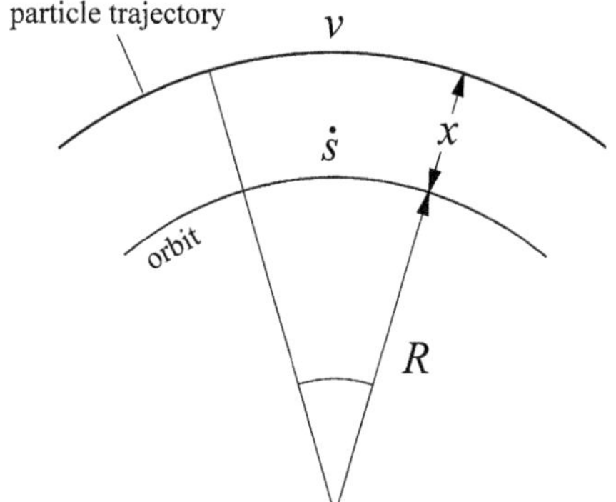

Fig. 3.4 Comparison between the velocity v of an individual particle and the velocity $\dot{s}$ of the co-moving coordinate system along the orbit.

$$\ddot{\boldsymbol{r}} = \frac{e}{m}(\dot{\boldsymbol{r}} \times \boldsymbol{B}) = \frac{e}{m} \begin{pmatrix} -\left(1+\frac{x}{R}\right)\dot{s}B_z \\ \left(1+\frac{x}{R}\right)\dot{s}B_x \\ x'\dot{s}B_z - z'\dot{s}B_x \end{pmatrix}. \tag{3.14}$$

Since the particles are moving with relativistic velocities, the effect of the magnetic field on their longitudinal velocity is negligible. In what follows we therefore need to consider only the transverse components x and z. Equations (3.12) and (3.14) then lead to the relations

$$\begin{aligned} x''\dot{s}^2 + x'\ddot{s} - \left(1+\frac{x}{R}\right)\frac{\dot{s}^2}{R} &= -\frac{e}{m}B_z\left(1+\frac{x}{R}\right)\dot{s} \\ z''\dot{s}^2 + z'\ddot{s} &= \frac{e}{m}B_x\left(1+\frac{x}{R}\right)\dot{s}. \end{aligned} \tag{3.15}$$

We can simplify these relations by assuming that that the velocity of the particles only varies very slowly as they pass through the magnetic field, and so $\ddot{s} \approx 0$. Furthermore we set $p = mv$. As may be seen from Fig. 3.4, $v \neq \dot{s}$. Instead simple geometry gives

$$v = \dot{s}\,\frac{R+x}{R} = \dot{s}\left(1+\frac{x}{R}\right). \tag{3.16}$$

Using this expression, (3.15) yields

$$\begin{aligned} x'' - \left(1+\frac{x}{R}\right)\frac{1}{R} &= -\frac{v}{\dot{s}}\frac{e}{p}B_z\left(1+\frac{x}{R}\right) \\ z'' &= \frac{v}{\dot{s}}\frac{e}{p}B_x\left(1+\frac{x}{R}\right). \end{aligned} \tag{3.17}$$

We now also assume that the particles have a well-defined momentum $p = p_0 + \Delta p$, where the momentum deviation Δp is very small compared to the nominal momentum p_0. This condition is very well satisfied in accelerators, with the relative momentum deviation $\Delta p/p$ in the beam generally being rather less than 1%. We are therefore justified in writing, to first order:

$$\frac{1}{p} = \frac{1}{p_0}\left(1 - \frac{\Delta p}{p_0}\right). \tag{3.18}$$

By analogy with (3.4) we describe the magnetic fields using the energy invariant dipole strength $1/R$ and quadrupole strength k, and consider only these linear terms in what follows. Let us now further assume that particles are only deflected in the horizontal plane, i.e. only horizontally acting dipoles fields are present. We then obtain

$$\frac{e}{p_0}B_z = \frac{1}{R} - k\,x \qquad\qquad \frac{e}{p_0}B_x = -k\,z. \tag{3.19}$$

The sign of the quadrupole strength k is arbitrary, and we choose a convention in which $k < 0$ if the quadrupole is **focusing** and $k > 0$, if it is **defocusing**. Substituting (3.18) and (3.19) into (3.17) and for simplicity writing p_0 as p, we obtain

$$\begin{aligned} x'' - \left(1+\frac{x}{R}\right)\frac{1}{R} &= -\left(1+\frac{x}{R}\right)^2\left(\frac{1}{R} - kx\right)\left(1 - \frac{\Delta p}{p}\right) \\ z'' &= -\left(1+\frac{x}{R}\right)^2 kz\left(1 - \frac{\Delta p}{p}\right). \end{aligned} \tag{3.20}$$

We multiply out the brackets on the right hand side and neglect all terms containing squares or products of x, z, and $\Delta p/p$. This is justified because $x \ll R$, $z \ll R$, and $\Delta p/p \ll 1$. With these simplifications we finally obtain the linear equations of motion for a particle travelling through the magnetic structure of an accelerator:

$$\boxed{\begin{aligned} x''(s) + \left(\frac{1}{R^2(s)} - k(s)\right)x(s) &= \frac{1}{R(s)}\frac{\Delta p}{p} \\ z''(s) + k(s)z(s) &= 0. \end{aligned}} \tag{3.21}$$

These equations form the basis of calculations in linear beam optics.

3.3 Beam steering magnets

Before turning to the solution of the trajectory equations (3.21) we first consider how the magnetic fields needed to bend and focus the beam are produced. We will discuss how magnets must be shaped in order to produce a particular field, such as a dipole field, quadrupole field and so on, with a strength $1/R$, k, m etc. We will only consider purely static fields here.

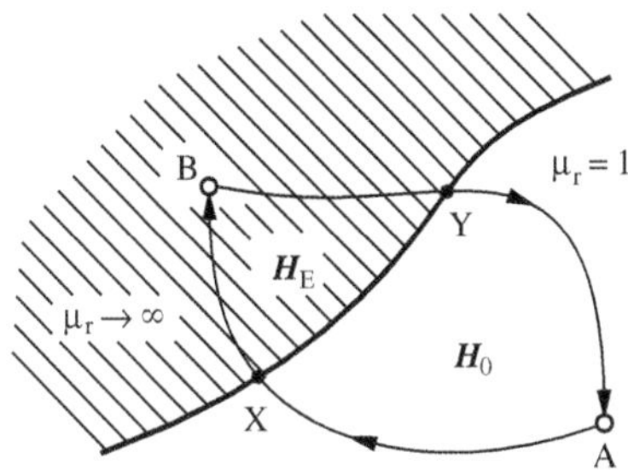

Fig. 3.5 Calculation of the magnetic potential at the boundary between a vacuum ($\mu_\mathrm{r} = 1$) and a highly permeable material with $\mu_\mathrm{r} \to \infty$.

3.3.1 Calculation of magnetic fields for beam steering

The basis of all calculations of static magnetic fields $\boldsymbol{H}$ is **Maxwell's equation**

$$\nabla \times \boldsymbol{H} = \boldsymbol{j}, \tag{3.22}$$

where $\boldsymbol{j}$ is the current density. In what follows, the field shape is important only in the vicinity of the beam, where no current is flowing, and so

$$\nabla \times \boldsymbol{H} = 0. \tag{3.23}$$

In this case $\boldsymbol{H}$ may be written in terms of a **scalar potential** φ as follows

$$\boldsymbol{H} = \nabla\varphi, \tag{3.24}$$

since $\nabla \times \nabla\varphi = 0$ always holds. Fixing the potential $\varphi(x, z, s)$ defines the field uniquely. For simplicity we will assume that the field only has transverse components, and that the structure of the field does not change along the beam axis s. This is true in general. The calculation of the magnetic field is thus reduced to solving a two-dimensional problem in the x-z plane. To produce a particular potential function $\varphi(x, z)$ it is necessary to fix the shape of the equipotentials $\varphi(x, z) = \text{const}$. Surfaces made of highly permeable materials are most suitable, especially iron, which has a relative permeability $\mu_\mathrm{r} > 1000$. If one moves along an arbitrary path from a point A outside the iron to a point B lying inside the iron, and then back along another arbitrary path to point A (Fig. 3.5), the total potential difference traversed is $\oint \boldsymbol{H} \cdot d\boldsymbol{s} = 0$. If the two points of entry into the surface of the iron are labelled X and Y, then one obtains

$$\oint \boldsymbol{H} \cdot d\boldsymbol{s} = \int_\mathrm{A}^\mathrm{X} \boldsymbol{H}_0 \cdot d\boldsymbol{s} + \int_\mathrm{X}^\mathrm{B} \boldsymbol{H}_\mathrm{Fe} \cdot d\boldsymbol{s} + \int_\mathrm{B}^\mathrm{Y} \boldsymbol{H}_\mathrm{Fe} \cdot d\boldsymbol{s} + \int_\mathrm{Y}^\mathrm{A} \boldsymbol{H}_0 \cdot d\boldsymbol{s} = 0. \tag{3.25}$$

Here $\boldsymbol{H}_0$ is the field in the vacuum and $\boldsymbol{H}_\mathrm{Fe}$ is that in the iron. The relation between the two at the surface is

$$|\boldsymbol{H}_\mathrm{Fe}| = \frac{1}{\mu_\mathrm{r}} |\boldsymbol{H}_0|. \tag{3.26}$$

Since μ_r is generally very large in iron, the contribution to the potential from $\boldsymbol{H}_{\text{Fe}}$ may be neglected. It then follows from (3.25) that

$$\int_{A}^{X} \boldsymbol{H}_0 \cdot d\boldsymbol{s} = \int_{A}^{Y} \boldsymbol{H}_0 \cdot d\boldsymbol{s}. \tag{3.27}$$

The points X and Y thus lie on an surface of equal potential. Since we have made no assumptions about the position of X and Y on the boundary surface, (3.27) tells us that all points on this surface are at the same potential. The surface of a highly permeable material such as iron therefore forms an **equipotential**. By carefully choosing the shape of the magnet poles it is possible to fix the potential $\varphi(x,z)$ and, as a consequence of (3.24), also the magnetic field $\boldsymbol{H}$.

In what follows it is more convenient to use the magnetic flux density $\boldsymbol{B} = \mu_r\mu_0\boldsymbol{H}$ rather than the magnetic field $\boldsymbol{H}$. By analogy with (3.24) we define a potential $\Phi(x,z) = \mu_r\mu_0\varphi(x,z)$, from which the magnetic flux density may be calculated according to

$$\boldsymbol{B} = \nabla\Phi. \tag{3.28}$$

Using the Maxwell equation $\nabla\boldsymbol{B} = 0$ the **Laplace equation** immediately follows,

$$\nabla^2\Phi = 0. \tag{3.29}$$

The equations (3.28) and (3.29) form the theoretical basis of the design of iron pole magnets. Here the first task is to fix the shape of the transverse field required for beam steering. It is sufficient to determine the form of just one field component along a particular axis, for example the vertical component $G_z(x)$ along the x-axis ($z \equiv 0$). This then uniquely determines the entire field distribution $\boldsymbol{B}(x,z)$. Taking the case of G_z, we may use the following general expression for the form of the z component of the field

$$B_z(x,z) = G_z(x) + f(z). \tag{3.30}$$

Here $f(z)$ is an unknown function, which only describes that part of the field which depends on the vertical coordinate z. The potential is then

$$\Phi(x,z) = \int B_z\,dz = G_z(x)\,z + \int f(z)\,dz. \tag{3.31}$$

The unknown function $f(z)$ can only be calculated by means of the Laplace equation (3.29). This gives

$$\nabla^2\Phi = \frac{\partial^2\Phi}{\partial x^2} + \frac{\partial^2\Phi}{\partial z^2} = \frac{d^2G_z(x)}{dx^2}\,z + \frac{df(z)}{dz} = 0, \tag{3.32}$$

from which it immediately follows that

$$f(z) = -\int \frac{d^2G_z(x)}{dx^2}\,z\,dz = -\frac{1}{2}\frac{d^2G_z(x)}{dx^2}\,z^2. \tag{3.33}$$

Substituting this expression into (3.31), we finally obtain the potential

$$\boxed{\Phi(x,z) = G_z(x)\,z - \frac{1}{6}\frac{d^2G_z(x)}{dx^2}\,z^3.} \tag{3.34}$$

From this we can calculate the final field distribution using

$$\boldsymbol{B}(x,z) = \begin{pmatrix} \dfrac{d\Phi(x,z)}{dx} \\ \dfrac{d\Phi(x,z)}{dz} \end{pmatrix}. \tag{3.35}$$

Using equations (3.34) and (3.35), the potential and the magnetic field in the entire x-z plane can be calculated for any field shape $G_z(x)$ along the x-axis.

3.3.2 Conventional ferromagnets

We will now use the procedure we have just developed to describe the most important types of conventional magnet, in which the poles are made of iron and the field is generated by current flowing through windings.

To bend charged particles around a circular path, **dipole magnets** are used. These have a constant field along the x-axis, which we determine as follows:

$$G_z(x) = B_0 = \text{const} \qquad \longrightarrow \qquad \frac{d^2G_z(x)}{dx^2} = 0. \tag{3.36}$$

The required potential then follows immediately from (3.34)

$$\Phi(x,z) = B_0\,z. \tag{3.37}$$

The equipotential line $\Phi(x,z) = \Phi_0 = \text{const}$ is therefore a line running parallel to the x-axis at a distance z from it. The dipole consists of two parallel iron poles of separation $h = 2z$ (Fig. 3.6).

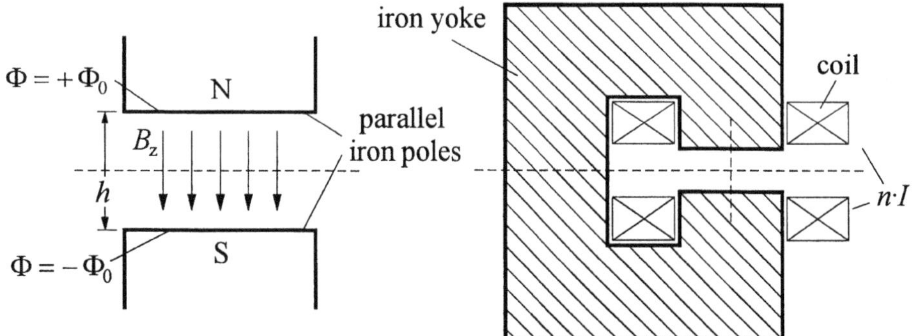

Fig. 3.6 A ferromagnet consisting of two parallel iron poles.

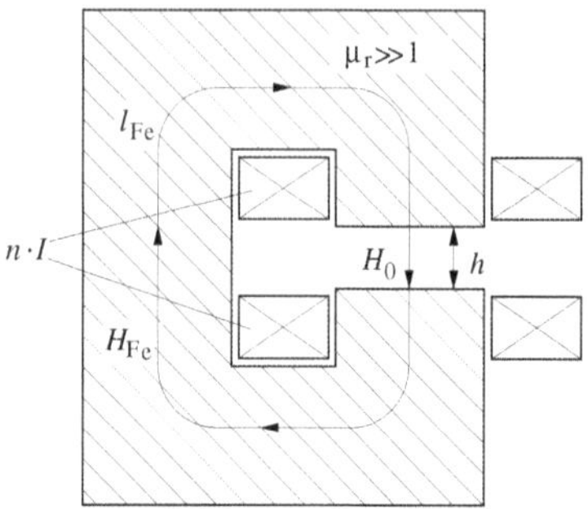

Fig. 3.7 Calculation of the relation between the current I and the magnetic field B of a dipole magnet.

The relationship between the current I and the resulting field B in the gap between the poles may be calculated most simply by using the Maxwell equation $\oint \boldsymbol{H} \cdot d\boldsymbol{s} = I_{\text{tot}}$. Let us assume that the coils have n windings in total, i.e. $I_{\text{tot}} = nI$, where I signifies the current through the coils. (Fig. 3.7). For simplicity we further assume that the field H_0 in the gap and H_{Fe} — in the iron are always constant. If we now notice that $H_{\text{Fe}} = H_0/\mu_{\text{r}}$ with $\mu_{\text{r}} \gg 1$, it follows that

$$nI = \oint \boldsymbol{H} \cdot d\boldsymbol{s} = H_{\text{Fe}} l_{\text{Fe}} + H_0 h \approx H_0 h. \tag{3.38}$$

The magnetic field in the gap is $B_0 = \mu_0 H_0$, since here $\mu_{\text{r}} = 1$. We thus obtain the following expression for the field of a dipole magnet:

$$B_0 = \mu_0 \frac{nI}{h}. \tag{3.39}$$

The dipole strength in (3.4) is then

$$\frac{1}{R} = \frac{e}{p} B_0 = \frac{e\mu_0}{p} \frac{nI}{h}. \tag{3.40}$$

In a realistic magnet there will be deviations from this ideal field. Firstly, a truly homogeneous field could only be produced by infinitely long poles and so is of course unattainable. Instead one finds that beyond a certain horizontal distance from the central axis of the magnet the field falls away, as the field lines at the edge of the poles are pushed outwards (Fig. 3.8). This restricts the size of the useful region of the field. The boundary of the field is generally taken to be the distance at which the field has decreased by a factor $\Delta B/B = 2 \times 10^{-4}$. The decrease in the field may be partly compensated for by fitting flat iron strips called shims to the ends of the poles, so increasing the useful region of the field for a given pole width. A further problem in ferromagnets is saturation, as can be seen from the excitation curve in Fig. 3.9. At low field strengths, i.e. $B < 1$ T, the relation between the magnetic field and the current is linear to a very good approximation. Above 1 T the field lags behind the current and levels off at a

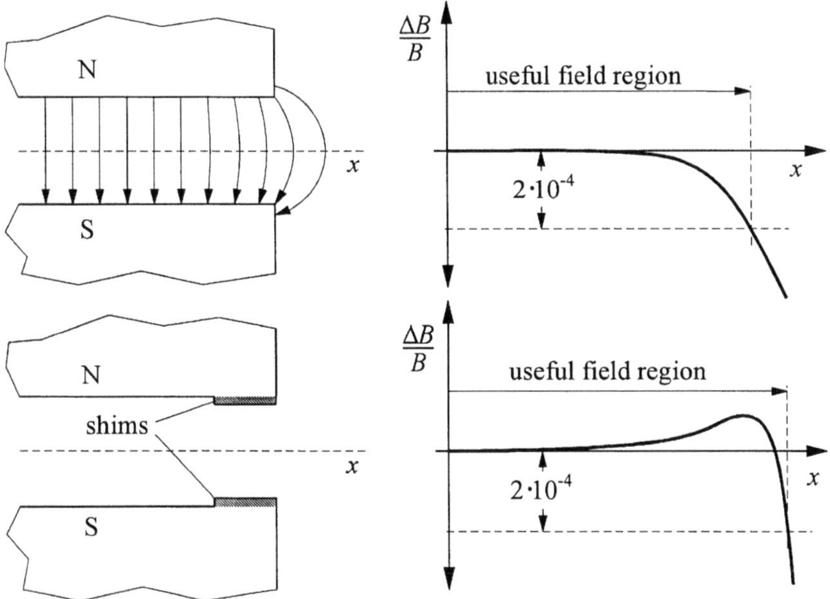

Fig. 3.8 Limitation of the useful field region due to the edge field. The useful field region can be extended somewhat by fitting iron shims to the edge of the poles.

constant value of around 2 T. It is then pointless to increase the current any further. 2 T thus represents the absolute upper limit for ferromagnets, although in practice they tend to be operated well below this value, not exceeding $B = 1.5$ T if possible. To focus the beams, **quadrupole fields** are used which, according to (3.4), disappear along the beam axis and increase linearly with transverse distance x. Their field shape may then be determined from the function

$$G_z(x) = gx \quad \text{with} \quad g = \frac{\partial B_z}{\partial x}. \tag{3.41}$$

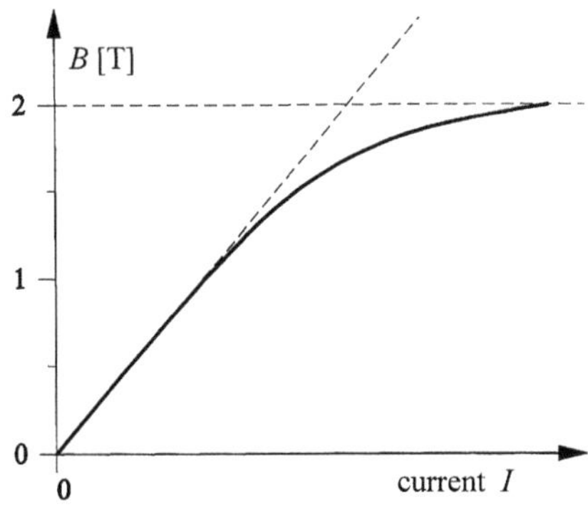

Fig. 3.9 Relation between magnetic field B and coil current I at high field strengths. Above $B = 1$ T saturation occurs, limiting the maximum field in conventional magnets to around 2 T.

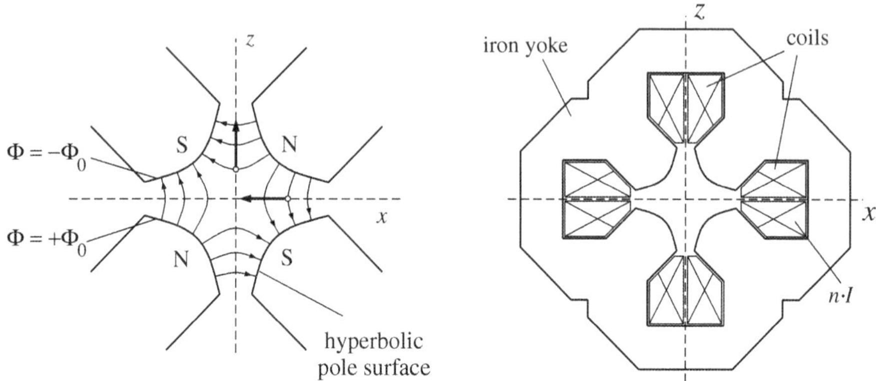

Fig. 3.10 A quadrupole magnet with hyperbolic pole surfaces.

The second derivative of G_z also disappears in this case. According to (3.34), the required potential is then

$$\Psi(x, z) \;=\; g\, x\, z. \tag{3.42}$$

It immediately follows that the equipotentials $z(x)$ for a given value of Ψ_0 are hyperbolae of the form

$$z(x) \;=\; \frac{\Psi_0}{gx}. \tag{3.43}$$

For this reason, a quadrupole consists of four poles with hyperbolic surfaces, arranged with alternating polarity North-South-North-South, as shown in Fig. 3.10. The four poles are excited by coils which surround them. The distribution of field lines between the poles causes a magnet which focuses in the horizontal plane to defocus the beam in the vertical direction. To properly focus the beam it is therefore necessary to use at least two quadrupoles, rotated through 90° relative to each other.

The relationship between the current I in the coils and the field gradient $g = \partial B_z/\partial x$ may easily be determined by again calculating the integral $\oint \boldsymbol{H} \cdot d\boldsymbol{s}$, choosing an appropriate closed integration contour around the conductor in the coil, as shown in Fig. 3.11. It begins on the beam axis (point 0) and runs through the saddle point of the pole (point 1), through the iron yoke to the x-axis (point 2), and from there back to the starting point 0.

The integral may then be broken up into pieces as follows:

$$\oint \boldsymbol{H} \cdot d\boldsymbol{s} \;=\; \int_0^1 \boldsymbol{H}_0 \cdot d\boldsymbol{s} \;+\; \int_1^2 \boldsymbol{H}_{\mathrm{Fe}} \cdot d\boldsymbol{s} \;+\; \int_2^0 \boldsymbol{H} \cdot d\boldsymbol{s} \;=\; nI. \tag{3.44}$$

The only non-zero contribution to the right-hand side of the equation comes from the integration from the beam axis to the pole $(0 \rightarrow 1)$. Within the iron yoke the integral vanishes because $\mu_{\mathrm{r}} \gg 1$ means that H_{Fe} is negligible. The integral

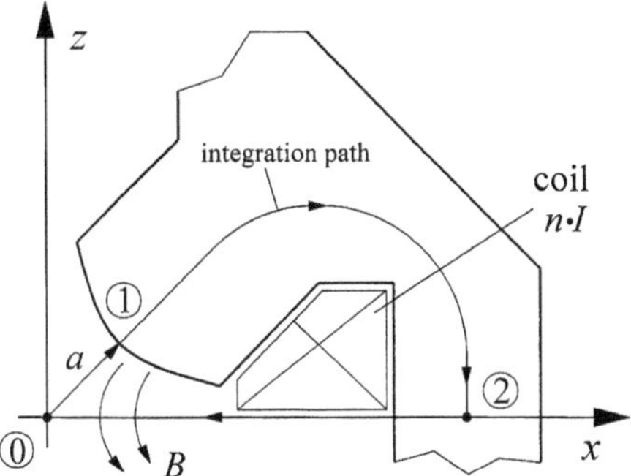

Fig. 3.11 Calculating the field gradient g of a quadrupole as a function of the current I in the coil.

along the x-axis is also zero, because here we always have $\boldsymbol{H} \perp \boldsymbol{s}$. We therefore only need to consider the field between the beam axis and the pole. This field may be obtained immediately from the expression for the potential (3.42) and is given by $B_x = gz$ and $B_z = gx$. The contribution of the magnetic field along the path of integration $0 \to 1$ is then given by

$$H = \frac{g}{\mu_0}\sqrt{x^2+z^2} = \frac{g}{\mu_0}\sqrt{2}x = \frac{g}{\mu_0}r. \tag{3.45}$$

Here we recognise that $x = z$ for this particular choice of path of integration and so r gives the line of integration from the origin. The apex of the pole is at $r = a$, allowing the integral (3.44) to be written as

$$\int_0^a H\,dr = \frac{g}{\mu_0}\int_0^a r\,dr = \frac{g}{\mu_0}\frac{a^2}{2} = nI. \tag{3.46}$$

From this it immediately follows that

$$g = \frac{2\mu_0 nI}{a^2}. \tag{3.47}$$

Since $g \propto 1/a^2$ it makes sense to keep the pole separation a as small as possible for high-strength quadrupoles, so that the current I and hence the power consumption of the magnet are reduced.

As a final example let us now consider the **sextupole magnet**, which is primarily used to compensate for chromatic abberation in strongly focusing magnetic structures. Along the x-axis this magnet has a field of the form

$$G_z(x) = \frac{1}{2}g'x^2 \qquad \text{with} \qquad \frac{d^2G_z(x)}{dx^2} = g'. \tag{3.48}$$

Notice that here the second derivative of $G_z(x)$ is not zero, as it is in cases of the dipole and quadrupole. Equation (3.33) then means that the function which describes the dependence of the field on the vertical coordinate z is non-zero, i.e. $f(z) \neq 0$. In the sextupole the particle motion in the horizontal plane is **coupled**

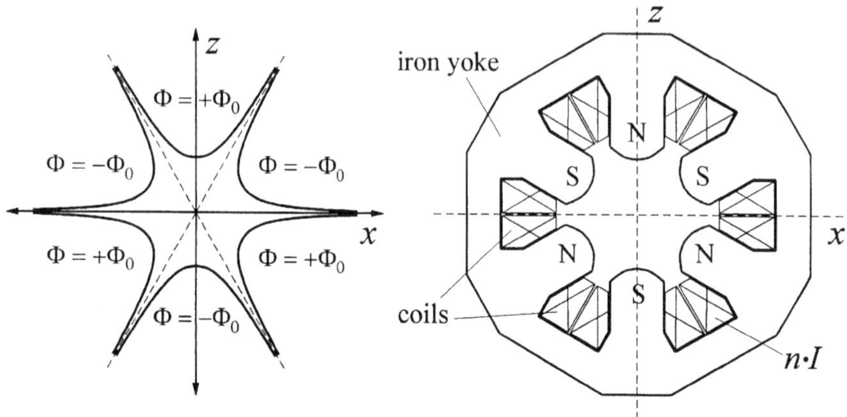

Fig. 3.12 Basic structure and equipotentials of a sextupole magnet.

to that in the vertical plane, and vice-versa. The same is true of all higher multipoles, whereas dipole and quadrupole magnets do not have this coupling. This means that for linear beam optics, in which only dipoles and quadrupoles are involved, we may calculate the particle motion in the two planes independently. Inserting the quadratic field (3.48) into the expression for the potential (3.34) gives

$$\Phi(x,z) = \frac{1}{2}g'\left[x^2 z - \frac{z^3}{3}\right]. \tag{3.49}$$

For a given constant potential Φ_0, solving this equation for $x(z)$ yields the equipotentials. In this case it is easier to fix the vertical coordinate z and to determine the associated value of x. For $z \neq 0$ it immediately follows that

$$x(z) = \sqrt{\frac{2\Phi_0}{g'z} + \frac{z^2}{3}}. \tag{3.50}$$

This expression determines the shape of the six poles, which are arranged with alternating polarity, each at an angle of 60° to the next. The basic structure of a sextupole magnet and the path of the equipotentials are shown in Fig. 3.12. The field strength along the beam axis of a sextupole with n turns per coil and a pole radius a is, for a given coil current I,

$$g' = \frac{\partial^2 B_z}{\partial x^2} = 6\mu_0 \frac{nI}{a^3}. \tag{3.51}$$

This expression is derived using the same procedure as in the case of the quadrupole, described above.

3.3.3 Superconducting magnets

The maximum field strength $B = 2$ T attainable with conventional iron magnets is not always sufficient for beam steering. In particular, proton accelerators at

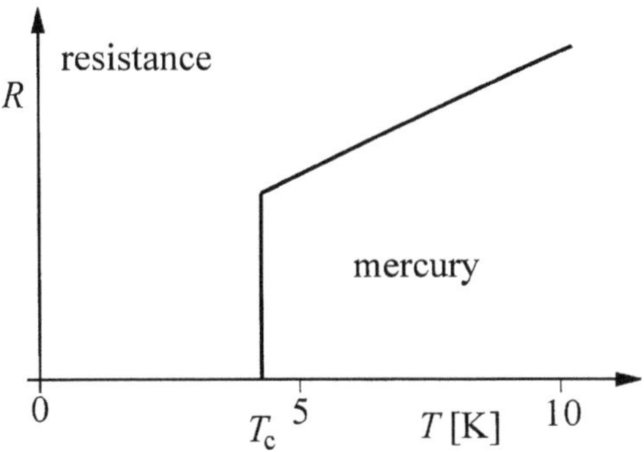

Fig. 3.13 Disappearance of ohmic resistance in mercury at very low temperatures.

energies above 100 GeV and synchrotrons producing extremely short wavelength radiation require fields of over 5 T. Such fields can only be produced by air core magnets, because the saturation effect means that iron may no longer be used. Here there is also the problem, however, that the conductors in the coils must carry an extremely high current density. If, for example, we want to produce a field of 5 T at a distance of 5 cm from a conductor, a current of $I = 1.25 \times 10^6$ A is required. If the conductor has a thickness of 3 cm this results in a current density $dI/da \approx 1700\ \mathrm{A\,mm^{-2}}$. Bearing in mind that even with very effective water cooling the maximum tolerable current density in copper is around $100\ \mathrm{A\,mm^{-2}}$, we immediately see that such a requirement cannot be satisfied using normal conductors. The problem lies in the ohmic resistance of the material, which above a certain current density results in such strong heating of the conductor that cooling cannot prevent it from melting.

The solution to this problem lies in the use of **superconductors**, which were discovered in 1911 by the Dutchman H. Kamerling Onnes. He observed that when mercury is cooled to very low temperatures its ohmic resistance suddenly disappears below a critical temperature $T_c = 4.2$ K (Fig. 3.13). In this state the current effectively flows without any losses. A whole series of materials were later discovered which also exhibited superconductivity, and with higher critical temperatures. However, it is striking that the best normal conductors such as copper and silver do not become superconducting, however low the temperature. On the other hand, some alloys which at normal temperatures are extremely bad conductors are excellent superconductors. A widely used alloy is niobium–titanium, which becomes superconducting below a critical temperature of $T_c \approx 10$ K.

Superconductivity does not just depend on the temperature. It is also influenced by external magnetic fields, as the curve in Fig. 3.14 shows. A superconductor cooled to below T_c placed in an increasing magnetic field suddenly passes back into the normal conducting state. Above a critical field strength $B_c(0)$, superconductivity cannot be achieved, even at absolute zero. The dependence of the critical field strength on temperature is given to a good approximation by

$$B_c(T) = B_c(0)\left[1 - \left(\frac{T}{T_c}\right)^2\right]. \tag{3.52}$$

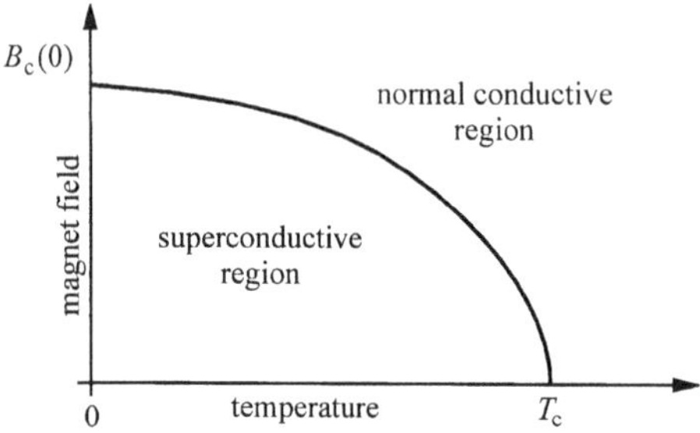

Fig. 3.14 Dependence of superconductivity on the temperature and the surrounding magnetic field.

As a result of the **Meissner-Ochsenfeld effect** the magnetic field is pushed out of the conductor at the onset of superconductivity. This means, however, that since $\nabla \times \boldsymbol{B} = \mu_0 \boldsymbol{j}$, the current itself can only flow near the surface of the conductor. Away from the surface the current density falls off exponentially, with a penetration depth of around 50 nm. In the conductor surface very high local current densities and magnetic fields result, and these fields are superimposed on the external field and push the conductor into the normal conducting state. This transition from the superconducting to normally conducting state, which occurs abruptly above a particular field strength, is known as 'quenching'. Quenching limits the maximum attainable field strength and is also one of the main problems in constructing superconducting magnets which may be operated safely. We will not go further into the physics of superconductivity here, see W. Buckel [56] for more details.

For a superconductor to produce extremely high field strengths it it necessary for the conductors to have a very large surface area. This is achieved by making the conductor out of a very large number of extremely fine filaments, as shown in Fig. 3.15. The individual niobium–titanium filaments have a thickness of around 10 μm. They are bundled into cords of several thousand filaments which are surrounded by a copper sleeve to provide mechanical strength and good cooling. As copper does not act as a superconductor, it does not contribute to the flow of current, and acts almost as an insulator. Several of these copper-sleeved cords of filaments are then bundled into a cable, usually roughly rectangular in shape, which is held together by kapton and glass ribbon. These cables, which are nowadays manufactured industrially, routinely allow current densities of $dI/da \approx 1500 \text{ A mm}^{-2}$ to be achieved in the superconductor itself, at an operating temperature of 4.6 K and within an external field of around 6 T. The overall current density in an assembly of superconducting strands with copper sleeves and insulators is of course considerably lower.

Using the superconductor technology described here it is possible to pass very high currents through comparatively thin cables and so produce the requisite high magnetic fields. As we have seen for conventional magnets in the previous section, the most important factor apart from the field strength is the field shape, i.e. the particular spatial distribution of the field in the region of the beam.

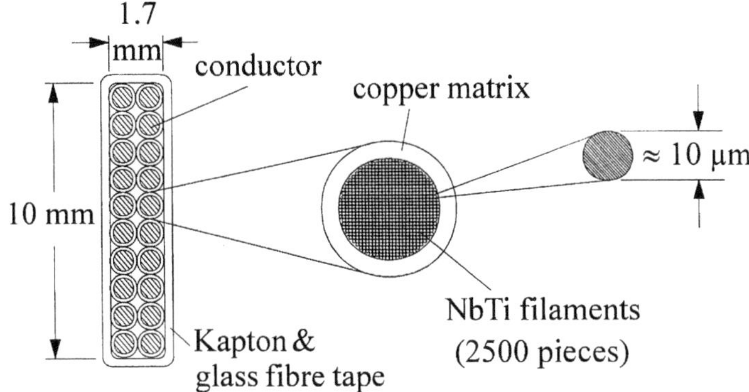

Fig. 3.15 Construction of a superconductor from niobium–titanium.

The challenge thus lies in arranging a particular number of current-carrying filaments without an iron core to produce, for example, a pure quadrupole field. We will begin by calculating the field around a single filament, assumed to be infinitely thin, and then use the principle of superposition to determine the field distribution resulting from an arrangement of many such filaments.

To calculate the static field around a conductor we start with the following fundamental equations:

$$\nabla \cdot \boldsymbol{B} = 0 \qquad \text{and} \qquad \nabla \times \boldsymbol{B} = \mu_0 \boldsymbol{j}. \tag{3.53}$$

The field $\boldsymbol{B}$ can be described by a vector potential $\boldsymbol{A}$ as

$$\boldsymbol{B} = \nabla \times \boldsymbol{A}, \tag{3.54}$$

which is consistent with the first equation in (3.53) since $\nabla \cdot (\nabla \times \boldsymbol{A}) = 0$ always holds. Let the conductor be infinitely long and run parallel to the beam axis s at a separation $a = |\boldsymbol{a}|$ (Fig. 3.16). The field $\boldsymbol{B}$ should be determined for any point P in the x-z plane, described by the position vector $\boldsymbol{r}$. Q is the point at which the conductor passes through the x-z plane. The separation between the conductor and the point P is given by the vector $\boldsymbol{R} = \boldsymbol{a} - \boldsymbol{r}$. The vectors $\boldsymbol{a}$, $\boldsymbol{r}$ and $\boldsymbol{R}$ all lie in the x-z plane perpendicular to the beam axis and the conductor. In the following we will also discuss the magnitudes of these vectors, which we will denote by a, r, and R, respectively. The vector potential $\boldsymbol{A}$ has only one component, in the s direction, since the field lines always run perpendicular to the beam axis. By integrating equation (3.54) and rearranging, we obtain the following form for the vector potential around a straight conductor carrying a current I

$$\boldsymbol{A} = \begin{pmatrix} 0 \\ 0 \\ A_s \end{pmatrix} \qquad \text{with} \qquad A_s = \frac{\mu_0 I}{2\pi} \ln R. \tag{3.55}$$

We must now find an appropriate expression for $\ln R$. Using the law of cosines we obtain the following expression for the magnitudes of the vectors:

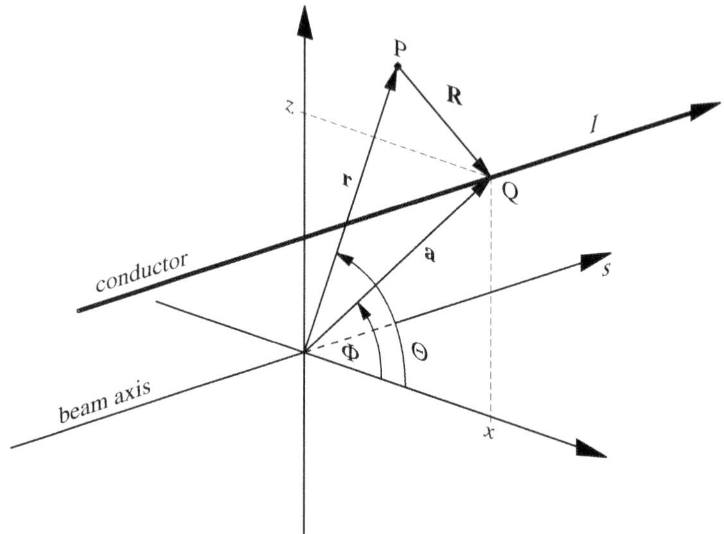

Fig. 3.16 Calculating the magnetic field around an infinitely thin, infinitely long conductor running parallel to the beam axis.

$$R^2 = a^2 + r^2 - 2ar\cos(\Phi - \Theta) = a^2\left[1 + \frac{r^2}{a^2} - 2\frac{r}{a}\cos(\Phi - \Theta)\right]. \tag{3.56}$$

Using the relation $\cos x = \cosh(ix) = \frac{1}{2}(\mathrm{e}^{ix} + \mathrm{e}^{-ix})$ this expression can be rearranged to give

$$R = a\sqrt{1 - \frac{r}{a}\mathrm{e}^{i(\Phi-\Theta)}}\sqrt{1 - \frac{r}{a}\mathrm{e}^{-i(\Phi-\Theta)}}. \tag{3.57}$$

The logarithm of this expression is

$$\ln R = \ln a + \frac{1}{2}\ln\left(1 - \frac{r}{a}\mathrm{e}^{i(\Phi-\Theta)}\right) + \frac{1}{2}\ln\left(1 - \frac{r}{a}\mathrm{e}^{-i(\Phi-\Theta)}\right). \tag{3.58}$$

It will be useful later when we come to calculate multipole fields to expand this logarithm as follows:

$$\ln(1 - z) \;=\; -\sum_{n=1}^{\infty}\frac{z^n}{n}. \tag{3.59}$$

This gives

$$\begin{aligned}
\ln R &= \ln a - \frac{1}{2}\sum_{n=1}^{\infty}\frac{1}{n}\left(\frac{r}{a}\mathrm{e}^{i\,(\Phi-\Theta)}\right)^n - \frac{1}{2}\sum_{n=1}^{\infty}\frac{1}{n}\left(\frac{r}{a}\mathrm{e}^{-i\,(\Phi-\Theta)}\right)^n \\
&= \ln a - \frac{1}{2}\sum_{n=1}^{\infty}\frac{1}{n}\left(\frac{r}{a}\right)^n\left(\mathrm{e}^{i\,n(\Phi-\Theta)} + \mathrm{e}^{-i\,n(\Phi-\Theta)}\right) \\
&= \ln a - \sum_{n=1}^{\infty}\frac{1}{n}\left(\frac{r}{a}\right)^n\cos\Big[n(\Phi-\Theta)\Big].
\end{aligned} \tag{3.60}$$

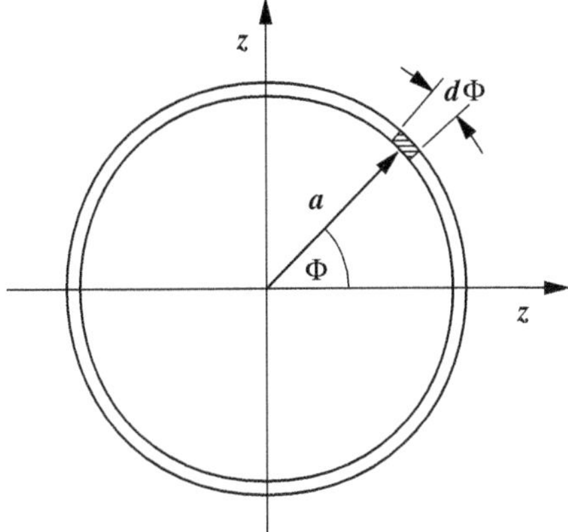

Fig. 3.17 Current along a cylindrical conductor with a magnitude dependent on the azimuthal angle Φ.

Inserting this into (3.55) finally yields the required vector potential in cylindrical coordinates

$$A_s(r,\Theta) = \frac{\mu_0 I}{2\pi}\sum_{n=1}^{\infty}\frac{1}{n}\left(\frac{r}{a}\right)^n \cos\Big[n(\Phi-\Theta)\Big]. \tag{3.61}$$

The constant term $\ln a$ is ignored here, since it does not contribute to the field. To calculate the field configuration resulting from an arrangement of a large number of conductors we will consider a conducting cylinder, in which current flows parallel to the axis. This axis coincides with the beam axis (Fig. 3.17). Let the current flowing along the cylinder surface have the following azimuthal angular distribution:

$$dI(\Phi) = I_0\cos(m\Phi)\, d\Phi \qquad m = 1,2,3,\ldots. \tag{3.62}$$

Inserting this current distribution into (3.61) and integrating over all current elements dI yields

$$A_s(r,\Theta) = \frac{\mu_0 I_0}{2\pi}\sum_{n=1}^{\infty}\frac{1}{n}\left(\frac{r}{a}\right)^n \int_0^{2\pi}\cos\Big[n(\Phi-\Theta)\Big]\cos(m\Phi)d\Phi. \tag{3.63}$$

To calculate the integral we use the sum rule for trigonometric functions and recall that $\int_0^{2\pi}\sin(n\Phi)\cos(m\Phi)d\Phi = 0$ and $\int_0^{2\pi}\cos(n\Phi)\cos(m\Phi)d\Phi = \pi$ when $n = m$ and $= 0$ otherwise. We thus obtain

$$A_s(r,\Theta) = \frac{\mu_0 I_0}{2}\frac{1}{m}\left(\frac{r}{a}\right)^m \cos(m\Theta). \tag{3.64}$$

This is an exact solution, and does not rely on any assumptions at all. We can immediately use it along with the relation (3.54) to calculate the field in polar coordinates:

$$\begin{aligned} \boldsymbol{B}(r,\Theta) &= \nabla \times \boldsymbol{A} = \left(\frac{1}{r}\frac{\partial A_s}{\partial \Theta}\,,\; -\frac{\partial A_s}{\partial r}\right) \\ &= -\frac{\mu_0 I_0}{2a}\left(\frac{r}{a}\right)^{m-1}\Big(\sin(m\Theta)\,,\;\cos(m\Theta)\Big). \end{aligned} \tag{3.65}$$

When considering this field distribution we only need to know its form along one axis, for example the x-axis, since this is enough to uniquely determine the distribution throughout all space. We must therefore find the component $B_z(x)$ for $z \equiv 0$. This is easily achieved by setting $\Theta = 0$, so that r becomes x and the only non-vanishing field component is B_z, giving

$$B_z(x) = -\frac{\mu_0 I_0}{2a^m}\, x^{m-1}. \tag{3.66}$$

The multipole may be specified by the choice of index m, as can be seen in Table 3.2.

Thus there is in general a field distribution $B_z \propto x^{m-1}$ and hence a $2m$ pole. To produce a dipole field in practice we need to form shells of current, whose thickness varies with azimuthal angle as $\cos(m\Phi)$ at constant current density. The left-hand diagram in Fig. 3.18 illustrates this for a dipole. It is, however, not possible to build these ideal shells of current using available superconductors, and so the $\cos(m\Phi)$ dependence must be approximated, as is shown in the right-hand diagram. As a result there will inevitably be deviations from a pure dipole field near to the conductors, but in the region of the beam these differences largely cancel out.

By arranging conductors in this way it is possible to build dipole, quadrupole, and sextupole magnets which, because they are superconducting, generate considerably higher fields than conventional magnets. Superconducting magnets are, however, a great deal more expensive, since they must be completely enclosed within a cryostat which cools them down to the operating temperature of 4.6 K using liquid helium. Here the electrical cables and mechanical support structures

Table 3.2 The multipoles produced by the azimuthal current distribution in a conducting cylinder.

index	field distribution	multipole
$m = 1$	$B_z = -\dfrac{\mu_0 I_0}{2a} = \text{const}$	dipole
$m = 2$	$B_z = -\dfrac{\mu_0 I_0}{2a^2} x$	quadrupole
$m = 3$	$B_z = -\dfrac{\mu_0 I_0}{2a^3} x^2$	sextupole
etc.		

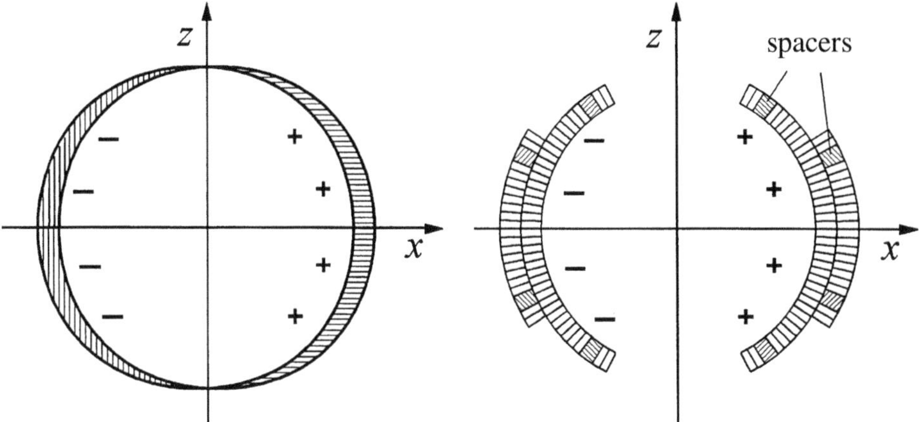

Fig. 3.18 Production of a pure dipole field by individual conductors with a current distribution proportional to $\cos(m\Phi)$.

pose a problem because they act as heat bridges. A further difficulty arises because the strong magnetic fields produce very strong forces of up to 10^6 N/m between the conductors. In addition, great care must be taken to ensure that in the event of a quench the entire coil immediately warms up and returns to the normally conducting state. This is to make certain that the energy stored in the magnet and converted to heat is evenly distributed over the entire coil rather than concentrated in the region of the quench, which would destroy the coil. Because they are so technically complex and difficult to operate, superconducting magnets are only used where it is really necessary to generate extremely high fields. If there is any doubt then conventional technology should be preferred for reasons of cost and reliability.

3.4 Particle trajectories and transfer matrices

We have shown in the preceding section how it is technically feasible to generate magnetic fields in the region of the beam with particular multipole strengths $1/R$, k, m, etc. We now turn to the solution of the trajectory equations (3.21). As we have seen, in dipole and quadrupole magnets there is no coupling between the horizontal and vertical motion of the particle. Hence it is sufficient to consider only one plane. We will choose the horizontal plane, i.e. the x-s plane.

To simplify the problem further we assume that the fields begin and end abruptly at the beginning and end of the magnets. Within the magnets we assume the fields are constant along the beam axis, i.e. $1/R$ and k are constant and independent of the s coordinate. The assumption of fields with this rectangular shape gives results which agree very well with measurements and so this **hard-edge model** is a very good approximation in calculations of the beam optics of complex structures of magnets. Now let us also assume that all particles have the nominal energy, i.e. $\Delta p/p = 0$. Using this set of assumptions we can solve the trajectory equation section by section, either within a magnet or within a so-called field-free **drift region**.

In principle it is possible to overlay different multipoles in one magnet. This is done in the classical synchrotron, for example, where a bending dipole field and a focusing quadrupole field are put together in a so-called **combined function** magnet. This very simple and cost-effective technique has the disadvantage, however, that the bending radius R and quadrupole strength k are strongly coupled and so the magnet structure may only be used for one fixed focus setting. In modern storage rings in particular, this restriction is unacceptable, and so the beam bending and focusing are performed by different **separated function** magnets. In what follows we will consider the two functions separately, for the sake of clarity.

We begin by solving for the particle trajectory in a quadrupole, which is characterized only by its strength k and length l. There is no bending of the beam ($1/R = 0$). The trajectory equation (3.21) thus simplifies to

$$x''(s) - kx(s) = 0 \qquad (k = \text{const}). \tag{3.67}$$

This homogeneous and linear second-order differential equation has the form of a normal oscillation equation which may be directly solved analytically.

In the case of a horizontally **defocusing** magnet with $k > 0$, we obtain the solution

$$\begin{aligned} x(s) &= A\cosh\sqrt{k}s + B\sinh\sqrt{k}s \\ x'(s) &= \sqrt{k}A\sinh\sqrt{k}s + \sqrt{k}B\cosh\sqrt{k}s. \end{aligned} \tag{3.68}$$

The constants of integration A and B are determined in the usual way by the initial conditions. We assume that at the start of the magnet $s = 0$ the particle trajectory has the displacement x_0 and gradient x_0' relative to the orbit. At this point the trajectory is thus defined by the trajectory vector

$$\boldsymbol{X}_0 = \begin{pmatrix} x_0 \\ x_0' \end{pmatrix} = \begin{pmatrix} x(0) \\ x'(0) \end{pmatrix}. \tag{3.69}$$

Inserting these initial conditions into the solution (3.68) immediately gives

$$\begin{aligned} x(s) &= x_0\cosh\sqrt{k}s + \frac{x_0'}{\sqrt{k}}\sinh\sqrt{k}s \\ x'(s) &= x_0\sqrt{k}\sinh\sqrt{k}s + x_0'\cosh\sqrt{k}s. \end{aligned} \tag{3.70}$$

These equations, which describe the evolution of the trajectory vector from the start of a magnet to a point s within the magnet, may also be more elegantly written in matrix notation:

$$\begin{pmatrix} x(s) \\ x'(s) \end{pmatrix} = \begin{pmatrix} \cosh\Omega & \dfrac{1}{\sqrt{k}}\sinh\Omega \\ \sqrt{k}\sinh\Omega & \cosh\Omega \end{pmatrix} \begin{pmatrix} x_0 \\ x_0' \end{pmatrix} \qquad \text{with} \qquad \Omega = \sqrt{k}s. \tag{3.71}$$

Equation (3.67) may be solved in the same way for a horizontally focusing quadrupole with $k < 0$ and for a zero-field drift region with $k = 0$. Depending on the choice of k we obtain the following transfer matrices, again with $\Omega = \sqrt{|k|}s$:

$$\mathbf{M} = \begin{cases} \begin{pmatrix} \cos\Omega & \dfrac{1}{\sqrt{|k|}}\sin\Omega \\ -\sqrt{|k|}\sin\Omega & \cos\Omega \end{pmatrix} & \text{if } k < 0 \quad \text{(focusing)} \\ \begin{pmatrix} 1 & s \\ 0 & 1 \end{pmatrix} & \text{if } k = 0 \quad \text{(drift section)} \\ \begin{pmatrix} \cosh\Omega & \dfrac{1}{\sqrt{k}}\sinh\Omega \\ \sqrt{k}\sinh\Omega & \cosh\Omega \end{pmatrix} & \text{if } k > 0 \quad \text{(defocusing)} \end{cases} \tag{3.72}$$

Calculating the determinant of these matrices yields

$$\det\mathbf{M} = 1 \tag{3.73}$$

in every case, and the same is true in general for all transfer matrices in linear beam optics.

We now turn to the calculation of the particle trajectory in a dipole magnet of constant bending radius R. We assume that the magnet has no gradient, namely $k = 0$. It follows immediately from (3.21) that by simply replacing $-k$ with $1/R^2$ in equation (3.67) we may calculate the trajectory function $x(s)$ in the same way as in a quadrupole. In this case the transfer matrix is

$$\mathbf{M}_{\text{dipole}} = \begin{pmatrix} \cos\dfrac{s}{R} & R\sin\dfrac{s}{R} \\ -\dfrac{1}{R}\sin\dfrac{s}{R} & \cos\dfrac{s}{R} \end{pmatrix}. \tag{3.74}$$

Comparing this matrix with the first matrix in (3.72), we immediately see that it describes beam focusing. This may at first seem surprising, since the dipole has no focusing field gradient. To understand this phenomenon let us consider the distribution of trajectories within a magnet with a total bending angle of 180°, as shown in Fig. 3.19. In this magnet all particle trajectories are semicircles with the same radius R. Consider a trajectory lying outside the curved orbit with a displacement $+x$ at the point where it enters the magnet (point A). The particle trajectory approaches the ideal orbit and crosses it at point B. It then runs inside the orbit and has an eventual displacement of $-x$ when it exits the magnet at point C. If we plot the trajectory as a function of the distance travelled along the ideal orbit, we immediately see that the particle is always bent in towards the orbit — in other words it is focused. Compared to specially-designed quadrupoles this focusing effect is, however, relatively weak, with $1/R^2 \ll k$ in general. For

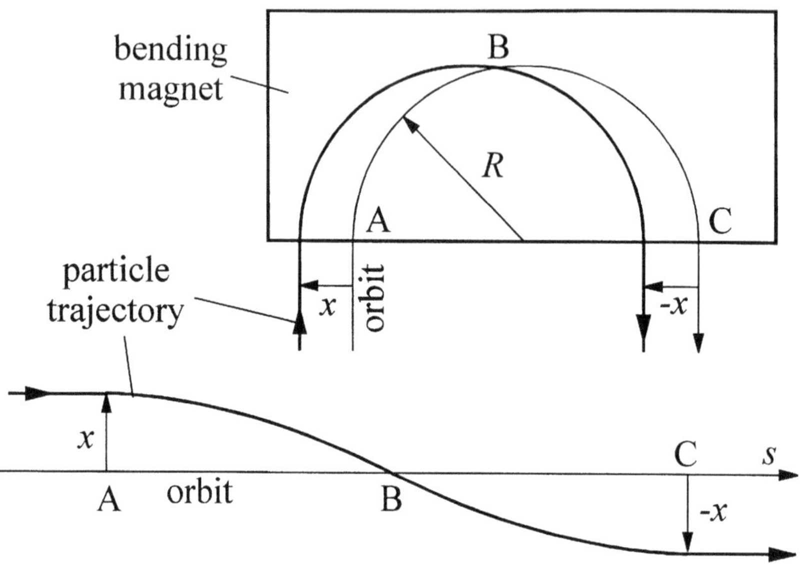

Fig. 3.19 Weak focusing in a homogeneous dipole magnet.

this reason, the focusing effect caused by dipole magnets alone is also called **weak focusing**.

Since the dipole magnets do not bend the beam in the vertical plane, they do not cause any focusing in this plane either. The particles continue undeflected, just as in a zero-field drift region. To describe the vertical evolution of the trajectory we thus use the matrix for $k = 0$ given in (3.72). In accelerators based on weak focusing, such as synchrocyclotrons, it is however possible to focus the beams in both planes simultaneously. By introducing a slight gradient in the field of the bending magnet, part of the weak focusing effect in the horizontal plane can be transferred to the vertical plane. Let us start with the equations of motion which follow from (3.21) for particles of ideal energy ($\Delta p/p = 0$):

$$\begin{aligned} x'' + \left(\frac{1}{R^2} - k\right) x &= 0 \\ z'' + kz &= 0. \end{aligned} \tag{3.75}$$

We define the **field index**

$$n \equiv R^2\, k \qquad \text{with} \qquad 0 < n < 1. \tag{3.76}$$

We may then replace $1/R^2$ with $k/n > 0$ in (3.75), i.e. we choose a gradient which focuses in the vertical plane and defocuses in the horizontal plane. The first equation in (3.75) then becomes

$$x'' + \left(\frac{k}{n} - k\right) x = x'' + \frac{1-n}{n}\, k\, x = 0. \tag{3.77}$$

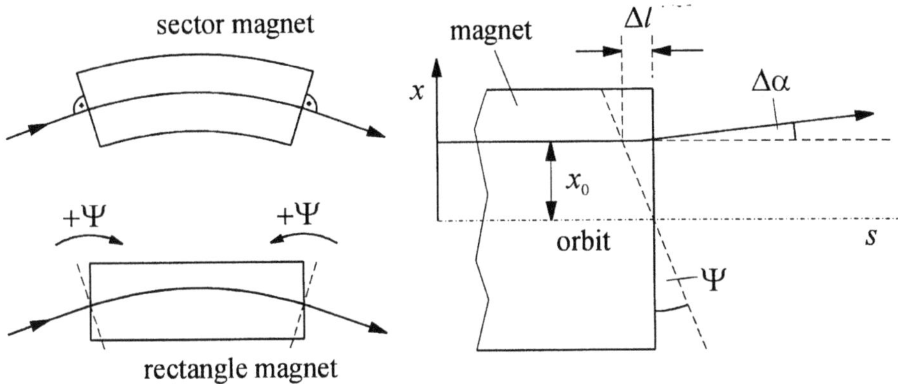

Fig. 3.20 Edge focusing by tilting the face of a dipole magnet around the perpendicular axis.

n is chosen such that $(1-n)/n > 0$ — in general $n \approx 0.5$. As a result the beam is also focused in the horizontal plane. This very simple principle of focusing in both planes at once is nowadays very rarely used, however, because the effect is relatively weak.

When considering bending magnets we have assumed that the field begins sharply at a particular point along the particle trajectory and ends sharply at another point. We assume this for the field distribution outside the orbit as well, i.e. the field begins and ends at the same time over the whole of the x-z plane. The faces of the bending magnets must therefore be perpendicular to the orbit, which due to the curvature of the trajectory requires the use of so-called **sector magnets**, like those shown in Fig. 3.20.

Because they are easier to manufacture, many **rectangular magnets** are nowadays used. These differ from sector magnets in that the magnet face is effectively tilted through an angle Ψ about the perpendicular axis (z-axis). The sign convention is chosen such that the change from a sector magnet to a rectangular magnet corresponds to a positive rotation (Fig. 3.20). The curvature of the orbit in the dipole is of course unaffected by this change, but the particles now cross the face of the magnet at an angle. This results in a special kind of focusing known as **edge focusing**, an effect illustrated in the right-hand diagram of Fig. 3.20. Consider a particle in the end region of the field, travelling parallel to the orbit, at a separation x_0 from it. Comparing the particle trajectory with that through a sector magnet, which does not cause any edge focusing, we find that the trajectory through the rectangular magnet is shorter by a distance

$$\Delta l = x_0 \tan\Psi. \tag{3.78}$$

As a result the particle is bent through a smaller angle, a difference of

$$\Delta\alpha = \frac{\Delta l}{R} = x_0 \frac{\tan\Psi}{R}, \tag{3.79}$$

which corresponds to a horizontal defocusing of the beam. For simplicity the end region of the field may be regarded as an infinitely thin lens, since Δl is generally a very short distance. The particle still crosses the end of the field at a transverse position x_0, but the angle of the trajectory is different before and after the field boundary. Hence the trajectory transformation is

$$\begin{aligned} x &= x_0 \\ x' &= x'_0 + x_0 \frac{\tan\Psi}{R}. \end{aligned} \tag{3.80}$$

This transformation may be rewritten in matrix form, namely

$$\mathbf{M}_{\text{edge}} = \begin{pmatrix} 1 & 0 \\ \dfrac{\tan\Psi}{R} & 1 \end{pmatrix}. \tag{3.81}$$

Let us now briefly consider the effect of edge focusing on the vertical motion of the particle. Recall that the field lines bend outwards at the end of the field, as is shown for example in Fig. 3.8. As a result the field in this region has components perpendicular to the magnet face. Along the orbit these components cancel out because of the symmetry of the magnet, but above and below the plane of the orbit they have finite values with opposite signs. In an ideal sector magnet these field components are parallel to the s-axis and so have no effect on the particle motion. The situation changes, however, if the magnet face is rotated by an angle Ψ. The field components at the end of the field are now at the same angle Ψ to the s-axis, and so may be separated into an s- and an x-component. This x-component, which increases with the angle of inclination, causes vertical focusing of the beam which may be described to a good approximation by the transfer matrix

$$\mathbf{M}_{\text{edge}} = \begin{pmatrix} 1 & 0 \\ -\dfrac{\tan\Psi}{R} & 1 \end{pmatrix}. \tag{3.82}$$

Comparing this with (3.81) shows that inclining the face of a bending magnet by an angle $+\Psi$ causes horizontal defocusing and vertical focusing of the beam.

In general a particle travelling through a magnet structure moves in the x-s as well as the z-s plane. As a result, four-dimensional trajectory vectors

$$\boldsymbol{X} = \begin{pmatrix} x \\ x' \\ z \\ z' \end{pmatrix} \tag{3.83}$$

are needed in order to completely describe the particle motion. This means that at least one 4×4 matrix is required to describe the particle transformation. We may immediately derive these matrices from the results in the preceding sections. We list them all here for reference, again with $\Omega = \sqrt{|k|}\, s$.

1. Horizontally focusing quadrupole ($k < 0$):

$$\mathbf{M}_{\mathrm{QF}} = \begin{pmatrix} \cos\Omega & \frac{1}{\sqrt{|k|}}\sin\Omega & 0 & 0 \\ -\sqrt{|k|}\sin\Omega & \cos\Omega & 0 & 0 \\ 0 & 0 & \cosh\Omega & \frac{1}{\sqrt{|k|}}\sinh\Omega \\ 0 & 0 & \sqrt{|k|}\sinh\Omega & \cosh\Omega \end{pmatrix} \tag{3.84}$$

2. Vertically focusing quadrupole ($k > 0$):

$$\mathbf{M}_{\mathrm{QD}} = \begin{pmatrix} \cosh\Omega & \frac{1}{\sqrt{k}}\sinh\Omega & 0 & 0 \\ \sqrt{k}\sinh\Omega & \cosh\Omega & 0 & 0 \\ 0 & 0 & \cos\Omega & \frac{1}{\sqrt{k}}\sin\Omega \\ 0 & 0 & -\sqrt{k}\sin\Omega & \cos\Omega \end{pmatrix} \tag{3.85}$$

3. Zero-field drift region ($k = 0$):

$$\mathbf{M}_{\mathrm{drift}} = \begin{pmatrix} 1 & s & 0 & 0 \\ 0 & 1 & 0 & 0 \\ 0 & 0 & 1 & s \\ 0 & 0 & 0 & 1 \end{pmatrix} \tag{3.86}$$

4. Dipole magnet ($k = 0, R > 0$):

$$\mathbf{M}_{\mathrm{dipole}} = \begin{pmatrix} \cos\frac{s}{R} & R\sin\frac{s}{R} & 0 & 0 \\ -\frac{1}{R}\sin\frac{s}{R} & \cos\frac{s}{R} & 0 & 0 \\ 0 & 0 & 1 & s \\ 0 & 0 & 0 & 1 \end{pmatrix} \tag{3.87}$$

5. Edge focusing:

$$\mathbf{M}_{\mathrm{edge}} = \begin{pmatrix} 1 & 0 & 0 & 0 \\ \frac{\tan\Psi}{R} & 1 & 0 & 0 \\ 0 & 0 & 1 & 0 \\ 0 & 0 & -\frac{\tan\Psi}{R} & 1 \end{pmatrix} \tag{3.88}$$

3.5 Calculation of a particle trajectory through a system of many beam steering magnets

So far we have only considered particle motion within one beam steering element, in which the field remains constant along the orbit. Using the transfer matrices (3.84) to (3.88), it is possible to evolve the trajectory vector $\boldsymbol{X}_0$ from the beginning to the end of the element, giving

$$\boldsymbol{X}_{\mathrm{E}} = \mathrm{M}\,\boldsymbol{X}_0. \tag{3.89}$$

By taking the evolved trajectory vector $\boldsymbol{X}_{\mathrm{E}}$ at the end of the element and using it as the starting vector for the next element, we can determine the particle motion through an arbitrary number of elements by successive multiplication of matrices. To illustrate this let us take as an example an arrangement of four quadrupoles, separated by zero-field drift sections. This arrangement is shown in Fig. 3.21.

At the beginning ($s = 0$) the trajectory vector is $\boldsymbol{X}_0 = (x_0, x_0')$. The trajectory then traverses the drift section D1. The vector at this point is obtained by a transfer matrix M_{D1} of the form of (3.86), in which $s = l$ is the length of the first drift section. The next section of the trajectory, through the focusing quadrupole Q1, is obtained using the matrix M_{Q1}. This has the form of (3.84) and is uniquely defined by the length of the magnet and the quadrupole strength. It is then followed by another drift section, then another magnet, and so on. The trajectory vector at the end of this arrangement of magnets is given by the product of a series of matrices

$$\boldsymbol{X}_{\mathrm{E}} = \mathrm{M}_{\mathrm{D5}} \cdot \mathrm{M}_{\mathrm{Q4}} \cdot \mathrm{M}_{\mathrm{D4}} \cdot \mathrm{M}_{\mathrm{Q3}} \cdot \mathrm{M}_{\mathrm{D3}} \cdot \mathrm{M}_{\mathrm{Q2}} \cdot \mathrm{M}_{\mathrm{D2}} \cdot \mathrm{M}_{\mathrm{Q1}} \cdot \mathrm{M}_{\mathrm{D1}} \cdot \boldsymbol{X}_0. \tag{3.90}$$

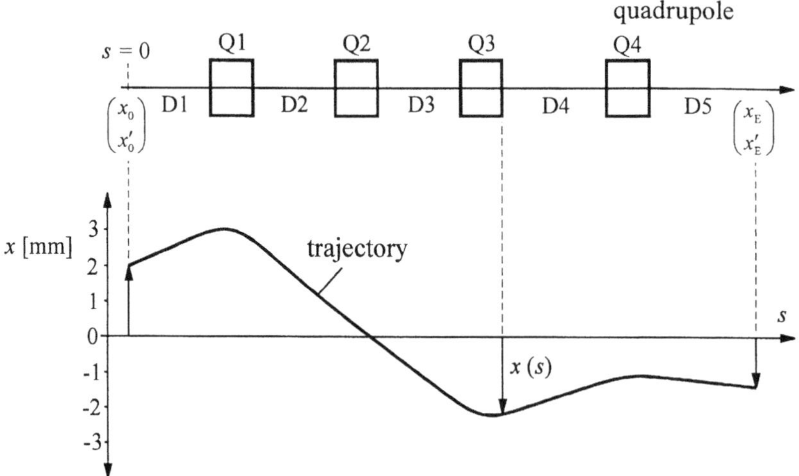

Fig. 3.21 Calculation of particle motion through a structure of multiple beam steering elements.

It is sometimes also necessary to determine the trajectory vector within an element, especially in relatively long elements. In this case it is helpful to divide the element into several identical shorter partial elements and to describe these by separate matrices. Calculating the particle trajectory through a long beam transport system thus involves a large number of individual matrices, ranging from a few hundred to sometimes more than a thousand. In such cases it is essential to perform the calculations by computer, and here the matrix formalism proves to be particularly effective.

Let us now use this principle to calculate the transfer matrix for a rectangular magnet. As Fig. 3.20 shows, a rectangular magnet may be described by taking a sector magnet and adding edge focusing at both ends. The matrix for a rectangular magnet is thus obtained by the matrix product

$$\mathbf{M}_{\text{rectangle}} = \mathbf{M}_{\text{edge}} \cdot \mathbf{M}_{\text{dipole}} \cdot \mathbf{M}_{\text{edge}}. \tag{3.91}$$

The partial matrices for the sector magnet and the edge focusing are given in (3.87) and (3.88). The bending angle of the dipole is determined by the ratio of the magnet length l and the bending radius R. The angle of inclination Ψ of the faces of a rectangular magnet compared to a sector magnet is exactly half the bending angle, i.e. $\Psi = l/2R$. For simplicity we will only consider bending magnets with relatively small bending angles, i.e. magnets with $l/R \ll 1$. It then follows immediately that

$$\tan\Psi \approx \Psi = \frac{l}{2R} \qquad \text{and} \qquad \cos\frac{l}{R} \approx 1 \quad \text{or} \quad \sin\frac{l}{R} \approx \frac{l}{R}. \tag{3.92}$$

If we insert these approximations into (3.87) or (3.88), with $\kappa = l/2R^2$, equation (3.91) then becomes

$$\mathbf{M}_{\text{rectangle}} = \begin{pmatrix} 1 & 0 & 0 & 0 \\ \kappa & 1 & 0 & 0 \\ 0 & 0 & 1 & 0 \\ 0 & 0 & -\kappa & 1 \end{pmatrix} \cdot \begin{pmatrix} 1 & l & 0 & 0 \\ -2\kappa & 1 & 0 & 0 \\ 0 & 0 & 1 & l \\ 0 & 0 & 0 & 1 \end{pmatrix} \cdot \begin{pmatrix} 1 & 0 & 0 & 0 \\ \kappa & 1 & 0 & 0 \\ 0 & 0 & 1 & 0 \\ 0 & 0 & -\kappa & 1 \end{pmatrix}. \tag{3.93}$$

After multiplying out the matrices we neglect all terms in $l^2/2R^2$, which according to approximation (3.92) are much smaller than 1. We finally obtain

$$\mathbf{M}_{\text{rectangle}} = \begin{pmatrix} 1 & l & 0 & 0 \\ 0 & 1 & 0 & 0 \\ 0 & 0 & 1 & l \\ 0 & 0 & -2\kappa & 1 \end{pmatrix}. \tag{3.94}$$

Comparing this with the matrix for a sector magnet we see that in rectangular magnets the weak focusing disappears in the horizontal plane and the magnet behaves like a drift region. However, the weak focusing in the vertical plane is increased by the same amount. Hence it is possible to transfer the weak focusing between the two planes by inclining the faces of the dipole magnet.

3.6 Dispersion and momentum compaction factor

Let us now consider the motion of off-momentum particles, i.e. particles with $\Delta p/p \neq 0$. Equation (3.21) tells us that the momentum deviation only has a significant effect on the trajectory of the particle if $1/R \neq 0$, and so we only need to solve the equations of motion inside the bending magnets. We again assume a homogeneous magnetic field with no gradient ($k = 0$). Equation (3.21) then becomes

$$x'' + \frac{1}{R^2}x = \frac{1}{R}\frac{\Delta p}{p}. \tag{3.95}$$

It is useful to define a special trajectory $D(s)$ for which $\Delta p/p = 1$. This function $D(s)$ is called a **dispersion function**. Equation (3.95) then takes the form

$$D''(s) + \frac{1}{R^2}D(s) = \frac{1}{R}. \tag{3.96}$$

This is an inhomogeneous differential equation, which we already solved in its homogeneous form when we considered the particle motion through a bending magnet. We therefore now only need to find a particular solution D_{p} of the inhomogeneous equation. Since the right hand side of (3.96) is a constant,

$$D_{\mathrm{p}} = C = \mathrm{const} \tag{3.97}$$

is a valid solution. Inserting this into (3.96) immediately yields

$$\frac{C}{R^2} = \frac{1}{R} \qquad \longrightarrow \qquad C = R. \tag{3.98}$$

The general solution of the trajectory equation for off-momentum particles is thus

$$\begin{aligned} D(s) &= A\cos\frac{s}{R} + B\sin\frac{s}{R} + R \\ D'(s) &= -\frac{A}{R}\sin\frac{s}{R} + \frac{B}{R}\cos\frac{s}{R}. \end{aligned} \tag{3.99}$$

The constants of integration A and B are determined by the initial conditions at $s = 0$

$$D(0) = D_0 \qquad \text{and} \qquad D'(0) = D_0'. \tag{3.100}$$

Inserting these into (3.99) we obtain

$$A = D_0 - R \qquad \text{and} \qquad B = RD_0'. \tag{3.101}$$

Using this result we may finally write the dispersion function as

$$\begin{aligned} D(s) &= D_0 \cos\frac{s}{R} + D_0' R \sin\frac{s}{R} + R\left(1 - \cos\frac{s}{R}\right) \\ D'(s) &= -\frac{D_0}{R}\sin\frac{s}{R} + D_0'\cos\frac{s}{R} + \sin\frac{s}{R}. \end{aligned} \tag{3.102}$$

This set of equations, restricted for the moment to the horizontal plane, can no longer be written as a 2×2 matrix as in (3.74). Hence we must switch to a 3×3 matrix, and from (3.102) obtain the transformation

$$\begin{pmatrix} D(s) \\ D'(s) \\ 1 \end{pmatrix} = \begin{pmatrix} \cos\frac{s}{R} & R\sin\frac{s}{R} & R\left(1-\cos\frac{s}{R}\right) \\ -\frac{1}{R}\sin\frac{s}{R} & \cos\frac{s}{R} & \sin\frac{s}{R} \\ 0 & 0 & 1 \end{pmatrix} \cdot \begin{pmatrix} D_0 \\ D_0' \\ 1 \end{pmatrix}. \tag{3.103}$$

Where the dispersion is non-zero, a particle with a momentum difference of $\Delta p/p$ has a transverse position of

$$x_{\mathrm{g}}(s) = x(s) + x_{\mathrm{D}}(s) = x(s) + D(s)\frac{\Delta p}{p}. \tag{3.104}$$

Here $x(s)$ is the transverse position that a particle of nominal momentum would have, and x_{D} is the displacement which results from the momentum difference. Hence the particle no longer follows the ideal orbit, instead following a dispersive trajectory determined by $\Delta p/p$. The general evolution of this trajectory through the magnet structure in the horizontal plane requires 3×3 matrices, by analogy with (3.103). To determine these matrices, we assume for the time being that the focusing is independent of the momentum deviation of the particle. Later we will take account of this dependence, the so-called **'chromaticity'**, via a perturbative calculation (see Section 3.16). We thus obtain the particle trajectory in the horizontal plane

$$\begin{pmatrix} x(s) \\ x'(s) \\ \Delta p/p \end{pmatrix} = \mathbf{M} \cdot \begin{pmatrix} x_0 \\ x_0' \\ \Delta p/p \end{pmatrix}. \tag{3.105}$$

Here $\mathbf{M}$ is a 3×3 matrix, which may be obtained for quadrupoles, drift sections, and edge focusing by extending the 2×2 matrices (3.72) and (3.81) or (3.82) as follows:

$$\mathbf{M} = \begin{pmatrix} a_{1,1} & a_{1,2} & 0 \\ a_{2,1} & a_{2,2} & 0 \\ 0 & 0 & 1 \end{pmatrix}. \tag{3.106}$$

$a_{i,k}$ are the elements of the respective 2×2 matrices. In order to again transform the particle motion in both planes simultaneously we must express the particle momenta in terms of a five-dimensional vector. The general transformation is then

$$\begin{pmatrix} x(s) \\ x'(s) \\ z(s) \\ z'(s) \\ \Delta p/p \end{pmatrix} = \mathbf{M} \cdot \begin{pmatrix} x_0 \\ x'_0 \\ z_0 \\ z'_0 \\ \Delta p/p \end{pmatrix}. \tag{3.107}$$

For quadrupoles, drift sections, and edge focusing we again have matrices of the form

$$\mathbf{M} = \begin{pmatrix} a_{1,1} & a_{1,2} & 0 & 0 & 0 \\ a_{2,1} & a_{2,2} & 0 & 0 & 0 \\ 0 & 0 & a_{3,3} & a_{3,4} & 0 \\ 0 & 0 & a_{4,3} & a_{4,4} & 0 \\ 0 & 0 & 0 & 0 & 1 \end{pmatrix}. \tag{3.108}$$

To evolve the trajectory in both planes through a dipole magnet, (3.103) with $\chi = s/R$ gives us

$$\mathbf{M}_{\text{dipole}} = \begin{pmatrix} \cos\chi & R\sin\chi & 0 & 0 & R(1-\cos\chi) \\ -(1/R)\sin\chi & \cos\chi & 0 & 0 & \sin\chi \\ 0 & 0 & 1 & s & 0 \\ 0 & 0 & 0 & 1 & 0 \\ 0 & 0 & 0 & 0 & 1 \end{pmatrix}. \tag{3.109}$$

Using these 5×5 matrices we can evolve all particle trajectories through any arbitrary structure of magnets, taking into account deviations in momentum. As a result, these matrices are the most important tool in calculations of linear beam optics. In what follows we will sometimes also use partial matrices extracted from them, in particular if we are only considering motion in one plane.

As we have seen, off-momentum particles follow dispersive trajectories $x_{\text{D}} = D(s)\Delta p/p$, which in general are not the same length as the ideal orbit. The path length is therefore a function of momentum. In circular accelerators this leads to a dependence of the revolution period on the particle momentum, which plays an important role the longitudinal phase focusing of the circulating particles. The ratio of the relative change in path length $\Delta L/L$ to the relative difference in momentum $\Delta p/p$ is termed the **momentum compaction factor**

$$\alpha = \frac{\Delta L/L}{\Delta p/p}. \tag{3.110}$$

It is easy to see that only bending magnets cause any significant change in the path length. A particle travelling at a separation x from the orbit covers an infinitesimal distance

$$d\tilde{s} = \frac{R+x}{R}ds, \tag{3.111}$$

if the origin of the co-moving coordinate system moves through the path element ds along the orbit. This relation follows from the simple geometry of similar triangles, as may be seen from Fig. 3.4. In quadrupoles the orbit is straight and any change in path length arises only through higher order effects, which may be neglected here. If the separation x in (3.111) is caused by the momentum difference of the particle then the total path length may be written as

$$L = L_0 + \Delta L = \oint \frac{R + x_{\mathrm{D}}}{R} ds = \oint ds + \frac{\Delta p}{p} \oint \frac{D(s)}{R(s)} ds. \tag{3.112}$$

Here it should be noted that the bending radius $R(s)$ may depend on the position s along the trajectory through the magnet structure. The length of the ideal orbit is $L_0 = \oint ds$, and so the change in length of the trajectory is given by

$$\Delta L = \frac{\Delta p}{p} \oint \frac{D(s)}{R(s)} ds. \tag{3.113}$$

Using the definition (3.110) we finally obtain the momentum compaction factor

$$\boxed{\alpha = \frac{1}{L_0} \oint \frac{D(s)}{R(s)} ds.} \tag{3.114}$$

3.7 Beta function and betatron oscillation

The matrix formalism developed so far allows us to calculate individual particle trajectories through an arbitrary structure of magnets and also to take into account variations in particle momentum. However, it does not yield any information about the properties of a beam of many particles. Since this is ultimately the decisive factor in the development of a system of beam optics, we must extend our techniques to describe the behaviour of an entire composite beam. To do this we go back to the fundamental equations (3.21) and assume only that $1/R = 0$ and $\Delta p/p = 0$. The quadrupole strength k is now a function of position s, however. We obtain Hill's differential equation of motion

$$x''(s) - k(s)\, x(s) \;=\; 0. \tag{3.115}$$

The trajectory function $x(s)$ describes a transverse oscillation about the orbit, known as a **betatron oscillation**, whose amplitude and phase depend on the position s along the orbit. We therefore solve this equation using the trial solution

$$x(s) \;=\; A\, u(s)\, \cos[\Psi(s) + \phi]. \tag{3.116}$$

The constant amplitude factor A and the phase ϕ are constants of integration which are fixed by the initial conditions. Inserting the solution (3.116) and its second derivative into (3.115) and writing $u = u(s)$ and $\Psi = \Psi(s)$, we obtain

$$A\Big[u'' - u\Psi'^2 - k(s)u\Big] \cos(\Psi + \phi) - A\Big[2u'\Psi' + u\Psi''\Big] \sin(\Psi + \phi) = 0. \tag{3.117}$$

Since the phase $\Psi(s)$ has a different value at every point around the orbit and the amplitude $A \neq 0$, equation (3.117) can only be satisfied if

$$u'' - u\Psi'^2 - k(s)u = 0 \tag{3.118}$$
$$2u'\Psi' + u\Psi'' = 0. \tag{3.119}$$

From (3.119) it follows that

$$2\frac{u'}{u} + \frac{\Psi''}{\Psi'} = 0. \tag{3.120}$$

This equation may be integrated directly, yielding

$$\Psi(s) = \int_0^s \frac{d\sigma}{u^2(\sigma)}. \tag{3.121}$$

Inserting this result into (3.118) gives

$$u'' - \frac{1}{u^3} - k(s)u = 0. \tag{3.122}$$

This non-linear differential equation has no general analytical solution and so may only be evaluated by numerical methods. This is hardly practical, however, in complicated magnet structures with many individual elements. In the following sections we will develop a matrix method to allow us to calculate the full beam optics in the same way as the particle trajectory, by means of matrix transformations. We thus introduce the **beta function** $\beta(s)$, also known as the **amplitude function**. It is defined as

$$\beta(s) \equiv u^2(s). \tag{3.123}$$

In addition we replace the amplitude factor A in (3.116) by $\sqrt{\varepsilon}$. The constant ε is termed the **emittance**, the meaning of which will be explained in the following sections. We finally write the solution of the trajectory equation (3.115) in the form

$$x(s) = \sqrt{\varepsilon}\sqrt{\beta(s)}\cos[\Psi(s) + \phi] \tag{3.124}$$

with

$$\Psi(s) = \int_0^s \frac{d\sigma}{\beta(\sigma)}. \tag{3.125}$$

Within the magnet structure, which has a net focusing effect, the particles perform betatron oscillations with a position-dependent amplitude given by

$$E(s) = \sqrt{\varepsilon\,\beta(s)}. \tag{3.126}$$

Here the beta function $\beta(s)$ depends on the beam focusing, which varies with position. It is a measure of the beam cross-section at that point, whereas ε

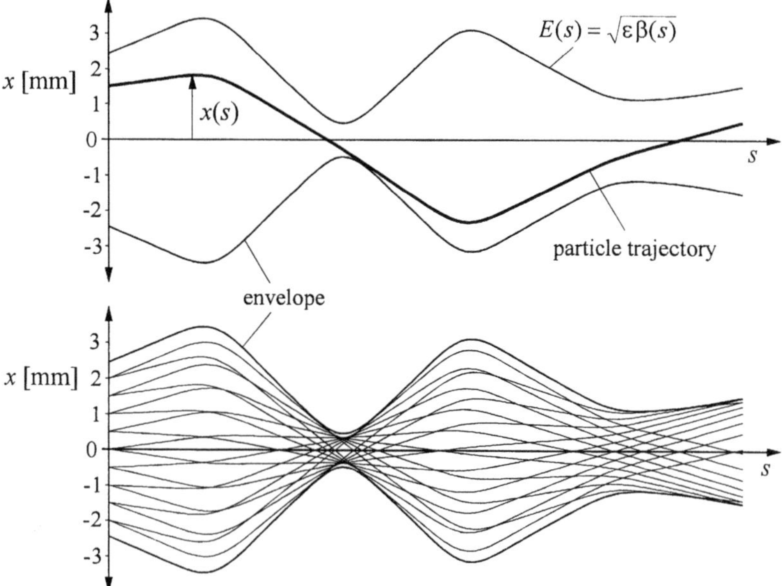

Fig. 3.22 Particle trajectories $x(s)$ within the envelope $E(s)$ of the beam. The upper figure shows a single trajectory, while the lower figure shows 18 different trajectories together. The beam is made up of a combination of all the individual trajectories.

remains constant throughout the whole beam transport system. We will prove this statement in the sections which follow. To be precise the particles undergo transverse motion about the orbit within a range marked out by the **envelope** $E(s)$, defined in (3.126). Since all particle trajectories lie inside this envelope, it defines the transverse size of the beam, as illustrated in Fig. 3.22. We now see the importance of the beta function $\beta(s)$. If it is possible to evolve $\beta(s)$ step-by-step through the magnet structure, in the same way as the particle trajectory $x(s)$, then for a given emittance ε we can determine the beam size at any point s.

Using (3.124) we calculate the second derivative of the trajectory function $x(s)$, which we will need several times in what follows. We write this in the form

$$x'(s) = -\frac{\sqrt{\varepsilon}}{\sqrt{\beta(s)}}\left[\alpha(s)\cos\Big(\Psi(s)+\phi\Big)+\sin\Big(\Psi(s)+\phi\Big)\right] \qquad (3.127)$$

with

$$\alpha(s) \equiv -\frac{\beta'(s)}{2}. \qquad (3.128)$$

This optical function should not be confused with the momentum compaction factor, which for historical reasons is also denoted by α. The functions $\beta(s)$ and $\alpha(s)$, along with the dispersion $D(s)$ and its derivative $D'(s)$, fully describe the linear beam optics in a plane. To these we add the phase $\Psi(s)$ of the beta function, calculated from (3.125).

3.8 The phase space ellipse and Liouville's theorem

As we have seen in the previous chapter, the general solution of the trajectory equation has the form

$$x(s) = \sqrt{\varepsilon}\sqrt{\beta(s)}\cos\Big(\Psi(s)+\phi\Big) \tag{3.129}$$

$$x'(s) = -\frac{\sqrt{\varepsilon}}{\sqrt{\beta(s)}}\Big[\alpha(s)\cos\Big(\Psi(s)+\phi\Big)+\sin\Big(\Psi(s)+\phi\Big)\Big]. \tag{3.130}$$

In order to arrive at an expression describing the particle motion in the x-x' phase space plane we must eliminate the terms which depend on the phase Ψ. From (3.129) we immediately obtain

$$\cos\Big(\Psi(s)+\phi\Big) = \frac{x}{\sqrt{\varepsilon}\sqrt{\beta(s)}}. \tag{3.131}$$

Substituting this into (3.130) and rearranging yields

$$\sin\Big(\Psi(s)+\phi\Big) = \frac{\sqrt{\beta(s)}x'}{\sqrt{\varepsilon}} + \frac{\alpha(s)x}{\sqrt{\varepsilon}\sqrt{\beta(s)}}. \tag{3.132}$$

If we now use the general relation $\sin^2\Theta + \cos^2\Theta = 1$ we obtain

$$\frac{x^2}{\beta(s)} + \left(\frac{\alpha(s)}{\sqrt{\beta(s)}}x + \sqrt{\beta(s)}x'\right)^2 = \varepsilon. \tag{3.133}$$

If we now introduce the definition

$$\gamma(s) \equiv \frac{1+\alpha^2(s)}{\beta(s)} \tag{3.134}$$

(3.133) gives

$$\boxed{\gamma(s)\; x^2(s) + 2\;\alpha(s)\; x(s)\; x'(s) + \beta(s)\; x'^2(s) = \varepsilon.} \tag{3.135}$$

This, as is shown in detail in appendix C, is the general equation of an ellipse in the x-x'-plane. It is plotted in Fig. 3.23. The emittance ε, introduced originally as a constant of integration, now has an obvious interpretation. It is, to within a factor π, the **area of the phase ellipse**, namely $\varepsilon = F/\pi$. Now according to **Louville's theorem**, which is deeply fundamental, every element of a volume of phase space is constant with respect to time if the particles obey the canonical equations of motion. This condition is generally satisfied in accelerators. This means that the area of the phase ellipse and hence the beam emittance are invariants of the particle motion. As the particle moves along the orbit the shape and position of the ellipse change according to the amplitude function $\beta(s)$, but the area remains constant. This result has important consequences for the calculation of linear beam optics.

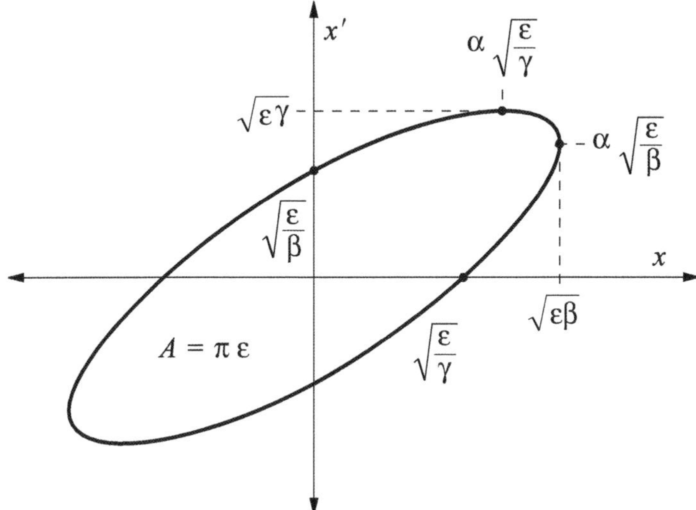

Fig. 3.23 The phase space ellipse of particle motion in the x-x'-plane.

3.9 Beam cross-section and emittance

So far we have only applied the expression for the emittance and the associated phase space ellipse to the motion of a single particle. If a single particle in a circular accelerator has the initial position $s = s_0$ and the trajectory $\boldsymbol{X}_0 = (x_0, x_0')$, then with every new revolution i the resulting trajectory vector $\boldsymbol{X}_i$ describes an ellipse in the $x - x'$-surface of phase space, as is shown in Fig. 3.23. The particle thus has a well-defined emittance associated with it, which is given directly by the area of the phase space ellipse.

There are, however, very many particles in a beam, moving with various amplitudes and each corresponding to a different ellipse in the phase space plane. This raises the question of what we mean by the average emittance of a beam consisting of an assembly of many particles. In particular, in the case of electrons the emission of synchrotron radiation results in stochastic fluctuations in the individual particle emittances. In order to provide a practical definition of the emittance of a particle beam, let us start with the equilibrium distribution of particles, which is constant over time. For electrons this is described to a very good approximation by a **Gaussian distribution**

$$\rho(x, z) = \frac{Ne}{2\pi\sigma_x\sigma_z} \exp\left(-\frac{x^2}{2\sigma_x^2} - \frac{z^2}{2\sigma_z^2}\right), \tag{3.136}$$

which gives the transverse charge density distribution. Here N is the number of particles of charge e in the beam. The horizontal distribution $\rho(x)$, which we obtain by setting $z = 0$, is shown in Fig. 3.24. σ_x and σ_z are the horizontal and vertical beam sizes, specifically the distances from the beam axis within which the charge density has decreased by $\exp(-\frac{1}{2}) = 0.607$. This corresponds to exactly one standard deviation of the statistical particle distribution. All particles which

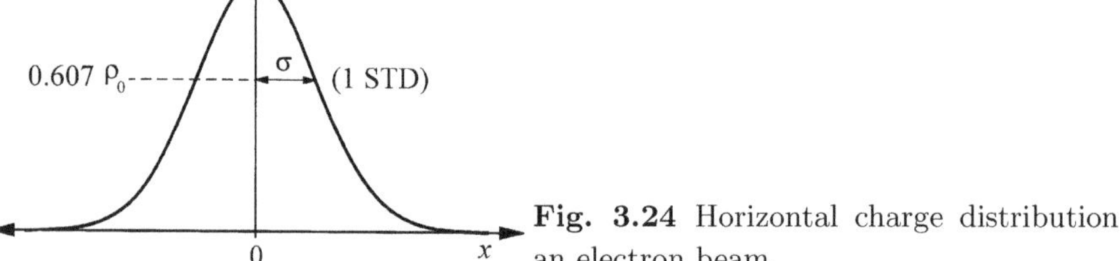

Fig. 3.24 Horizontal charge distribution in an electron beam.

lie exactly one standard deviation σ from the beam axis may be assigned a precise emittance $\varepsilon_{\mathrm{STD}}$ via the relation

$$\sigma(s) = \sqrt{\varepsilon_{\mathrm{STD}}\,\beta(s)}. \tag{3.137}$$

The value defined in this way

$$\varepsilon_{\mathrm{STD}} = \frac{\sigma^2(s)}{\beta(s)} \tag{3.138}$$

will henceforth be used to represent the emittance of the entire beam. We will omit the index STD and always use ε to denote the emittance at one standard deviation.

When operating an accelerator it is important to ensure that the beam has sufficient room available in the transverse phase space. Even particles undergoing extremely large betatron oscillations can then still circulate stably. This raises the question of how large the phase space ellipse of a particle is allowed to be before it collides with the wall of the vacuum chamber and is lost. This limiting case is plotted in Fig. 3.25. Here we must keep in mind that the width of the beam is proportional to $\sqrt{\beta(s)}$. This means that the aperture d of the vacuum chamber is not a direct measure of the space available to the beam. As can be seen from Fig. 3.22, the beam is indeed very narrow at positions where the

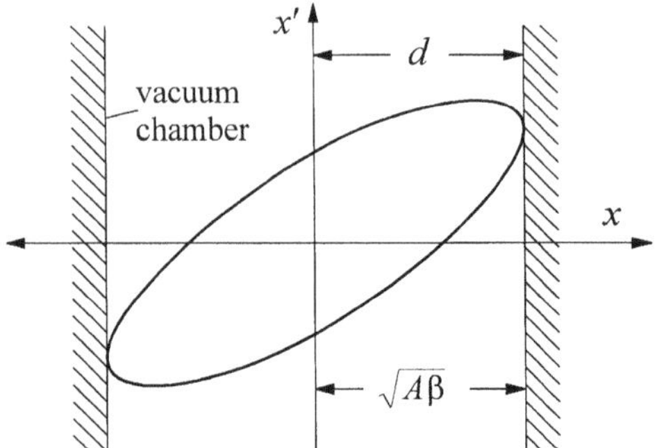

Fig. 3.25 The largest possible phase space ellipse for a stably circulating particle. The limit is generally determined by the size of the vacuum chamber.

beta function is small, so here even a relatively narrow vacuum chamber does not cause any restriction. Ultimately the decisive quantity is $d/\sqrt{\beta(s)}$, which varies around the orbit. The narrowest regions of the beam optics coincide with the points around the accelerator at which $d/\sqrt{\beta(s)}$ is smallest, resulting in a reduction in the transverse phase space available to the beam. Of course similar restrictions exist in both planes, but normally both do not occur at the same point.

By analogy with the beam emittance we define the **transverse acceptance of the accelerator**, which corresponds to the emittance of the particle with the largest possible phase space ellipse. It is given by

$$A = \left(\frac{d^2}{\beta}\right)_{\min}, \tag{3.139}$$

where d and β are the values at the narrowest optical position. In order for the beam to circulate without any losses the acceptance of the machine must be definitely larger than the beam emittance. In the case of storage rings a particularly large value of A/ε is required. If this value is not large enough then over many revolutions too many particles with large oscillation amplitudes will hit the wall of the vacuum chamber and be lost. In electron beams, synchrotron radiation can cause the betatron amplitude of all the particles to vary stochastically and so each one has a certain probability of having a very large amplitude. Ultimately this means that if the acceptance of the accelerator is too small then each particle has a chance of being lost within a finite time, leading to very short beam lifetimes. Consequently the narrowest point in the vacuum chamber of an electron storage ring must be at least seven times the the beam width, and for trouble-free operation considerably more space is in fact required, i.e. $A > 50\varepsilon$.

3.10 Evolution of the beta function through the magnet structure

As we have seen in the preceding chapters, the beta function is an important quantity in linear beam optics. From it we may for example calculate the beam dimensions using (3.137) and the phase of the betatron oscillations using (3.125). Here we will describe how to determine the evolution of the beta function itself through the course of the magnet structure. We assume that the value of this function at a given initial position $s = s_0$ is known. Starting from this initial value the beta function may then be calculated step by step along the structure of beam steering magnets by the use of an appropriate transformation. We will introduce two different transformation methods.

3.10.1 Method 1

The trajectory of the particle may be described by a trajectory vector $\boldsymbol{X}$, which travels around the phase space ellipse during the motion of the particle around an orbit. At the beginning of the magnet structure, i.e. at $s = 0$, it has the value $\boldsymbol{X} = \boldsymbol{X}_0$. This vector may be formally expressed as a 2×1 matrix, namely

$$\mathbf{X}_0 = \begin{pmatrix} x_0 \\ x_0' \end{pmatrix} \quad \text{with the transposed matrix} \quad \mathbf{X}_0^{\mathrm{T}} = (x_0 \; x_0'). \tag{3.140}$$

In addition we define the **beta matrix**

$$\mathbf{B}_0 \equiv \begin{pmatrix} \beta_0 & -\alpha_0 \\ -\alpha_0 & \gamma_0 \end{pmatrix}, \tag{3.141}$$

with the determinant $\det \mathbf{B}_0 = 1$. We then calculate the product

$$\begin{aligned} \mathbf{X}_0^{\mathrm{T}} \cdot \mathbf{B}_0^{-1} \cdot \mathbf{X}_0 &= (x_0 \; x_0') \cdot \begin{pmatrix} \gamma_0 & \alpha_0 \\ \alpha_0 & \beta_0 \end{pmatrix} \cdot \begin{pmatrix} x_0 \\ x_0' \end{pmatrix} \\ &= \gamma_0 \, x_0^2 + 2\alpha_0 \, x_0 \, x_0' + \beta_0 \, {x_0'}^2 \\ &= \varepsilon. \end{aligned} \tag{3.142}$$

This product corresponds to the beam emittance, according to (3.135), and so is position invariant. The trajectory vector may be calculated at any position s using the known transfer matrices and the relation

$$\boldsymbol{X}_1 = \mathbf{M} \, \boldsymbol{X}_0. \tag{3.143}$$

Here the matrix $\mathbf{M}$, which describes the transformation from a position s_0 to another position s_1, is in general the product of a large number of matrices. From (3.142) and using $\mathbf{M}^{-1}\mathbf{M} = 1$ and $\mathbf{M}^{\mathrm{T}}(\mathbf{M}^{\mathrm{T}})^{-1} = 1$ we obtain the relation

$$\begin{aligned} \varepsilon &= \mathbf{X}_0^{\mathrm{T}} \cdot \mathbf{B}_0^{-1} \cdot \mathbf{X}_0 \\ &= \mathbf{X}_0^{\mathrm{T}}(\mathbf{M}^{\mathrm{T}}(\mathbf{M}^{\mathrm{T}})^{-1})\mathbf{B}_0^{-1}(\mathbf{M}^{-1}\mathbf{M})\mathbf{X}_0 \\ &= \mathbf{X}_0^{\mathrm{T}}\mathbf{M}^{\mathrm{T}}((\mathbf{M}^{\mathrm{T}})^{-1}\mathbf{B}_0^{-1}\mathbf{M}^{-1})\mathbf{M}\mathbf{X}_0 \end{aligned} \tag{3.144}$$

with

$$\mathbf{A}^{\mathrm{T}}\mathbf{B}^{\mathrm{T}} = (\mathbf{B}\mathbf{A})^{\mathrm{T}} \qquad \text{and} \qquad \mathbf{A}^{-1}\mathbf{B}^{-1} = (\mathbf{B}\mathbf{A})^{-1}. \tag{3.145}$$

It follows that

$$\begin{aligned} \varepsilon &= \mathbf{X}_0^{\mathrm{T}}\mathbf{M}^{\mathrm{T}}((\mathbf{M}^{\mathrm{T}})^{-1}(\mathbf{M}\mathbf{B}_0)^{-1})\mathbf{M}\mathbf{X}_0 \\ &= \mathbf{X}_0^{\mathrm{T}}\mathbf{M}^{\mathrm{T}}(\mathbf{M}\mathbf{B}_0\mathbf{M}^{\mathrm{T}})^{-1}\mathbf{M}\mathbf{X}_0 \\ &= (\mathbf{M}\mathbf{X}_0)^{\mathrm{T}}(\mathbf{M}\mathbf{B}_0\mathbf{M}^{\mathrm{T}})^{-1}\mathbf{M}\mathbf{X}_0. \end{aligned} \tag{3.146}$$

Using the transformation relation for the trajectory vector (3.143), we have

$$\mathbf{X}_1^{\mathrm{T}} = (\mathbf{M}\mathbf{X}_0)^{\mathrm{T}}.$$

Applying this to (3.146) we then obtain

$$\varepsilon = \mathbf{X}_1^{\mathrm{T}}(\mathbf{M}\mathbf{B}_0\mathbf{M}^{\mathrm{T}})^{-1}\mathbf{X}_1. \tag{3.147}$$

At the point $s = s_1$ the particle trajectory is given by the trajectory vector $\boldsymbol{X}_1$ and the beta matrix is $\mathbf{B}_1$. From (3.142) it then follows that

$$\varepsilon = \mathbf{X}_1^{\mathrm{T}} \mathbf{B}_1^{-1} \mathbf{X}_1. \tag{3.148}$$

Comparing this expression with (3.147) we finally obtain

$$\boxed{\mathbf{B}_1 = \mathbf{M} \cdot \mathbf{B}_0 \cdot \mathbf{M}^{\mathrm{T}}.} \tag{3.149}$$

This is the relation we need to transform the known beta matrix $\mathbf{B}_0$ at the position s_0 into the matrix $\mathbf{B}_1$ at the point s_1. The 2×2 matrices $\mathbf{M}$ are the same as those used to describe the evolution of the particle trajectory.

Let us now consider a simple example to illustrate the evolution of the beta function. We will calculate the form of the beta function about a symmetry point in a field-free drift section. This point, which we may arbitrarily take to be $s = 0$, is the point at which the gradient of the beta function is zero, i.e. $\alpha^* = 0$. Let the beta function itself have the value β^*. Using the transfer matrix for the drift section (3.72) it follows from (3.149) that

$$\begin{aligned} \mathbf{B}_1(s) &= \begin{pmatrix} 1 & s \\ 0 & 1 \end{pmatrix} \cdot \begin{pmatrix} \beta^* & 0 \\ 0 & 1/\beta^* \end{pmatrix} \cdot \begin{pmatrix} 1 & 0 \\ s & 1 \end{pmatrix} \\ &= \begin{pmatrix} \beta^* + \dfrac{s^2}{\beta^*} & \dfrac{s}{\beta^*} \\ \dfrac{s}{\beta^*} & \dfrac{1}{\beta^*} \end{pmatrix}. \end{aligned} \tag{3.150}$$

From this expression we immediately obtain the beta function around a point of symmetry in a drift region

$$\begin{aligned} \beta(s) &= \beta^* + \frac{s^2}{\beta^*} \\ \alpha(s) &= -\frac{s}{\beta^*}, \end{aligned} \tag{3.151}$$

which is plotted in Fig. 3.26 for various values of β^*. The beta function grows quadratically with the distance s from the symmetry point and grows faster the smaller the value of β^*. This is a direct consequence of Liouville's theorem. Since the area of the phase space ellipse cannot be reduced by focusing, any decrease in the transverse size of the beam entails a corresponding increase in the angular divergence. This fact is illustrated in Fig. 3.27.

3.10.2 Method 2

Let us now develop a second method of transforming the beta function , starting with two trajectory vectors at different positions s_0 and s, which we write as

$$\boldsymbol{X}_0 = \begin{pmatrix} x_0 \\ x_0' \end{pmatrix} \quad \text{and} \quad \boldsymbol{X} = \begin{pmatrix} x \\ x' \end{pmatrix}. \tag{3.152}$$

The optical functions at the two points are described by the quantities β_0, α_0, and γ_0 or β, α, and γ respectively. For both vectors we can insert these quantities

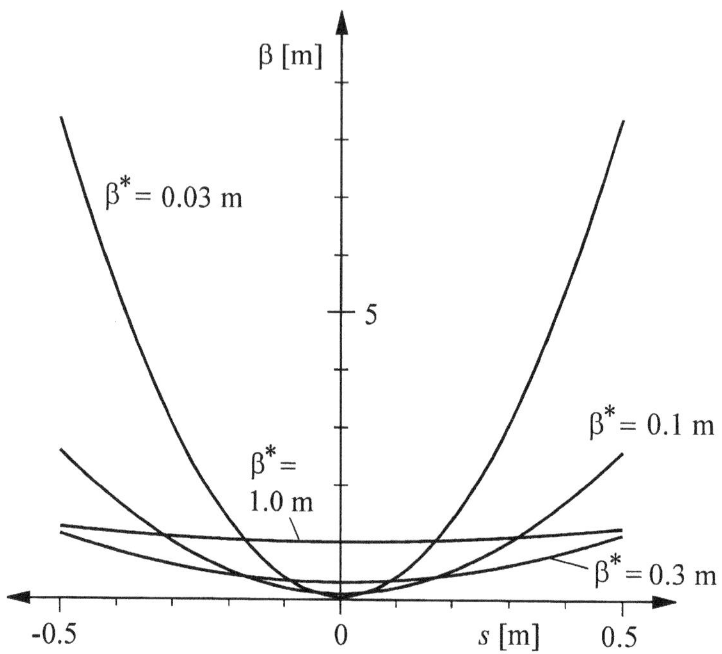

Fig. 3.26 Form of the beta function $\beta(s)$ around a symmetry point of the beam optics. The parameter β^* is the beta function at the symmetry point, $s = 0$.

into the function describing the phase space ellipse (3.135). Since the emittance is invariant we then obtain the equation

$$\varepsilon = \beta\, x'^2 + 2\,\alpha\, x\, x' + \gamma\, x^2 = \beta_0\, x_0'^{\,2} + 2\,\alpha_0\, x_0\, x_0' + \gamma_0\, x_0^2. \tag{3.153}$$

The two trajectory vectors are then related by the transfer matrix $\mathbf{M}$ as follows

$$\begin{pmatrix} x \\ x' \end{pmatrix} = \mathbf{M} \begin{pmatrix} x_0 \\ x_0' \end{pmatrix} \quad \text{with} \quad \mathbf{M} = \begin{pmatrix} m_{11} & m_{12} \\ m_{21} & m_{22} \end{pmatrix}. \tag{3.154}$$

Solving for $\boldsymbol{X}_0$ gives

$$\begin{pmatrix} x_0 \\ x_0' \end{pmatrix} = \mathbf{M}^{-1} \begin{pmatrix} x \\ x' \end{pmatrix} \quad \text{with} \quad \mathbf{M}^{-1} = \begin{pmatrix} m_{22} & -m_{12} \\ -m_{21} & m_{11} \end{pmatrix} \tag{3.155}$$

or equivalently

$$\begin{aligned} x_0 &= m_{22}\, x - m_{12}\, x' \\ x_0' &= -m_{21}\, x + m_{11}\, x'. \end{aligned} \tag{3.156}$$

Inserting this result into (3.153), we obtain

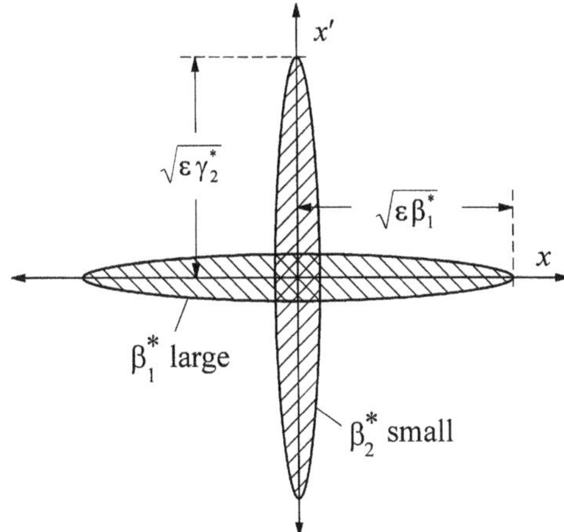

Fig. 3.27 Phase space ellipses at a symmetry point for a large and a small value of β^*. As a consequence of Liouville's theorem, the angular divergence grows with decreasing β^*.

$$
\begin{aligned}
\varepsilon &= \beta\, x'^2 + 2\,\alpha\, x\, x' + \gamma\, x^2 \\
&= \beta_0(-m_{21}x + m_{11}x')^2 \\
&\qquad +2\alpha_0(m_{22}x - m_{12}x')(-m_{21}x + m_{11}x') \\
&\qquad +\gamma_0(m_{22}x - m_{12}x')^2 \\
&= (\beta_0 m_{21}^2 - 2\alpha_0 m_{22}m_{21} + \gamma_0 m_{22}^2)x^2 \\
&\qquad +2(-\beta_0 m_{21}m_{11} + \alpha_0(m_{22}m_{11} + m_{12}m_{21}) - \gamma_0 m_{22}m_{12})xx' \\
&\qquad +(\beta_0 m_{11}^2 - 2\alpha_0 m_{12}m_{11} + \gamma_0 m_{12}^2)x'^2.
\end{aligned} \tag{3.157}
$$

This relation must hold for any values of x and x', so we may compare coefficients to obtain

$$
\begin{aligned}
\beta &= m_{11}^2\beta_0 - 2m_{12}m_{11}\alpha_0 + m_{12}^2\gamma_0 \\
\alpha &= -m_{21}m_{11}\beta_0 + (m_{22}m_{11} + m_{12}m_{21})\alpha_0 - m_{22}m_{12}\gamma_0 \\
\gamma &= m_{21}^2\beta_0 - 2m_{22}m_{21}\alpha_0 + m_{22}^2\gamma_0
\end{aligned} \tag{3.158}
$$

or in matrix notation

$$
\boxed{\begin{pmatrix} \beta \\ \alpha \\ \gamma \end{pmatrix} = \begin{pmatrix} m_{11}^2 & -2m_{11}m_{12} & m_{12}^2 \\ -m_{11}m_{21} & m_{11}m_{22} + m_{12}m_{21} & -m_{22}m_{12} \\ m_{21}^2 & -2m_{22}m_{21} & m_{22}^2 \end{pmatrix} \begin{pmatrix} \beta_0 \\ \alpha_0 \\ \gamma_0 \end{pmatrix}.} \tag{3.159}
$$

This second form of the transfer matrix may also be used to calculate the beta function within any structure of magnets. It too uses the matrix elements which were developed to describe the transformation of the particle trajectory. The results of the two methods presented here are of course equivalent, and both require similar amounts of computing power when used in a calculation.

3.11 Determination of the transfer matrix from the beta function

As we showed in the previous section, the elements of the transfer matrix **M** can be used to uniquely determine the values of the optical functions β, α, and γ at the end of a magnet structure from the known initial values. The converse is also true, namely if the values of β, α, and γ at the beginning and end of a magnetic structure are known then the transfer matrix is uniquely defined. The elements of this matrix must therefore be expressible in terms of the values of the optical functions. This has the advantage that many of the characteristics of a beam transport system can be discussed without a detailed knowledge of the magnet structure, and allows us to make some very general statements about beam optics.

We again start with the solution of the trajectory equation (3.129) and (3.130). Using the rules of addition of trigonometric functions we cast this into the form

$$\begin{aligned} x(s) &= \sqrt{\varepsilon}\sqrt{\beta(s)}\Big[\cos\Psi(s)\cos\phi - \sin\Psi(s)\sin\phi\Big] \\ x'(s) &= -\frac{\sqrt{\varepsilon}}{\sqrt{\beta(s)}}\Big[\alpha(s)\cos\Psi(s)\cos\phi - \alpha(s)\sin\Psi(s)\sin\phi \\ &\qquad + \sin\Psi(s)\cos\phi + \cos\Psi(s)\sin\phi\Big]. \end{aligned} \tag{3.160}$$

With the initial conditions $x(0) = x_0$, $x'(0) = x'_0$, $\beta(0) = \beta_0$, $\alpha(0) = \alpha_0$, and $\Psi(0) = 0$ it immediately follows that

$$\begin{aligned} \cos\phi &= \frac{x_0}{\sqrt{\varepsilon\,\beta_0}} \\ \sin\phi &= -\frac{1}{\sqrt{\varepsilon}}\left(x'_0\sqrt{\beta_0} + \frac{\alpha_0 x_0}{\sqrt{\beta_0}}\right). \end{aligned} \tag{3.161}$$

Inserting this into (3.160), we obtain

$$\begin{aligned} x(s) &= \sqrt{\frac{\beta(s)}{\beta_0}}\Big[\cos\Psi(s) + \alpha_0\sin\Psi(s)\Big]\,x_0 + \sqrt{\beta(s)\beta_0}\sin\Psi(s)\,x'_0 \\ x'(s) &= \frac{1}{\sqrt{\beta(s)\beta_0}}\Big[\big(\alpha_0 - \alpha(s)\big)\cos\Psi(s) - \big(1+\alpha_0\alpha(s)\big)\sin\Psi(s)\Big]\,x_0 \\ &\qquad + \sqrt{\frac{\beta_0}{\beta(s)}}\Big[\cos\Psi(s) - \alpha(s)\sin\Psi(s)\Big]\,x'_0. \end{aligned} \tag{3.162}$$

These equations may again be expressed by a transfer matrix. Using the shorthand $\beta = \beta(s)$, $\alpha = \alpha(s)$, and $\Psi = \Psi(s)$ we obtain the trajectory vector at the point s. It has the form

$$\begin{pmatrix} x(s) \\ x'(s) \end{pmatrix} = \mathbf{M}\begin{pmatrix} x_0 \\ x'_0 \end{pmatrix} \tag{3.163}$$

with

$$\mathbf{M} = \begin{pmatrix} \sqrt{\frac{\beta}{\beta_0}}(\cos\Psi + \alpha_0 \sin\Psi) & \sqrt{\beta\beta_0}\sin\Psi \\ \frac{(\alpha_0 - \alpha)\cos\Psi - (1+\alpha_0\alpha)\sin\Psi}{\sqrt{\beta\beta_0}} & \sqrt{\frac{\beta_0}{\beta}}(\cos\Psi - \alpha\sin\Psi) \end{pmatrix}. \tag{3.164}$$

Notice that as well as the values of the optical functions at the beginning and end of the structure, the advance in phase Ψ of the betatron oscillation between these two points must also be included in this matrix.

3.12 Matching of beam optics

In the development of a system of beam optics, the values of the optical functions at the beginning of a beam transport system are often fixed in advance. They may be defined by a pre-accelerator upstream or constrained by the requirements of an experiment. The evolution of the optical functions through the magnet structure may then be calculated using the matrix equations derived above. In addition, it is usually necessary for the optical functions to be tuned to specific values at the end of the structure. For example, it is extremely important to ensure that the beta function stays within reasonable limits, otherwise the beam dimensions will exceed the aperture of the vacuum chamber. A major task in developing a system of beam optics therefore consists of choosing the strengths k of the quadrupole magnets so that the optical functions have the desired form and, above all, have the required final values at the end of the magnet structure. This procedure is called **matching of beam optics**. Figure 3.28 illustrates this problem. Given the values β_0, α_0, and γ_0 at the beginning of the structure, the quadrupole strengths k_1, $k_2 \ldots k_m$ must be carefully chosen to produce the values β_{E}, α_{E}, and γ_{E} at the end. To achieve this we need to know the relationship between the quadrupole strength k_j and the values of the optical functions at an arbitrary point s. We already have all the mathematical tools we need to hand. The quadrupole strengths determine the matrix elements of the transfer matrices

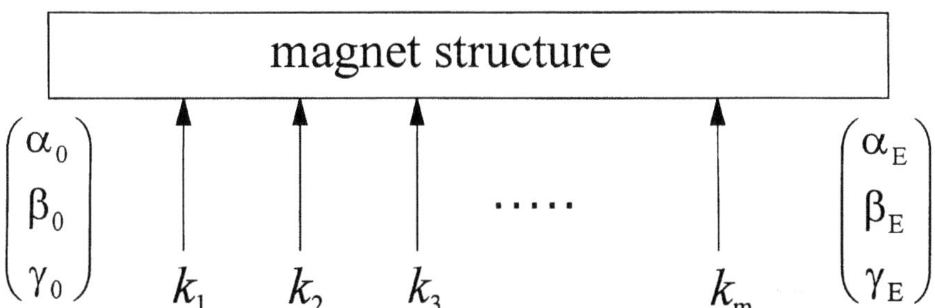

Fig. 3.28 Matching of the optical function values β_{E}, α_{E}, and γ_{E} at the end of a magnet structure by appropriate choice of the quadrupole strengths k_1, $k_2 \ldots k_m$.

(3.84) and (3.85), and these in turn give the beta function and its gradient via (3.149) or (3.159). The dispersion $D(s)$ and its gradient are also determined, like the particle trajectory, by the quadrupole strengths. If we wish for example to adjust the beta function $\beta(s)$ at the point s to a particular value, we need to know the relation

$$\beta(s) = f(k_j), \tag{3.165}$$

where k_j is the strength of the jth quadrupole. $f(k_j)$ may be any non-linear function, and so it is not in general possible to find the solution of (3.165) with respect to k_j by analytical methods. Instead a method must be developed which allows the solution to be found by numerical iteration.

3.12.1 The one-dimensional case

We first of all assume that only one function must be matched, for example β_{E} to the value β_{ideal}. We choose one particular quadrupole, which initially has the arbitrary strength k_0. At the end of the magnet structure the beta function has the resulting value $\beta_{\mathrm{E}}(k_0)$. In general this will not equal the required value. Let us denote the quadrupole strength needed to give the correct form of the beta function by k. The difference between the ideal and actual values of the beta function may then be expanded as a function of the difference in the quadrupole strength as follows

$$\beta_{\mathrm{ideal}} - \beta_{\mathrm{E}}(k_0) = \frac{d\beta_{\mathrm{E}}}{dk}(k - k_0) + \frac{1}{2}\frac{d^2\beta_{\mathrm{E}}}{dk^2}(k - k_0)^2 + \dots\ . \tag{3.166}$$

The differential of the beta function with respect to k is always evaluated at $k = k_0$. Of course this equation still cannot be solved exactly with respect to k so we linearize it, neglecting all quadratic and higher-order terms. It then follows that

$$\beta_{\mathrm{ideal}} - \beta_{\mathrm{E}}(k_0) \approx \frac{d\beta_{\mathrm{E}}}{dk}(k - k_0) \tag{3.167}$$

and the required quadrupole strength is approximately given by

$$k = k_0 + \frac{\beta_{\mathrm{ideal}} - \beta_{\mathrm{E}}(k_0)}{\left(\dfrac{d\beta_{\mathrm{E}}}{dk}\right)_{k=k_0}}. \tag{3.168}$$

This solution is of course not exact, since all the higher order terms in the expansion have been neglected, but in general it will yield a better value than the initial quadrupole strength k_0. This improved value may then be inserted back into equation (3.168) in place of k_0, and the same process repeated to give an even better value. We have thus found a procedure which takes an inexact solution k_p and yields an improved solution k_{p+1}, namely

$$\boxed{k_{p+1} = k_p + \frac{\beta_{\mathrm{ideal}} - \beta_{\mathrm{E}}(k_p)}{\left(\dfrac{d\beta_{\mathrm{E}}}{dk}\right)_{k=k_p}}}. \tag{3.169}$$

Table 3.3 Optical functions which can or must be matched by varying the quadrupole strengths.

optical function	horizontal	vertical
beta function	$\beta_x(s)$	$\beta_z(s)$
gradient of the beta function	$\alpha_x(s)$	$\alpha_z(s)$
betatron phase	$\Psi_x(s)$	$\Psi_z(s)$
dispersion	$D_x(s)$	
gradient of the dispersion	$D'_x(s)$	

This iterative process may be repeated as many times as necessary until the required precision is reached. The starting value k_0 is obtained by trial-and-error or by a rough analytical estimate. The derivative of the beta function is calculated by taking the value of the beta function at the end of the magnet structure for a quadrupole strength k_p, namely $\beta_{\mathrm{E}}(k_p)$, and then considering a slight variation Δk in strength to determine the value $\beta_{\mathrm{E}}(k_p + \Delta k)$. Using this value one obtains

$$\left(\frac{d\beta_{\mathrm{E}}}{dk}\right)_{k=k_p} = \frac{\beta_{\mathrm{E}}(k_p + \Delta k) - \beta_{\mathrm{E}}(k_p)}{\Delta k}. \tag{3.170}$$

For complicated structures of magnets this procedure involves considerable numerical computation and the use of computers becomes essential.

3.12.2 The n-dimensional case

In practice it is very rare that just one optical function has to be matched. What is more, with a single quadrupole it is not even possible to adjust only one optical function, since varying the quadrupole strength will in general change all the optical functions. When developing a system of beam optics it is almost always necessary to match all the optical functions to particular values at the end of the magnet structure, and to treat both planes simultaneously. Table 3.3 lists all the optical functions which must be taken into account in a complete beam optics calculation. Only certain of the optical functions in the table may need to be optimized at once, depending on the situation. Sometimes it is not necessary to pay particular attention to the dispersion. Alternatively, in a beam transport system the form of the betatron phase is not critical and need not be considered. We must develop a procedure for matching n arbitrarily chosen functions at the end of a magnet structure. We denote these n functions by $f_1,\ f_2, \dots f_i, \dots f_n$, as is shown in Fig. 3.29. To adjust the values of these functions we have m quadrupoles available, of strengths $k_1,\ k_2, \dots k_j, \dots k_m$, where we require that $m \geq n$. In the calculation we always choose exactly n of the m quadrupoles, usually taking the most effective ones.

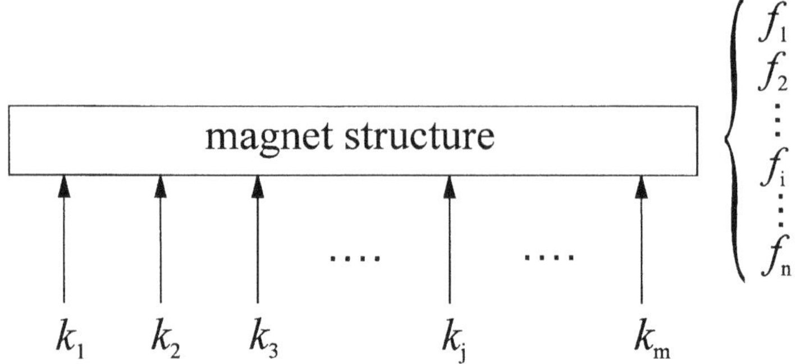

Fig. 3.29 Matching the n optical functions f_i using the m quadrupole strengths k_j

Each of the n optical functions f_i depends on the strengths of all n of the chosen quadrupoles, namely

$$f_i = f_i(k_1,\ k_2, \ldots k_j, \ldots k_n). \tag{3.171}$$

We write the arbitrary starting values of the quadrupole strengths together in the vector $\boldsymbol{K}_0 = (k_{1,0},\ k_{2,0}, \ldots k_{n,0})$. At the end of the magnet structure the resulting values of the functions are then $\boldsymbol{F}_0 = (f_{1,0},\ f_{2,0}, \ldots f_{n,0})$. The ideal values of the optical functions are written $\boldsymbol{F}_{\text{ideal}} = (f_{1,s},\ f_{2,s}, \ldots f_{n,s})$. We again expand to first order the difference between the ideal and current values of the functions and obtain

$$\begin{aligned}
f_{1,s} - f_{1,0} &= \frac{\partial f_1}{\partial k_1}(k_1 - k_{1,0}) + \frac{\partial f_1}{\partial k_2}(k_2 - k_{2,0}) + \cdots + \frac{\partial f_1}{\partial k_n}(k_n - k_{n,0}) \\
f_{2,s} - f_{2,0} &= \frac{\partial f_2}{\partial k_1}(k_1 - k_{1,0}) + \frac{\partial f_2}{\partial k_2}(k_2 - k_{2,0}) + \cdots + \frac{\partial f_2}{\partial k_n}(k_n - k_{n,0}) \\
&\vdots \\
f_{n,s} - f_{n,0} &= \frac{\partial f_n}{\partial k_1}(k_1 - k_{1,0}) + \frac{\partial f_n}{\partial k_2}(k_2 - k_{2,0}) + \cdots + \frac{\partial f_n}{\partial k_n}(k_n - k_{n,0}).
\end{aligned} \tag{3.172}$$

Writing these equations in matrix notation we have

$$\begin{pmatrix} f_{1,s} \\ f_{2,s} \\ \vdots \\ f_{n,s} \end{pmatrix} - \begin{pmatrix} f_{1,0} \\ f_{2,0} \\ \vdots \\ f_{n,0} \end{pmatrix} = \begin{pmatrix} \frac{\partial f_1}{\partial k_1} & \frac{\partial f_1}{\partial k_2} & \cdots & \frac{\partial f_1}{\partial k_n} \\ \frac{\partial f_2}{\partial k_1} & \frac{\partial f_2}{\partial k_2} & \cdots & \frac{\partial f_2}{\partial k_n} \\ \vdots & \vdots & \ddots & \vdots \\ \frac{\partial f_n}{\partial k_1} & \frac{\partial f_n}{\partial k_2} & \cdots & \frac{\partial f_n}{\partial k_n} \end{pmatrix} \cdot \begin{pmatrix} k_1 - k_{1,0} \\ k_2 - k_{2,0} \\ \vdots \\ k_n - k_{n,0} \end{pmatrix} \tag{3.173}$$

The matrix only contains elements of the form $\partial f_i / \partial k_j$ and shows how sensitively each optical function depends on each of the quadrupole strengths. It is known as the **response matrix A**. (3.173) may also be written in simplified form as

$$\boldsymbol{F}_{\text{ideal}} - \boldsymbol{F}_0 = \mathbf{A}(\boldsymbol{K} - \boldsymbol{K}_0) \tag{3.174}$$

or, solving for the quadrupole strength k_j we wish to determine

$$\boldsymbol{K} = \boldsymbol{K}_0 + \mathbf{A}^{-1}(\boldsymbol{F}_{\text{ideal}} - \boldsymbol{F}_0). \tag{3.175}$$

From this relation we immediately obtain the required iteration

$$\boxed{\boldsymbol{K}_{p+1} = \boldsymbol{K}_p + \mathbf{A}_p^{-1}(\boldsymbol{F}_{\text{ideal}} - \boldsymbol{F}_p).} \tag{3.176}$$

The matrix elements of $\mathbf{A}$ are determined piece by piece. We choose the jth quadrupole, whose strength is set to the value k_j. We then use numerical methods to calculate the values f_1 to f_n of all the optical functions. After varying the quadrupole strength by a small amount Δk, all the optical functions are recalculated. The elements of the jth partial vector are then given by

$$\frac{\partial f_i}{\partial k_j} = \frac{f_i(k_1,\ k_2, \ldots, k_j + \Delta k, \ldots, k_n) - f_i(k_1,\ k_2, \ldots, k_j, \ldots, k_n)}{\Delta k}, \tag{3.177}$$

where the index i runs from 1 to n. It should be noted that this procedure does not always yield a useful solution. The high degree of non-linearity can result in the iterative process 'getting stuck' in a local minimum, i.e. the calculation does not converge to a limiting value but instead oscillates around this local minimum. Alternatively it may converge to a value that is not physically meaningful or is not technically achievable. For example, the calculation may sometimes yield unfeasibly high quadrupole strengths, or such huge values for the beta function that the beam would no longer fit within the aperture of the vacuum chamber. In such cases it is necessary to try again using better initial values for the quadrupole strengths $k_{j,0}$. If better values cannot be found then the magnet structure must be modified.

3.13 Periodicity conditions in circular accelerators

In circular accelerators there is no definite beginning or end to the magnet structure. This means that in general there are no predefined values that the optical functions must match at particular points along the orbit. On the other hand, the orbit in a circular accelerator is closed, i.e. the optical functions must repeat continuously after a complete orbit. Another common constraint comes from the introduction of symmetry points through which the optical functions are reflected. Both types of condition allow the optical functions to be defined at particular points around the ring and hence all the way around the orbit.

3.13.1 The periodic solution

Let us choose an arbitrary point s_0 along the orbit, at which the optical functions are initially unknown. We require that after a full revolution, i.e. at the point $s_0 + L$, where L is the circumference of the accelerator, the functions must have the same values as at the starting point s_0. This results in the following **periodicity conditions** for the optical functions:

$$\begin{aligned}
\beta(s_0 + L) &= \beta(s_0) = \beta_0 \\
\alpha(s_0 + L) &= \alpha(s_0) = \alpha_0 \\
D(s_0 + L) &= D(s_0) = D_0 \\
D'(s_0 + L) &= D'(s_0) = D'_0.
\end{aligned} \tag{3.178}$$

Let us denote the transfer matrix for one full revolution by $\mathbf{M}_{\text{rev}}$. We begin by calculating the beta function, and define a beta matrix $\mathbf{B}_0$ at the point s_0, by analogy with (3.141). Applying the periodicity conditions to the transformation equation (3.149) gives us the following relation

$$\mathbf{B}_0 = \mathbf{M}_{\text{rev}} \cdot \mathbf{B}_0 \cdot \mathbf{M}_{\text{rev}}^{\text{T}} \tag{3.179}$$

or explicitly

$$\begin{pmatrix} \beta_0 & -\alpha_0 \\ -\alpha_0 & \gamma_0 \end{pmatrix} = \begin{pmatrix} m_{11} & m_{12} \\ m_{21} & m_{22} \end{pmatrix} \cdot \begin{pmatrix} \beta_0 & -\alpha_0 \\ -\alpha_0 & \gamma_0 \end{pmatrix} \cdot \begin{pmatrix} m_{11} & m_{21} \\ m_{12} & m_{22} \end{pmatrix}. \tag{3.180}$$

Multiplying out we immediately obtain the equations determining β_0, α_0, and γ_0

$$\begin{aligned}
\beta_0 &= \frac{2\, m_{12}}{\sqrt{2 - m_{11}^2 - 2m_{12}m_{21} - m_{22}^2}} \\
\alpha_0 &= \frac{m_{11} - m_{22}}{2m_{12}} \beta_0 \\
\gamma_0 &= \frac{1 + \alpha_0^2}{\beta_0}.
\end{aligned} \tag{3.181}$$

It is clear that only one solution exists which satisfies these conditions, namely when the expression under the square root is greater than zero, i.e.

$$2 - m_{12}^2 - 2m_{12}m_{21} - m_{22}^2 > 0. \tag{3.182}$$

The calculation of the periodic solution for the dispersion proceeds along the same lines except that this time we use a 3×3 matrix to describe a full revolution. The elements are determined in standard fashion by multiplying all the partial matrices of the magnet structure, as shown in Section 3.6 in the calculation of the evolution of the dispersive trajectory. In this case the periodicity condition gives

$$\begin{pmatrix} D_0 \\ D'_0 \\ 1 \end{pmatrix} = \begin{pmatrix} m_{11} & m_{12} & m_{13} \\ m_{21} & m_{22} & m_{23} \\ 0 & 0 & 1 \end{pmatrix} \cdot \begin{pmatrix} D_0 \\ D'_0 \\ 1 \end{pmatrix}. \tag{3.183}$$

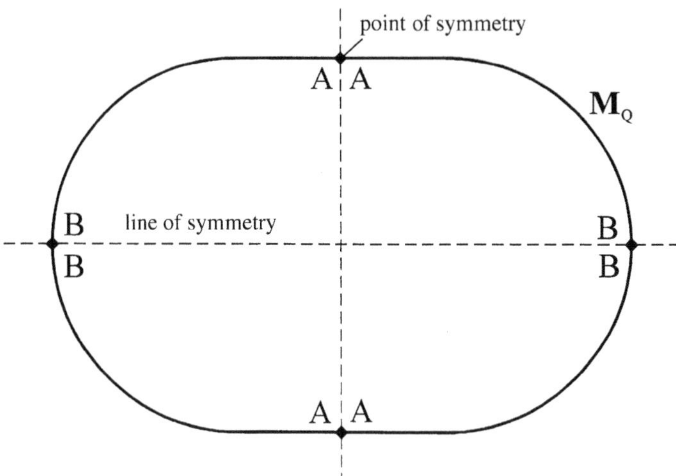

Fig. 3.30 Division of a circular accelerator into four identical quadrants reflected about the two lines of symmetry.

Solving for D and D' yields

$$
\begin{aligned}
D_0' &= \frac{m_{21}m_{13} + m_{23}(1 - m_{11})}{2 - m_{11} - m_{22}} \\
D_0 &= \frac{m_{12}D_0' + m_{13}}{1 - m_{11}}.
\end{aligned}
\tag{3.184}
$$

Here there is almost always a periodic solution, ignoring the cases where $m_{11} + m_{22} = 2$ or $m_{11} = 1$.

3.13.2 The symmetric solution

In circular accelerators attempts are always made to arrange the magnets according to certain symmetries in order simplify the calculation of the beam optics and to reduce the number of different magnet circuits. An example is shown in Fig. 3.30 in which the accelerator is divided into four identical quadrants reflected about the symmetry points A and B. The transfer matrix for one quadrant is $\mathbf{M}_\mathrm{Q}$. In this case it is only necessary to calculate the beam optics for one such quadrant - the optics of the other quadrants are mirror-symmetric. In this arrangement the beginning A and end B of each quadrant are symmetry points. The derivatives of the optical functions must therefore vanish at these points, namely

$$
\begin{aligned}
\alpha_\mathrm{A} &= 0 \\
\alpha_\mathrm{B} &= 0 \\
D_\mathrm{A}' &= 0 \\
D_\mathrm{B}' &= 0.
\end{aligned}
\tag{3.185}
$$

If $\mathbf{B}_\mathrm{A}$ is the beta matrix at the beginning of each quadrant and $\mathbf{B}_\mathrm{B}$ that at the end, then the transformation equation (3.149) again yields

$$\mathbf{B}_\mathrm{B} = \mathbf{M}_\mathrm{Q} \cdot \mathbf{B}_\mathrm{A} \cdot \mathbf{M}_\mathrm{Q}^\mathrm{T}. \tag{3.186}$$

Imposing the symmetry condition (3.185) gives

$$\begin{pmatrix} \beta_\mathrm{B} & 0 \\ 0 & 1/\beta_\mathrm{B} \end{pmatrix} = \begin{pmatrix} m_{11} & m_{12} \\ m_{21} & m_{22} \end{pmatrix} \cdot \begin{pmatrix} \beta_\mathrm{A} & 0 \\ 0 & 1/\beta_\mathrm{A} \end{pmatrix} \cdot \begin{pmatrix} m_{11} & m_{21} \\ m_{12} & m_{22} \end{pmatrix}. \tag{3.187}$$

Multiplying out the product of matrices on the right-hand side and comparing the resulting matrix elements with those of the matrix on the left-hand side, we obtain

$$\begin{aligned} \beta_\mathrm{B} &= \beta_\mathrm{A} m_{11}^2 + \frac{m_{12}^2}{\beta_\mathrm{A}} \\ 0 &= \beta_\mathrm{A} m_{11} m_{21} + \frac{m_{12} m_{22}}{\beta_\mathrm{A}}. \end{aligned} \tag{3.188}$$

These equations determine the beta function at the beginning and end of the quadrant. Solving them yields the values

$$\begin{aligned} \beta_\mathrm{A} &= \sqrt{-\frac{m_{12} m_{22}}{m_{21} m_{11}}} \\ \beta_\mathrm{B} &= -\frac{1}{\beta_\mathrm{A}} \frac{m_{12}}{m_{21}}. \end{aligned} \tag{3.189}$$

Recall that $\det \mathbf{M}_\mathrm{Q} = m_{11} m_{22} - m_{12} m_{21} = 1$. From these expressions we immediately obtain the conditions for the existence of a symmetric solution. Since the beta function is always positive definite, we must have

$$\frac{m_{12}}{m_{21}} < 0 \tag{3.190}$$

and since the square root should only give real solutions we require

$$\frac{m_{22}}{m_{11}} > 0. \tag{3.191}$$

To find the symmetric solution for the dispersion we again take the 3×3 matrix and form the transformation equation

$$\begin{pmatrix} D_\mathrm{B} \\ 0 \\ 1 \end{pmatrix} = \begin{pmatrix} m_{11} & m_{12} & m_{13} \\ m_{21} & m_{22} & m_{23} \\ 0 & 0 & 1 \end{pmatrix} \cdot \begin{pmatrix} D_\mathrm{A} \\ 0 \\ 1 \end{pmatrix}, \tag{3.192}$$

which gives us the expressions

$$\begin{aligned} D_\mathrm{B} &= m_{11} D_\mathrm{A} + m_{13} \\ 0 &= m_{21} D_\mathrm{A} + m_{23}. \end{aligned} \tag{3.193}$$

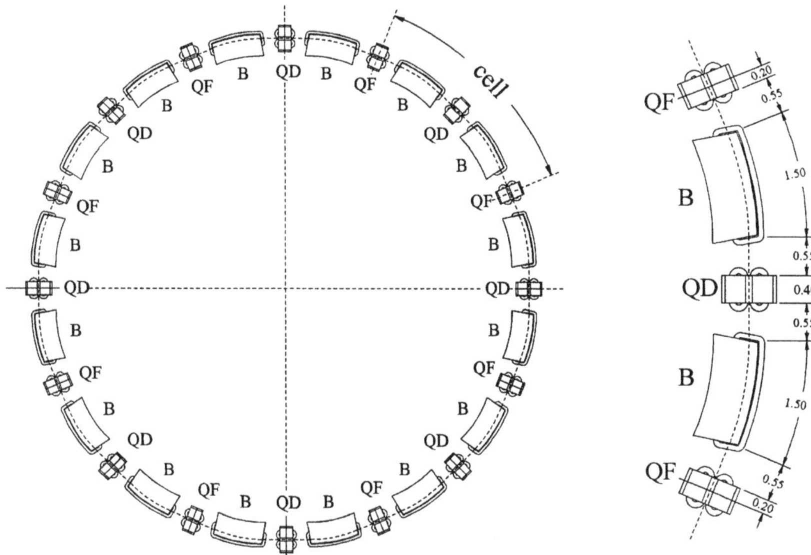

Fig. 3.31 Example of a circular accelerator employing a FODO structure. The ring consists of a number of identical cells, each consisting of two bending magnets, with quadrupoles arranged with alternating polarity between them.

The values for the dispersion at the beginning and end of the quadrant are then

$$\begin{aligned} D_{\mathrm{A}} &= -\frac{m_{23}}{m_{21}} \\ D_{\mathrm{B}} &= -\frac{m_{11}m_{23}}{m_{21}} + m_{13}. \end{aligned} \tag{3.194}$$

Equations (3.189) and (3.194) allow us to evaluate the beta function and dispersion at the beginning and end of a quadrant, or indeed of any generalized magnet structure with symmetry points, provided that it satisfies equations (3.190) and (3.191). If they are not satisfied then no solution exists and the quadrupole strengths must be altered until a solution can be found.

3.13.3 Worked example: beam optics of a circular accelerator with a FODO structure

It is often necessary to transport beams over a long distance without the beam cross-section becoming too large. This is the case, for example, in the machine section which connects a pre-accelerator to a storage ring, or within a storage ring, in order to steer the beam in a controlled way from one interaction point to another. The simplest arrangement of magnets for this task is the **FODO structure**, illustrated in Fig. 3.31. Since quadrupoles always focus only in one plane while defocusing in the other plane, the magnets are arranged one after another with alternating polarity. In between each pair is a bending magnet, which bends the beam around the circular path. The quadrupoles and dipoles are usually placed a certain distance apart, so that field-free drift regions exist between them. From the point of view of beam focusing, this arrangement consists of a series of focusing and defocusing quadrupoles with a section between them, indicated by 0, where there is no focusing (ignoring weak and edge focusing.) We thus have an arrangement of the form: F 0 D 0 etc., hence the name of this magnet structure.

The example shown in Fig. 3.31 consists of 16 dipoles in total plus an equal number of quadrupoles. Rectangular magnets are used in the dipoles, so we must take account of the edge focusing at both ends. Together with the 32 drift sections this gives a total of 96 transfer matrices. This is already a large number, just for this small example ring, but they do not all need to be calculated because the FODO structure means that the ring is symmetric. It consists of a total of eight **cells**, each of which begins in the middle of a focusing quadrupole (QF) and ends in the middle of the next QF. In between there are two bending magnets, the defocusing quadrupole QD and four identical drift sections. One such cell is shown in Fig. 3.31, labelled with the physical dimensions relating to this example. The middle of the quadrupole is always a symmetry point. Hence we only need to calculate the beam optics for one such cell, which satisfies the symmetry conditions outlined in the preceding sections. The optics of the complete ring are obtained by simply chaining together the appropriate number of **FODO cells**. The actual number of cells making up a whole ring does not matter. The symmetry conditions thus considerably reduce the amount of computation needed.

Let us now quantitatively calculate the optics of one of the FODO cells which makes up the simple circular accelerator shown in Fig. 3.31. We will assume the following values for the individual elements of the structure:

Quadrupoles QF and QD :

QF		QD	
k_{QF}	$=$ - 1.20 m^{-2}	k_{QD}	$=$ 1.20 m^{-2}
l_{QF}	$=$ 0.20 m	l_{QD}	$=$ 0.40 m

Dipole magnet B with edge focusing EB :

B		EB	
R_{B}	$=$ 3.8197 m	Ψ_{B}	$=$ 11.25 °
l_{B}	$=$ 1.50 m		$=$ 0.1964 rad

Drift section D : length $l_{\mathrm{D}} = 0.55$ m

Using relations (3.84) to (3.88) and (3.109) we can explicitly calculate the transfer matrices of the individual elements, based on these parameters:

Focusing quadrupole:

$$\mathbf{M}_{\mathrm{QF}} = \begin{pmatrix} 0.9761 & 0.1984 & 0 & 0 & 0 \\ -0.2381 & 0.9761 & 0 & 0 & 0 \\ 0 & 0 & 1.0241 & 0.2106 & 0 \\ 0 & 0 & 0.2419 & 1.0241 & 0 \\ 0 & 0 & 0 & 0 & 1 \end{pmatrix} \qquad (3.195)$$

Defocusing quadrupole:

$$\mathbf{M}_{\mathrm{QD}} = \begin{pmatrix} 1.0975 & 0.4129 & 0 & 0 & 0 \\ 0.4955 & 1.0975 & 0 & 0 & 0 \\ 0 & 0 & 0.9055 & 0.3873 & 0 \\ 0 & 0 & -0.4648 & 0.9055 & 0 \\ 0 & 0 & 0 & 0 & 1 \end{pmatrix} \qquad (3.196)$$

Dipole:

$$\mathbf{M}_{\mathrm{B}} = \begin{pmatrix} 0.9239 & 1.4617 & 0 & 0 & 0.2908 \\ -0.1002 & 0.9239 & 0 & 0 & 0.3827 \\ 0 & 0 & 1 & 1.5000 & 0 \\ 0 & 0 & 0 & 1 & 0 \\ 0 & 0 & 0 & 0 & 1 \end{pmatrix} \qquad (3.197)$$

Edge focusing:

$$\mathbf{M}_{\mathrm{EB}} = \begin{pmatrix} 1 & 0 & 0 & 0 & 0 \\ 0.0521 & 1 & 0 & 0 & 0 \\ 0 & 0 & 1 & 0 & 0 \\ 0 & 0 & -0.0521 & 1 & 0 \\ 0 & 0 & 0 & 0 & 1 \end{pmatrix} \qquad (3.198)$$

Drift section:

$$\mathbf{M}_{\mathrm{D}} = \begin{pmatrix} 1 & 0.5500 & 0 & 0 & 0 \\ 0 & 1 & 0 & 0 & 0 \\ 0 & 0 & 1 & 0.5500 & 0 \\ 0 & 0 & 0 & 1 & 0 \\ 0 & 0 & 0 & 0 & 1 \end{pmatrix} \qquad (3.199)$$

Finally we obtain the transfer matrix for the whole cell from these individual matrices according to

$$\begin{aligned} \mathbf{M}_{\mathrm{Z}} &= \mathbf{M}_{\mathrm{QF}}\mathbf{M}_{\mathrm{D}}\mathbf{M}_{\mathrm{EB}}\mathbf{M}_{\mathrm{B}}\mathbf{M}_{\mathrm{EB}}\mathbf{M}_{\mathrm{D}}\mathbf{M}_{\mathrm{QD}}\mathbf{M}_{\mathrm{D}}\mathbf{M}_{\mathrm{EB}}\mathbf{M}_{\mathrm{B}}\mathbf{M}_{\mathrm{EB}}\mathbf{M}_{\mathrm{D}}\mathbf{M}_{\mathrm{QF}} \\ &= \begin{pmatrix} 0.0808 & 9.7855 & 0 & 0 & 3.1424 \\ -0.1015 & 0.0808 & 0 & 0 & 0.3471 \\ 0 & 0 & -0.4114 & 1.1280 & 0 \\ 0 & 0 & -0.7365 & -0.4114 & 0 \\ 0 & 0 & 0 & 0 & 1 \end{pmatrix} . \end{aligned} \qquad (3.200)$$

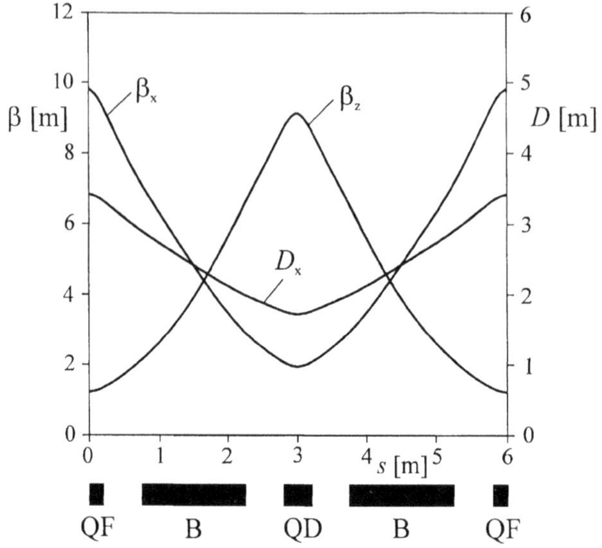

Fig. 3.32 Optical functions of the FODO cell in the example ring.

Using equations (3.189), these matrix elements allow us to calculate the beta functions in the x- and z-planes at the beginning and end of the cell, namely

$$\begin{aligned} \beta_{\mathrm{A,x}} &= \sqrt{-\frac{m_{12}m_{22}}{m_{21}m_{11}}} = 9.8176 \text{ m} \\ \beta_{\mathrm{B,x}} &= -\frac{1}{\beta_{\mathrm{A,x}}}\frac{m_{12}}{m_{21}} = 9.8176 \text{ m} \end{aligned} \tag{3.201}$$

and

$$\begin{aligned} \beta_{\mathrm{A,z}} &= \sqrt{-\frac{m_{34}m_{44}}{m_{43}m_{33}}} = 1.2376 \text{ m} \\ \beta_{\mathrm{B,z}} &= -\frac{1}{\beta_{\mathrm{A,z}}}\frac{m_{34}}{m_{43}} = 1.2376 \text{ m}. \end{aligned} \tag{3.202}$$

From (3.194) we obtain the dispersion

$$\begin{aligned} D_{\mathrm{A,x}} &= -\frac{m_{25}}{m_{21}} = 3.4187 \text{ m} \\ D_{\mathrm{B,x}} &= -\frac{m_{11}m_{25}}{m_{21}} + m_{15} = 3.4187 \text{ m}. \end{aligned} \tag{3.203}$$

Owing to the symmetrical cell structure, the beta function and dispersion have the same values at the beginning and end of the cell. As a result they can be easily chained together, without changing the optics. The form of the beta function within a cell can be calculated using the transformation relation (3.149), while the dispersion is simply treated just like a particle trajectory using (3.105). The result of these calculations for the FODO cell used in this example ring is shown in in Fig. 3.32.

3.14 Tune and optical resonances

In circular accelerators such as synchrotrons or storage rings the beam repeatedly encounters the same magnet structure, once every full revolution. It therefore undergoes forces which repeat periodically. Since the focusing magnets can cause the beam particles to undergo transverse oscillations, under certain conditions the magnetic structure may cause the circulating beam to resonate. This makes the oscillation amplitude of the particles grow rapidly, causing the beam to blow up or in extreme cases to be lost. This phenomenon is known as **optical resonance**, and we will study it further in the following sections.

3.14.1 Periodic solution of Hill's differential equation

In circular accelerators, as in all beam transport systems, the trajectory of particles with nominal momentum, i.e. $\Delta p/p = 0$, is given by Hill's differential equation

$$x''(s) + K(s)x(s) = 0. \tag{3.204}$$

In a circular machine the focusing function $K(s) = 1/R(s) - k(s)$ is periodic with the circumference of the ring:

$$K(s + L) = K(s). \tag{3.205}$$

Using **Floquet's theorem**, we obtain the same solution as before

$$x(s) = \sqrt{\varepsilon}\sqrt{\beta(s)}\cos\Big[\Psi(s) + \phi\Big], \tag{3.206}$$

but this time the beta function $\beta(s)$ is also periodic, with the same period as $K(s)$. The resonant behaviour depends crucially on the betatron phase $\Delta\Psi = \Psi(s + L) - \Psi(s)$ over one complete revolution. We thus define the **tune** or Q **value** of a circular accelerator as

$$Q \equiv \frac{\Delta\Psi}{2\pi} = \frac{1}{2\pi}\int_0^L \frac{ds}{\beta(s)} = \frac{1}{2\pi}\oint \frac{ds}{\beta(s)}. \tag{3.207}$$

Owing to the periodicity of $\beta(s)$, Q is independent of position s. The tune tells us how many betatron oscillations a particle undergoes per revolution. The periodicity conditions $\beta(s + L) = \beta(s)$ and $\alpha(s + L) = \alpha(s)$ allow us to simplify the transfer matrix for a full revolution (3.164), namely

$$\mathbf{M}_{s\to s+L} = \begin{pmatrix} \cos\mu + \alpha(s)\sin\mu & \beta(s)\sin\mu \\ -\gamma(s)\sin\mu & \cos\mu - \alpha(s)\sin\mu \end{pmatrix} \tag{3.208}$$

with $\mu = 2\pi Q$. If we move an infinitesimal distance from the point s to $s + ds$, (3.208) gives us

$$\mathbf{M}_{s+ds\to s+ds+L} = \begin{pmatrix} \cos\mu + \left(\alpha + \dfrac{d\alpha}{ds}ds\right)\sin\mu & \left(\beta + \dfrac{d\beta}{ds}ds\right)\sin\mu \\ -\left(\gamma + \dfrac{d\gamma}{ds}ds\right)\sin\mu & \cos\mu - \left(\alpha + \dfrac{d\alpha}{ds}ds\right)\sin\mu \end{pmatrix}$$

$$= \mathbf{M}_{s\to s+L} + \begin{pmatrix} \alpha'(s)\sin\mu & \beta'(s)\sin\mu \\ -\gamma'(s)\sin\mu & -\alpha'(s)\sin\mu \end{pmatrix} ds. \quad (3.209)$$

This matrix can also be expressed in another form. Let us consider two points s_1 and s_2 along the orbit of a circular accelerator, chosen such that $s_2 > s_1$. Using Fig. 3.33 we immediately see that

$$\begin{aligned} \mathbf{M}_{s_1\to s_2+L} &= \mathbf{M}_{s_2\to s_2+L}\,\mathbf{M}_{s_1\to s_2} \\ &= \mathbf{M}_{s_1\to s_2}\,\mathbf{M}_{s_1\to s_1+L}. \end{aligned} \quad (3.210)$$

From this we obtain

$$\begin{aligned} \mathbf{M}_{s_2\to s_2+L} &= \mathbf{M}_{s_2\to s_2+L}\,\mathbf{M}_{s_1\to s_2}\,\mathbf{M}^{-1}_{s_1\to s_2} \\ &= \mathbf{M}_{s_1\to s_2}\,\mathbf{M}_{s_1\to s_1+L}\,\mathbf{M}^{-1}_{s_1\to s_2}. \end{aligned} \quad (3.211)$$

We may thus describe the transformation from the position $s + ds$ to the end of the next full revolution by the matrix

$$\mathbf{M}_{s+ds\to s+ds+L} = \mathbf{M}_{s\to s+ds}\,\mathbf{M}_{s\to s+L}\,\mathbf{M}^{-1}_{s\to s+ds}. \quad (3.212)$$

The transformation along the path element ds can be described by the matrix for an infinitely thin lens

$$\mathbf{M}_{s\to s+ds} = \begin{pmatrix} 1 & ds \\ -K(s)ds & 1 \end{pmatrix} \quad (3.213)$$

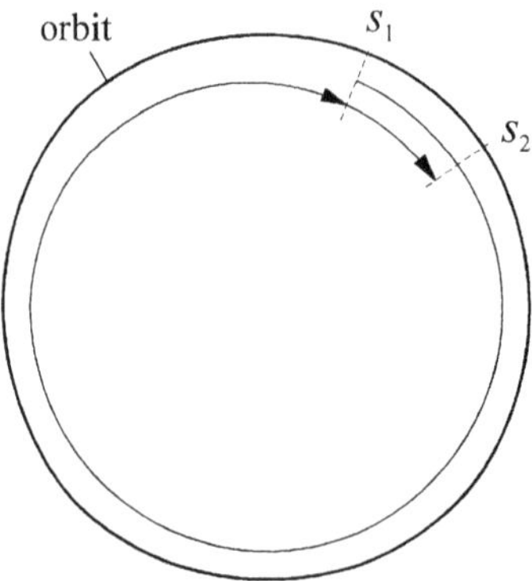

Fig. 3.33 Calculating the transfer matrix in a circular accelerator.

which is easily derived from one of the relations (3.84) to (3.87) by setting the variable s to the very small value ds. Inserting the matrices (3.213) and (3.208) into (3.212), we obtain

$$
\begin{aligned}
&\mathbf{M}_{s+ds\to s+ds+L} \\
&= \begin{pmatrix} 1 & ds \\ -K(s)ds & 1 \end{pmatrix} \begin{pmatrix} \cos\mu + \alpha\sin\mu & \beta\sin\mu \\ -\gamma\sin\mu & \cos\mu - \alpha\sin\mu \end{pmatrix} \begin{pmatrix} 1 & -ds \\ K(s)ds & 1 \end{pmatrix} \\
&= \begin{pmatrix} \cos\mu + \alpha sin\mu + (\beta K(s)-\gamma)\sin\mu\, ds & \beta\sin\mu - 2\alpha\sin\mu\, ds \\ -\gamma\sin\mu - 2K(s)\alpha\sin\mu\, ds & \cos\mu - \alpha\sin\mu - (K(s)\beta-\gamma)\sin\mu\, ds \end{pmatrix} \\
&= \begin{pmatrix} \cos\mu + \alpha\sin\mu & \beta\sin\mu \\ -\gamma\sin\mu & \cos\mu - \alpha\sin\mu \end{pmatrix} \\
&\qquad + \begin{pmatrix} (\beta K(s)-\gamma)\sin\mu & -2\alpha\sin\mu \\ -2K(s)\alpha\sin\mu & -(\beta K(s)-\gamma)\sin\mu \end{pmatrix} ds.
\end{aligned}
\tag{3.214}
$$

Comparing this relation with (3.209) yields

$$
\begin{aligned}
\alpha'(s) &= \beta(s)K(s) - \gamma(s) \\
\beta'(s) &= -2\alpha(s) \\
\gamma'(s) &= 2\alpha(s)K(s).
\end{aligned}
\tag{3.215}
$$

3.14.2 Floquet's transformation

When considering optical resonances it is sometimes helpful to use a form of the equation of motion (3.204) which is periodic with the machine circumference. We therefore introduce the variable

$$
\phi(s) \equiv \frac{\Psi(s)}{Q} = \frac{1}{Q}\int \frac{ds}{\beta(s)} \tag{3.216}
$$

which changes value by exactly 2π per revolution. Its derivative is $\dfrac{d\phi}{ds} = \dfrac{1}{Q\beta(s)}$. We also replace the transverse particle amplitude $x(s)$ by the normalized quantity

$$
\eta(s) \equiv \frac{x(s)}{\sqrt{\beta(s)}}. \tag{3.217}
$$

Its derivatives with respect to the new variable ϕ are

$$\begin{aligned} \frac{d\eta}{d\phi} &= \frac{d\eta}{ds}\frac{ds}{d\phi} = \frac{d}{ds}\left(\frac{x(s)}{\sqrt{\beta(s)}}\right) Q\beta(s) \\ &= \left(\frac{\alpha(s)}{\sqrt{\beta(s)}}\, x(s) + \sqrt{\beta(s)}\; x'(s)\right) Q \end{aligned} \tag{3.218}$$

and

$$\begin{aligned} \frac{d^2\eta}{d\phi^2} &= \frac{d}{d\phi}\left(\frac{d\eta}{d\phi}\right) = \frac{d}{ds}\left(\frac{d\eta}{d\phi}\right) Q\beta(s) \\ &= \left\{\beta^{\frac{3}{2}}(s)\; x''(s) + \frac{\alpha^2(s)}{\sqrt{\beta(s)}}\, x(s) + \alpha'(s)\sqrt{\beta(s)}\; x(s)\right\} Q^2. \end{aligned} \tag{3.219}$$

Using (3.215) it follows that

$$\begin{aligned} \frac{d^2\eta}{d\phi^2} &= \left\{\beta^{\frac{3}{2}}(s)x''(s) + \alpha^2(s)\frac{x(s)}{\sqrt{\beta(s)}} + \left[K(s)\beta(s) - \frac{1+\alpha^2(s)}{\beta(s)}\right]\sqrt{\beta(s)}x(s)\right\} Q^2 \\ &= \left\{\beta^{\frac{3}{2}}(s)\Big[x''(s) + K(s)x(s)\Big] - \frac{x(s)}{\sqrt{\beta(s)}}\right\} Q^2. \end{aligned} \tag{3.220}$$

We can use these relations to obtain a convenient transformed version of the equation of motion, which is also known as **Floquet's transformation**. We multiply the equation of motion (3.204) by $\beta^{\frac{3}{2}}(s)Q^2$, yielding

$$\begin{aligned} 0 &= \Big[x''(s) + K(s)x(s)\Big]\beta^{\frac{3}{2}}(s)Q^2 \\ &= \underbrace{\left\{\beta^{\frac{3}{2}}\left[x''(s) + K(s)x(s)\right] - \frac{x(s)}{\sqrt{\beta(s)}}\right\} Q^2}_{= d^2\eta/d\phi^2} + \frac{x(s)}{\sqrt{\beta(s)}}Q^2. \end{aligned} \tag{3.221}$$

Using the definition (3.217) we finally arrive at the transformed equation of motion

$$\boxed{\frac{d^2\eta}{d\phi^2} + Q^2\eta = 0.} \tag{3.222}$$

3.14.3 Optical resonances

In an ideal linear machine the magnetic field produced around the beam by the magnet structure may be written according to (3.4) in the form

$$\frac{e}{p}B_z(x,s) = \frac{e}{p}B_{z0}(s) + \frac{e}{p}\frac{dB_z(s)}{dx}x(s) = \frac{1}{R(s)} + k(s)x(s) \tag{3.223}$$

where $\dfrac{e}{p} = \dfrac{1}{R\,B_{z0}}$. By an ideal machine we mean a circular accelerator in which the magnetic fields agree exactly with the theoretical predictions. In realistic accelerators, however, this is of course impossible. First of all the relative field strengths vary from magnet to magnet by at least $\Delta B/B > 10^{-4}$, and secondly the magnets have fringe fields which are not sufficiently well described by the hard-edge model upon which the beam optics calculations are based. Under certain conditions these field errors can induce resonant betatron oscillations of the particles, leading to beam instabilities. In order to study the effect of field errors more closely let us introduce a realistic field, consisting of the ideal theoretical field with field errors superimposed upon it. We assume a field distribution of the form

$$\begin{aligned} \frac{e}{p}\tilde{B}_z(x,s) &= \frac{1}{R(s)} + k(s)x(s) + \frac{e}{p}\Delta B(x,s) \\ &= \frac{1}{R(s)} + k(s)x(s) + \frac{\Delta B(x,s)}{R\ B_{z0}} \end{aligned} \tag{3.224}$$

where $\Delta B(x,s)$ describes the field errors. For particles of nominal momentum this becomes the inhomogeneous form of Hill's differential equation (3.204)

$$x''(s) + K(s)x(s) = \frac{\Delta B(x,s)}{R\ B_{z0}}. \tag{3.225}$$

Applying Floquet's transformation immediately yields

$$\boxed{\frac{d^2\eta}{d\phi^2} + Q^2\eta = \beta^{\frac{3}{2}}Q^2\frac{\Delta B}{R\ B_{z0}}.} \tag{3.226}$$

Using Floquet's theorem we then obtain the periodic solution

$$\eta(\phi) = \frac{Q}{2\sin\pi Q}\int_{\phi}^{\phi+2\pi}\beta^{\frac{3}{2}}\frac{\Delta B}{R\ B_{z0}}\cos\Big[Q(\pi+\phi+\vartheta)\Big]\ d\vartheta. \tag{3.227}$$

We immediately see that the amplitude of the particle oscillations grows without limit as the tune Q of the accelerator approaches a whole number. Clearly the beam cannot circulate stably under such conditions, and so the region close to an integer value of Q is called an **integer stopband**. This integer resonance condition may be visualized with the help of Fig. 3.34. For simplicity it is assumed that all but one of the magnets are free of errors. A particle starts off exactly along the orbit, i.e. $\boldsymbol{x} = 0$, and in an ideal machine would continue along this orbit. In this case, however, it encounters the defective magnet and is deflected onto a new trajectory, at a non-zero angle $\Delta\alpha$ to the orbit. This angle is proportional to the strength of the field error. The focusing system bends the trajectory of the particle back towards the orbit and the particle continues, performing betatron oscillations about the orbit. At the end of a full revolution the particle arrives back at the faulty magnet and once again undergoes the same angular deflection $\Delta\alpha$. If the tune Q is a whole number then this change in the trajectory angle

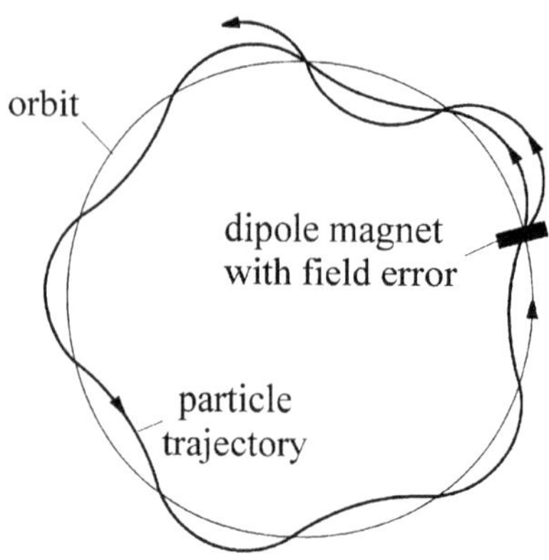

Fig. 3.34 Resonant excitation of betatron oscillations by a dipole field error at an integer value of Q (integer stopband).

always occurs at the same value of the betatron phase and so the angle increases with each revolution. The oscillation amplitude thus grows linearly with the number of revolutions until the particles hit the wall of the vacuum chamber.

This simple example illustrates the principle of optical resonance, but in reality things are rather more complicated. Firstly, all magnets have field errors which may be stronger or weaker depending on their relative position and on the particular beam optics. Furthermore, the focusing magnets cause the beam cross-section to vary around the orbit, which changes the effect of the field error. In order to study this problem in more detail, let us expand the field error in terms of its multipoles:

$$\Delta B(x) = \Delta B_0 + \frac{d\Delta B}{dx}x + \frac{1}{2}\frac{d^2\Delta B}{dx^2}x^2 + \frac{1}{3}\frac{d^3\Delta B}{dx^3}x^3 + \ldots. \tag{3.228}$$

Recalling that $\frac{d}{d\eta} = \frac{d}{dx}\frac{dx}{d\eta}$ with $\frac{dx}{d\eta} = \sqrt{\beta}$, it follows from (3.228) that

$$\Delta B(\eta) = \Delta B_0 + \beta^{\frac{1}{2}}\frac{d\Delta B}{d\eta}\eta + \frac{1}{2}\beta^{\frac{2}{2}}\frac{d^2\Delta B}{d\eta^2}\eta^2 + \ldots. \tag{3.229}$$

Inserting this field expansion into (3.226), we obtain

$$\frac{d^2\eta}{d\phi^2} + Q^2\eta = \frac{Q^2}{R\,B_{z0}}\left(\beta^{\frac{3}{2}}\Delta B_0 + \beta^{\frac{4}{2}}\frac{d\Delta B}{d\eta}\eta + \frac{1}{2}\beta^{\frac{5}{2}}\frac{d^2\Delta B}{d\eta^2}\eta^2 + \ldots\right). \tag{3.230}$$

If, for $Q = p$ (p = integer), the product $\beta^{\frac{3}{2}}\Delta B$ contains a non-vanishing term of the pth harmonic of a full revolution, the betatron oscillations will be resonantly excited. This is the case of the integer stopband, which we have already seen. For half-integer Q values, i.e. $Q = n + \frac{1}{2}$ with n an integer, $2Q = p$. In this case the pth harmonic of $\beta^{\frac{4}{2}}\frac{d\Delta B}{d\eta}\eta$ can excite resonances. The third term describes the effect of sextupole fields. Here the resonance condition is satisfied for $3Q = p$ if $\beta^{\frac{5}{2}}\frac{d^2\Delta B}{d\eta^2}\eta^2$ contains components of the pth harmonic. In the same way, higher multipole fields can also induce resonances.

Let us now look in more detail at the mechanism of resonant excitation of betatron oscillations by higher multipoles of the beam steering fields. We will look explicitly at the first three multipole components. We begin with the general solution of the equations of motion (3.129) and (3.130), which we write with $\chi = \Psi + \phi$ in the form

$$\begin{aligned} x(s) &= \sqrt{\varepsilon}\sqrt{\beta(s)}\cos\chi(s) \\ x'(s) &= -\frac{\sqrt{\varepsilon}}{\sqrt{\beta(s)}}\Big[\alpha(s)\cos\chi(s)+\sin\chi(s)\Big]. \end{aligned} \tag{3.231}$$

Let $a = x(0)$ be the amplitude of the betatron oscillation at the position $s = 0$. Writing $\beta_0 = \beta(0)$ we have $\sqrt{\varepsilon} = a/\sqrt{\beta_0}$. We can thus write the trajectory function in the form

$$x(s) = a\sqrt{\frac{\beta(s)}{\beta_0}}\cos\chi(s). \tag{3.232}$$

From the second equation of (3.231) it follows that

$$\begin{aligned} \beta(s)x'(s) &= -a\sqrt{\frac{\beta(s)}{\beta_0}}\Big[\alpha(s)\cos\chi(s)+\sin\chi(s)\Big] \\ &= \underbrace{-a\sqrt{\frac{\beta(s)}{\beta_0}}\cos\chi(s)}_{=x(s)}\ \alpha(s) - a\sqrt{\frac{\beta(s)}{\beta_0}}\sin\chi(s). \end{aligned} \tag{3.233}$$

We now take this rearranged form and replace $x'(s)$ by the new variable

$$y(s) \equiv \beta(s)x'(s) + \alpha(s)x(s) = -a\sqrt{\frac{\beta(s)}{\beta_0}}\sin\chi(s). \tag{3.234}$$

Using this choice of variables, the particle traces out an exact circle in the x-y phase plane, with a radius equal to the amplitude a. Each field error $\Delta B(x,s)$, which acts over a distance Δs, induces a change of angle of

$$\Delta x'(s) = -\frac{e}{p}\Delta B(x,s)\ \Delta s = -\frac{\Delta B(x,s)}{B_0\ R}\ \Delta s \tag{3.235}$$

in the trajectory of the particle. The resulting change in the variable $y(s)$ is obtained by substitution into equation (3.234), yielding

$$\Delta y(s) = -\beta(s)\frac{\Delta B(x,s)}{B_0\ R}\ \Delta s. \tag{3.236}$$

Alternatively the change in the variables x and y may be written in the general form

$$\Delta x = \frac{dx}{da}\Delta a + \frac{dx}{d\chi}\Delta\chi = \sqrt{\frac{\beta}{\beta_0}}\left[\cos\chi\ \Delta a - a\sin\chi\ \Delta\chi\right] = 0$$

$$\Delta y = \frac{dy}{da}\Delta a + \frac{dy}{d\chi}\Delta\chi = -\sqrt{\frac{\beta}{\beta_0}}\left[\sin\chi\ \Delta a + a\cos\chi\ \Delta\chi\right] = -\beta\frac{\Delta B}{B_0\ R}\ \Delta s. \quad (3.237)$$

When the particle traverses a short magnetic field the angle of its motion changes but its position does not, since $\Delta x = 0$ in the first equation. Both these equations are linear in Δa and $\Delta\chi$ and may be directly solved with respect to these quantities. We set $\chi(s) = \vartheta + \Psi(s) = \vartheta + Q\ \phi(s)$, where the phase $\phi(s) = \int \frac{ds}{Q\ \beta(s)}$ again increases by 2π during each revolution. It then follows that

$$\Delta a = \sqrt{\beta_0\beta(s)}\frac{\Delta B(x,s)}{B_0\ R}\sin\left[\vartheta + Q\ \phi(s)\right]\ \Delta s$$

$$\Delta\chi = \frac{\sqrt{\beta_0\beta(s)}}{a}\frac{\Delta B(x,s)}{B_0\ R}\cos\left[\vartheta + Q\ \phi(s)\right]\ \Delta s. \quad (3.238)$$

From the first equation we may directly calculate the change per revolution in the betatron oscillation amplitude by integrating Δa over a full revolution. We obtain

$$\boxed{\frac{da}{dn} = \frac{\sqrt{\beta_0}}{B_0\ R}\oint\sqrt{\beta(s)}\Delta B(x,s)\sin\left[\vartheta + Q\ \phi(s)\right]ds,} \quad (3.239)$$

where $\Delta B(x,s)$ describes an arbitrary field distribution. If $da/dn \neq 0$ the amplitude of the betatron oscillations increases steadily with each revolution, i.e. the oscillation is resonantly excited. In the following sections we explicitly study this process for the three most important multipole fields.

The integer resonance $Q = n$

To begin with let us assume a dipole field error which is constant in the horizontal plane and only varies along the beam axis, namely

$$\Delta B(x,s) = \Delta B_0(s). \quad (3.240)$$

Inserting this into (3.239), we obtain

$$\frac{da}{dn} = \frac{\sqrt{\beta_0}}{B_0\ R}\oint\sqrt{\beta(s)}\Delta B_0(s)\sin\left[\vartheta + Q\phi(s)\right]ds$$

$$= \frac{\sqrt{\beta_0}}{B_0\ R}\oint\sqrt{\beta(s)}\Delta B_0(s)\left[\sin\vartheta\cos Q\phi(s) + \cos\vartheta\sin Q\phi(s)\right]ds. \quad (3.241)$$

The function $F(s) = \sqrt{\beta(s)}\Delta B_0(s)$ is periodic, with a period of one revolution, and may always be written as a Fourier series. Since the phase $\phi(s)$ ranges from $0 - 2\pi$ per revolution, it makes sense to expand $F(s)$ as a function of ϕ, namely

$$F(s) = F_0 + \sum_{p=1}^{\infty} \Big[a_p \cos(p\phi(s)) + b_p \sin(p\phi(s)) \Big]. \tag{3.242}$$

We insert this expansion into (3.241) and multiply out the trigonometric functions, recalling that $\int_0^{2\pi} \cos mx \sin nx \; dx = 0$. This yields

$$\begin{aligned} \frac{da}{dn} &= \frac{\sqrt{\beta_0}}{B_0 \; R} \Big\{ \oint F_0 \Big[\sin\vartheta \cos Q\phi + \cos\vartheta \sin Q\phi \Big] \; ds \\ &+ \oint \sum_{p=1}^{\infty} \Big[a_p \sin\vartheta \cos p\phi \cos Q\phi + b_p \cos\vartheta \sin p\phi \sin Q\phi \Big] \; ds \Big\}. \end{aligned} \tag{3.243}$$

The first integral, with the constant term F_0, makes no contribution on average to the change in amplitude of the betatron oscillation. Since the integral

$$\int_0^{2\pi} \cos mx \cos nx dx \qquad \text{and} \qquad \int_0^{2\pi} \sin mx \sin nx dx$$

is non-zero only when $m = n$, only terms with $p = Q$ contribute in (3.243). If the tune of the circular accelerator is is a whole number, i.e.

$$Q = p \qquad \text{with } p = \text{integer}, \tag{3.244}$$

the change in amplitude

$$\frac{da}{dn} = \frac{\sqrt{\beta_0}}{B_0 \; R} \oint \left(a_Q \sin\vartheta \cos^2 Q\phi + b_Q \cos\vartheta \sin^2 Q\phi \right) \; ds \tag{3.245}$$

has a non-zero value. We have already encountered this integer resonance, caused by a dipole magnet error. It is the strongest of all optical resonances, and if the machine is to operate stably then the tune must be chosen to be sufficiently far from any integer value.

The half-integer resonance $Q = n + \frac{1}{2}$

Let us now consider the resonances which can be caused by errors in quadrupole fields, given by

$$\Delta B(x, s) \;\; = \;\; g(s) \; x(s). \tag{3.246}$$

The trajectory function $x(s)$ may be determined from (3.232) by again setting the phase to $\vartheta + Q\phi(s)$:

$$x(s) = a \sqrt{\frac{\beta(s)}{\beta_0}} \cos\big[\vartheta + Q\phi(s)\big]. \tag{3.247}$$

Using (3.239) we obtain

$$\frac{da}{dn} = \frac{a}{B_0\ R}\oint \beta(s)g(s)\cos\Big[\vartheta + Q\phi(s)\Big]\ \sin\Big[\vartheta + Q\phi(s)\Big]\ ds$$

$$= \frac{a}{2B_0\ R}\oint \beta(s)g(s)\Big[\sin 2\vartheta\cos 2Q\phi(s) + \cos 2\vartheta\sin 2Q\phi(s)\Big]\ ds. \tag{3.248}$$

$\beta(s)g(s)$ is again a periodic function, with a period of one revolution, and so in general also contains a term of the pth harmonic, where $p = 2Q$. This term relates to the quadrupole field. Multiplied by $\sin 2Q\phi$ or $\cos 2Q\phi$ in the integral, it again leads to a non-vanishing change of amplitude. Hence quadrupoles can induce half-integer resonances.

The one-third integer resonance $Q = n + \frac{1}{3}$

As a final example let us study the effect of a sextupole magnet on the resonance condition for a circulating particle. The sextupole field has the form

$$\Delta B(x,s) = \frac{1}{2}g'(s)x^2(s), \tag{3.249}$$

or, substituting into the trajectory (3.232)

$$\Delta B(x,s) = \frac{1}{2}g'(s)a^2\frac{\beta(s)}{\beta_0}\cos^2\Big[\vartheta + Q\phi(s)\Big]. \tag{3.250}$$

Under these conditions the change in amplitude is given by

$$\frac{da}{dn} = \frac{a^2}{2B_0\ R\sqrt{\beta_0}}\oint \beta^{\frac{3}{2}}(s)g'(s)\cos^2\Big[\vartheta + Q\phi(s)\Big]\ \sin\Big[\vartheta + Q\phi(s)\Big]\ ds. \tag{3.251}$$

Using $\cos^2 x\sin x = \frac{1}{4}(\sin 3x + \sin x)$ it follows that

$$\frac{da}{dn} = \frac{a^2}{8B_0\ R\sqrt{\beta_0}}\Big\{\oint \beta^{\frac{3}{2}}(s)g'(s)\Big[\sin\vartheta\cos Q\phi(s) + \cos\vartheta\sin Q\phi(s)\Big]\ ds$$

$$+ \oint \beta^{\frac{3}{2}}(s)g'(s)\Big[\sin 3\vartheta\cos 3Q\phi(s) + \cos 3\vartheta\sin 3Q\phi(s)\Big]\ ds\Big\} \tag{3.252}$$

The first integral again gives an integer resonance, which we have already discussed in the preceding sections. Let us now consider the second integral. If the tune fulfills the condition $3Q = p$ the beam can resonate with the pth harmonic of the function $\beta^{\frac{3}{2}}(s)g'(s)$. This condition is met when Q has a one-third integer value, i.e. $Q = n + \frac{1}{3}$. Sextupoles thus induce third-integer resonances.

Table 3.4 Optical resonances and the multipole fields which drive them.

driving field	condition
dipole	$Q = p$
quadrupole	$2Q = p$
sextupole	$3Q = p$
octupole	$4Q = p$
etc.	$\vdots$

The Q_x-Q_z diagram (tune diagram)

Multipole fields of even higher order induce correspondingly higher resonances, for example an octupole field can produce a fourth order betatron oscillation resonance. The calculation proceeds along the same lines as before, using (3.239). In reality, however, it is very difficult to determine the strength of the resonances exactly, since the field errors are usually known only very imprecisely, if at all. This means that the Fourier coefficients in (3.242) are usually not known. However it is a general rule that the strength of a resonance decreases sharply with the order. Table 3.4 summarizes the optical resonances and the multipole fields which drive them.

The finite mechanical tolerance and the limited pole width of the magnets mean that in principle all possible multipole fields are present in an accelerator. As a result there are always resonances when $mQ = p$, where m and p are integers. We must bear in mind that every accelerator has two tune values, namely Q_x in the horizontal and Q_z in the vertical plane. In general these are different, and so there are resonance conditions in both planes. For higher multipole fields the strenth in one plane depends on the beam position in the other, which leads to coupling between between the betatron oscillations in the two planes and hence to **coupled resonances**. The condition for optical resonance in both planes may thus be expressed as

$$\boxed{m\,Q_x + n\,Q_z = p \qquad (m, n, p = \text{integers}).} \tag{3.253}$$

The sum $|m| + |n|$ is called the **order** of the resonance. For stable operation a pair of values for Q_x and Q_z must be chosen which avoid optical resonances. This pair of values is termed the **working point**. The strength of the resonance decreases rapidly with the order, and so when choosing a working point we are generally only concerned about resonances up to 5th order.

The optical resonances given by (3.253) are shown up to third order for both planes as lines in the diagram in Fig. 3.35. Care must be taken to choose a working point that is sufficiently far from these lines. This is particularly important in storage rings, because the long beam lifetimes mean that even relatively weak multipole fields can cause significant resonant beam disturbances.

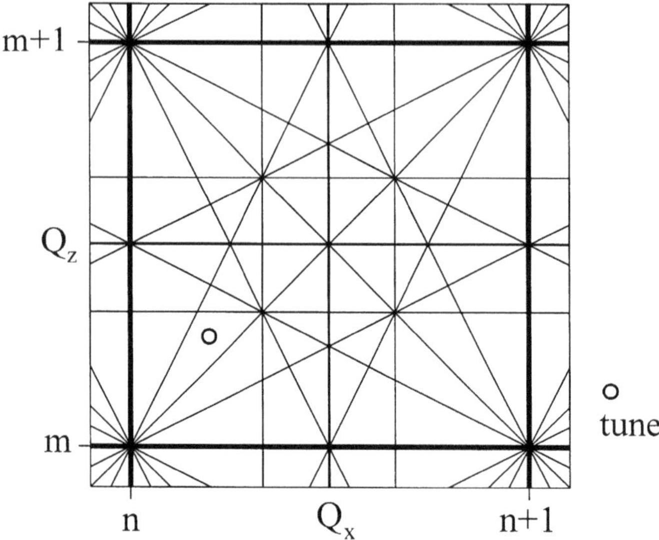

Fig. 3.35 Optical resonances up to 3rd order in both planes of oscillation. A possible choice of working point is indicated in the diagram.

Hence the working point must be carefully chosen in advance when the beam optics are designed.

3.15 The effect of magnetic field errors on beam optics

Ideal magnets, which agree exactly with the hard-edge model used in calculations of linear optics, cannot be achieved in practice. Because of finite manufacturing tolerances and the effects of pole ends there are always non-negligible deviations from the ideal field, which we will call **field errors**. In the previous section on optical resonances we met an important example of the effect of field errors on the stability of the particle dynamics. In the following sections we will consider the effect of dipole and quadrupole field errors on the beam optics.

3.15.1 Effect of dipole kicks

We assume a dipole field error of strength ΔB acts over a length l. This field changes the angle of the particle trajectory by the amount

$$\Delta x' = \frac{e}{p}\,\Delta B\, l. \tag{3.254}$$

If l is not too long, the disturbance may be described by a localized angular kick right in the middle of the disturbing field at $l/2$. We may therefore approximate it to an infinitesimally short field disturbance, which considerably simplifies the calculation. Consider a particle travelling exactly along the orbit in front of the disturbing field, i.e. with the trajectory vector $(x, x') = (0, 0)$. This particle therefore has zero emittance. Immediately after the field disturbance the particle

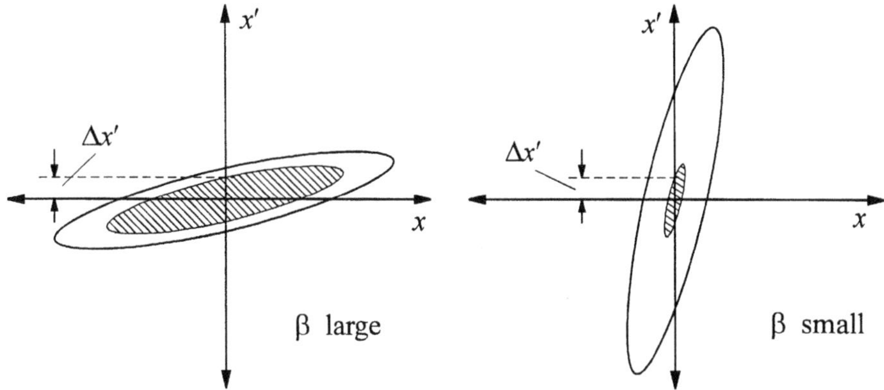

Fig. 3.36 The effect of a constant field error on the beam at two points with different beta functions. The change in angle $\Delta x'$ is the same in both cases. When the beta function is large (*left*) the angular acceptance is relatively small, and so as a consequence of Liouville's theorem the angular deviation $\Delta x'$ gives rise to a relatively large phase space ellipse. When the beta function is small (*right*) the corresponding ellipse is small.

now travels at an angle $\Delta x'$ with respect to the orbit. The trajectory vector is now $\boldsymbol{x} = (0, \Delta x')$. This deflection will lead to betatron oscillations, as a result of the beam focusing. Inserting this trajectory vector into the ellipse equation (3.135) gives a non-zero emittance

$$\varepsilon_{\text{error}} = \beta \, \Delta x'^2. \tag{3.255}$$

It should be noted that for a given field error the increase in emittance is proportional to the beta function at the point of the disturbance. This fact is illustrated in Fig. 3.36. It is a fundamental property of beam optics that the effect of a field error increases with the beta function. This means that special care must be taken in the construction of the magnets to be used at points where the beta function is large. In a circular accelerator the beam passes through the same field disturbance again during each revolution, and so is deflected through the same angle each time. After many revolutions a stable equilibrium is established, resulting in a new **distorted orbit**, consisting of a static betatron oscillation about the unperturbed ideal orbit, with a phase or angular shift at the point of the error, as Fig. 3.37 shows. In equilibrium the perturbed orbit has a displacement x from the ideal orbit at the point s_0. The angle is $x' - \Delta x'$ immediately before the disturbance and x' immediately after it. Hence the trajectory begins at the point s_0 with the trajectory vector (x, x'). After a full revolution it again passes through the field disturbance, assumed to be infinitesimally short, this time with the trajectory vector $(x, x' - \Delta x')$. The distorted orbit may therefore be evolved using

$$\begin{pmatrix} x \\ x' - \Delta x' \end{pmatrix} = \mathbf{M}_{\text{rev}} \begin{pmatrix} x \\ x' \end{pmatrix}. \tag{3.256}$$

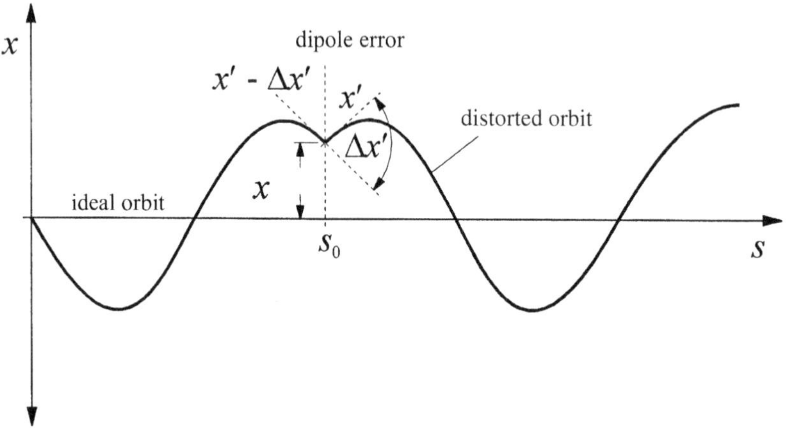

Fig. 3.37 The distorted orbit resulting from a dipole kick at the point s_0. The static trajectory is bent through an angle $\Delta x'$ at the point s_0.

Here $\mathbf{M}_{\mathrm{rev}}$ is the transfer matrix for a full revolution. This matrix can be easily determined from (3.164) by replacing the betatron phase for a revolution by $\Psi = 2\pi Q$ where Q is the tune, and by realizing that the periodicity conditions imply that the beta function and its derivative are the same at the beginning and end of a revolution, i.e. $\beta(s_0) = \beta_0$ and $\alpha(s_0) = \alpha_0$. It then follows that

$$\mathbf{M}_{\mathrm{rev}} = \begin{pmatrix} \cos 2\pi Q + \alpha(s_0) \sin 2\pi Q & \beta(s_0) \sin 2\pi Q \\ -\gamma(s_0) \sin 2\pi Q & \cos 2\pi Q - \alpha(s_0) \sin 2\pi Q \end{pmatrix}. \tag{3.257}$$

Using this matrix, (3.256) yields the equations which determine the components x and x' of the distorted orbit vectors immediately after the disturbance

$$\begin{aligned} \left[\cos 2\pi Q + \alpha(s_0) \sin 2\pi Q - 1\right] x \; + \; \beta(s_0) \sin 2\pi Q \; x' &= 0 \\ -\gamma(s_0) \sin 2\pi Q \; x \; + \; \left[\cos 2\pi Q - \alpha(s_0) \sin 2\pi Q - 1\right] x' &= -\Delta x'. \end{aligned} \tag{3.258}$$

After a little manipulation of the trigonometric functions, the solution of these two equations finally yields the required result

$$\boxed{\begin{aligned} x &= \Delta x' \frac{\beta(s_0)}{2 \; \tan \pi Q} \\ x' &= \frac{\Delta x'}{2} \left(1 - \frac{\alpha(s_0)}{\tan \pi Q}\right). \end{aligned}} \tag{3.259}$$

From this initial vector we may calculate the distorted orbit at any point around the ring, using the known transfer matrices. We immediately see that the orbit

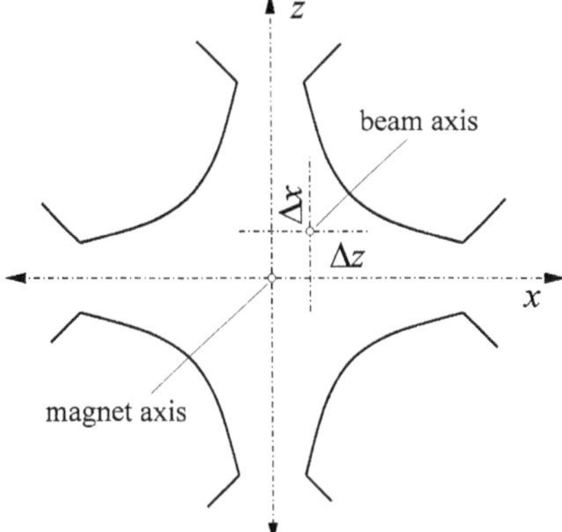

Fig. 3.38 Orbit distortion caused by a transversely misaligned quadrupole. The quadrupole axis lies parallel to the orbit but has a displacement $\Delta \boldsymbol{x} = (\Delta x, \Delta z)$.

distortion grows without bound if the tune Q approaches an integer. This is the result of the integer resonance, which we have already encountered. In general there are many such disturbances around the ring, which must be individually calculated and then summed together. This can lead to considerable shifts in the orbit, which must be compensated for by specially arranged small correcting magnets.

3.15.2 Effect of quadrupole field errors

The simplest case of a field error caused by a quadrupole is a transverse misalignment, in which the axis of the quadrupole correctly lies parallel to the line of the orbit but has a displacement, as shown in Fig. 3.38. If the quadrupole has a gradient $g = \partial B_z/\partial x$ then at the orbit there is a field

$$\begin{pmatrix} \Delta B_x \\ \Delta B_z \end{pmatrix} = g \begin{pmatrix} \Delta z \\ \Delta x \end{pmatrix} . \tag{3.260}$$

This leads to a deflection in both planes of

$$\begin{pmatrix} \Delta x' \\ \Delta z' \end{pmatrix} = \frac{e}{p}\, l \begin{pmatrix} \Delta B_z \\ \Delta B_x \end{pmatrix} = \frac{e}{p}\, g\, l \begin{pmatrix} \Delta x \\ \Delta z \end{pmatrix} = k\, l \begin{pmatrix} \Delta x \\ \Delta z \end{pmatrix} . \tag{3.261}$$

Like a dipole field error, this misalignment thus causes an angular deflection of the trajectory in both planes. This again leads to an orbit distortion, which may be calculated by inserting the angular deflection from (3.261) into (3.259). The orbit distortion is proportional to the size of the misalignment and also proportional to the beta function at the quadrupole.

An error in the gradient g in a quadrupole changes the focusing and hence the tune in a circular accelerator. It also changes the beta function around the ring. We will study this problem using a perturbative calculation in which the gradient error Δg is assumed to be very small compared to the gradient itself.The

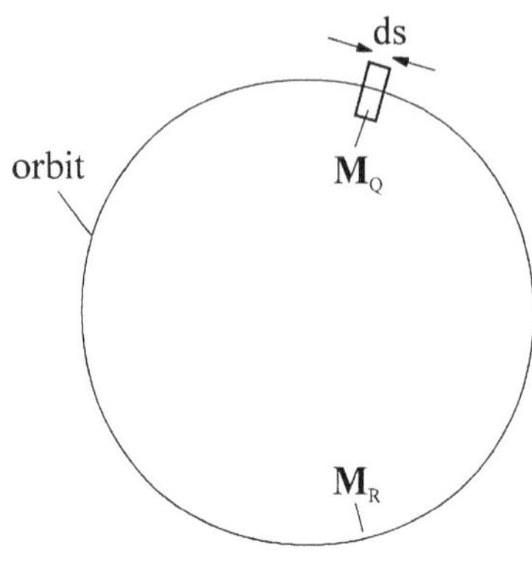

Fig. 3.39 Circular accelerator with a separate infinitesimal quadrupole section.

gradient deviation or alternatively the deviation in quadrupole strength are then written in the form

$$\begin{aligned} g &= g_{\text{ideal}} + \Delta g \quad \text{with} \quad \Delta g \ll g \\ k &= k_{\text{ideal}} + \Delta k \quad \text{with} \quad \Delta k \ll k. \end{aligned} \tag{3.262}$$

In the undistorted case, i.e. when $\Delta k = 0$, the transfer matrix for a full revolution is

$$\mathbf{M}_{\text{rev}} = \begin{pmatrix} \cos 2\pi Q + \alpha_0 \sin 2\pi Q & \beta_0 \sin 2\pi Q \\ -\gamma_0 \sin 2\pi Q & \cos 2\pi Q - \alpha_0 \sin 2\pi Q \end{pmatrix}, \tag{3.263}$$

where Q is the tune, $\Psi = 2\pi Q$ is the betatron phase shift per revolution, and β_0 and α_0 are the optical functions at the point s_0.

To determine the effect of the gradient error we cut an infinitesimal quadrupole section of length ds out of the magnet structure at the point s_0, as shown in Fig. 3.39. We will assume, without any loss of generality, that this segment is a focusing quadrupole. $\mathbf{M}_\text{Q}$ is the transfer matrix for the quadrupole section and $\mathbf{M}_\text{R}$ that for the rest of the ring. The revolution matrix may thus be written in the form

$$\mathbf{M}_{\text{rev}} = \mathbf{M}_\text{Q}\mathbf{M}_\text{R}. \tag{3.264}$$

For a short error-free quadrupole section of length ds it immediately follows from (3.84) that

$$\mathbf{M}_\text{Q} = \begin{pmatrix} 1 & ds \\ -k\ ds & 1 \end{pmatrix}. \tag{3.265}$$

When there is a quadrupole error this becomes

$$\mathbf{M}_\text{Q}^* = \begin{pmatrix} 1 & ds \\ -(k + \Delta k)\ ds & 1 \end{pmatrix}. \tag{3.266}$$

The matrix product

$$\begin{pmatrix} 1 & 0 \\ -\Delta k\ ds & 1 \end{pmatrix} \cdot \begin{pmatrix} 1 & ds \\ -k\ ds & 1 \end{pmatrix} = \begin{pmatrix} 1 & ds \\ -(k + \Delta k)\ ds & 1 - \Delta k\ ds^2 \end{pmatrix} \tag{3.267}$$

gives the distorted matrix $\mathbf{M}_\mathrm{Q}^*$ since $|\Delta k\ ds^2| \ll 1$, and so the distorted matrix may be written as

$$\mathbf{M}_\mathrm{Q}^* = \begin{pmatrix} 1 & 0 \\ -\Delta k\ ds & 1 \end{pmatrix} \mathbf{M}_\mathrm{Q}. \tag{3.268}$$

From (3.264), the perturbation to the revolution matrix resulting from the gradient error is thus

$$\begin{aligned} \mathbf{M}_\mathrm{rev}^* &= \begin{pmatrix} 1 & 0 \\ -\Delta k\ ds & 1 \end{pmatrix} \mathbf{M}_\mathrm{rev} \\ &= \begin{pmatrix} \cos 2\pi Q + \alpha_0 \sin 2\pi Q & \beta_0 \sin 2\pi Q \\ -\Delta k\, ds\, (\cos 2\pi Q - \alpha_0 \sin 2\pi Q) - \gamma_0 \sin 2\pi Q & -\Delta k\, ds\, \beta_0 \sin 2\pi Q + \cos 2\pi Q - \alpha_0 \sin 2\pi Q \end{pmatrix}. \end{aligned} \tag{3.269}$$

The quadrupole error Δk causes a change in the beam focusing and hence a tune shift dQ. The distorted revolution matrix may also be expressed in terms of this shift by replacing Q by $Q + dQ$ in the undistorted revolution matrix (3.263). Using $\chi = 2\pi(Q + dQ)$ gives

$$\mathbf{M}_\mathrm{rev}(Q + dQ) = \begin{pmatrix} \cos\chi + \alpha_0 \sin\chi & \beta_0 \sin\chi \\ -\gamma_0 \sin\chi & \cos\chi - \alpha_0 \sin\chi \end{pmatrix}. \tag{3.270}$$

The matrices $\mathbf{M}_\mathrm{rev}^*$ and $\mathbf{M}_\mathrm{rev}(Q + dQ)$ are two different representations of the same physical quantities. Since they do not necessarily use the same coordinate system, the simple matrix identity $\mathbf{M}_\mathrm{rev}^* = \mathbf{M}_\mathrm{rev}(Q + dQ)$, i.e. equating the individual matrix elements, does not lead to the desired result. We require a matrix identity which describes the same spatial transformation for any choice of coordinate system. The identity we need is that of **similarity** of matrices. The rules of linear algebra state that matrices are **similar** if their traces (the sum of the diagonal elements) are the same. Imposing this requirement one obtains

$$\text{trace } \mathbf{M}_\mathrm{rev}^* = \text{trace } \mathbf{M}_\mathrm{rev}(Q + dQ). \tag{3.271}$$

Using (3.269) and (3.270) it then follows immediately that

$$2\ \cos 2\pi Q - \Delta k\, \beta_0\ \sin 2\pi Q\ ds = 2\ \cos 2\pi(Q + dQ). \tag{3.272}$$

Because the shift in Q is very small, $\cos 2\pi dQ \approx 1$ and $\sin 2\pi dQ \approx 2\pi dQ$, and (3.272) then reduces to

$$4\pi\ dQ = \Delta k\ \beta_0\ ds. \tag{3.273}$$

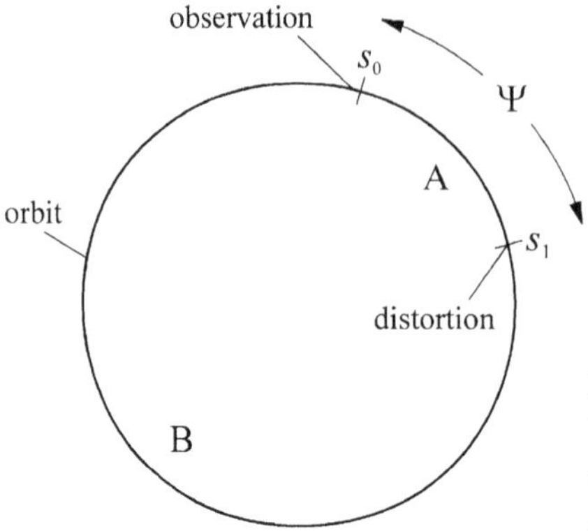

Fig. 3.40 Calculating the deviation in the beta function at the point s_0 due to a quadrupole error at the point s_1 in a circular accelerator.

From this we may calculate the tune shift ΔQ due to a quadrupole of finite length l with an error Δk by simply integrating over the length of the magnet:

$$\boxed{\Delta Q = \frac{1}{4\pi} \int_{s_0}^{s_0+l} \Delta k \ \beta(s) \ ds.} \tag{3.274}$$

It is important to note that this formula is only valid for very small quadrupole errors Δk.

As well as causing a tune shift, a quadrupole error also has an effect on the beta function. Let us calculate its value at the point s_0 in the ring when a quadrupole error, again assumed to be infinitesimally short, is acting at another point s_1 (Fig. 3.40). Between s_0 and s_1 the betatron phase changes by the amount Ψ. The matrix $\mathbf{A}$ describes the magnet structure between the points s_0 and s_1, and the matrix $\mathbf{B}$ describes the structure from s_1 to s_0. The unperturbed revolution matrix at the point s_0 is

$$\mathbf{M}_{\text{rev}}(s_0) = \begin{pmatrix} m_{11} & m_{12} \\ m_{21} & m_{22} \end{pmatrix} = \mathbf{B} \cdot \mathbf{A} \tag{3.275}$$

with

$$\mathbf{A} = \begin{pmatrix} a_{11} & a_{12} \\ a_{21} & a_{22} \end{pmatrix} \qquad \mathbf{B} = \begin{pmatrix} b_{11} & b_{12} \\ b_{21} & b_{22} \end{pmatrix} \tag{3.276}$$

As in (3.268), we again introduce the perturbation by multiplying by the distortion matrix, giving

$$\mathbf{M}^*_{\text{rev}}(s_0) = \begin{pmatrix} m^*_{11} & m^*_{12} \\ m^*_{21} & m^*_{22} \end{pmatrix} = \mathbf{B} \cdot \begin{pmatrix} 1 & 0 \\ -\Delta k \ ds & 1 \end{pmatrix} \cdot \mathbf{A}. \tag{3.277}$$

In what follows it is sufficient to consider only the matrix elements m_{12} or m^*_{12}. Using the relation (3.164) and writing $\beta_0 = \beta(s_0)$, we may write m_{12} in the form

$$m_{12} = \beta_0 \ \sin 2\pi Q. \tag{3.278}$$

The quadrupole error changes the tune by a small amount and also induces a small change in the beta function. By analogy with (3.278), the distorted matrix element m_{12}^* may thus be written in the general form

$$m_{12}^* = (\beta_0 + d\beta)\ \sin 2\pi(Q + dQ). \tag{3.279}$$

Alternatively we may obtain m_{12}^* from the matrix product (3.277)

$$\begin{aligned} m_{12}^* &= \underbrace{b_{11}a_{12} + b_{12}a_{22}}_{=m_{12}} - a_{12}b_{12}\Delta k\ ds \\ &= \beta_0\ \sin 2\pi Q - a_{12}b_{12}\Delta k\ ds. \end{aligned} \tag{3.280}$$

If we now set (3.279) and (3.280) to be equal and again use the approximations $\cos 2\pi dQ \approx 1$ and $\sin 2\pi dQ \approx 2\pi dQ$ we obtain

$$(\beta_0 + d\beta)[\sin 2\pi Q + 2\pi dQ \cos 2\pi Q] = \beta_0 \sin 2\pi Q - a_{12}b_{12}\Delta k\ ds. \tag{3.281}$$

As dQ and $d\beta$ are infinitesimally small quantities we may neglect terms containing products $dQd\beta$, giving

$$2\pi dQ\beta_0 \cos 2\pi Q + d\beta \sin 2\pi Q = -a_{12}b_{12}\Delta k\ ds. \tag{3.282}$$

Using the expression for the tune shift (3.273) we may further rearrange this expression to give

$$\frac{1}{2}\beta_0\beta(s_1)\Delta k\ ds\ \ \cos 2\pi Q + d\beta \sin 2\pi Q = -a_{12}b_{12}\Delta k\ ds. \tag{3.283}$$

Solving with respect to $d\beta$ finally yields

$$d\beta = -\frac{1}{2\sin 2\pi Q}\left[2a_{12}b_{12} + \beta_0\beta(s_1)\cos 2\pi Q\right]\Delta k\ ds. \tag{3.284}$$

We still have to calculate the matrix elements a_{12} and b_{12}. Here we use the values of the beta function at the points s_0 and s_1 and denote the advance in phase along the section (s_0, s_1) by Ψ and that along the section (s_1, s_0) by $2\pi Q - \Psi$. Inserting these terms into (3.164) gives

$$\begin{aligned} a_{12} &= \sqrt{\beta_0\ \beta(s_1)}\sin\Psi \\ b_{12} &= \sqrt{\beta_0\ \beta(s_1)}\sin(2\pi Q - \Psi). \end{aligned} \tag{3.285}$$

Using (3.284) it follows that

$$d\beta = -\frac{\beta_0\ \beta(s_1)}{2\sin 2\pi Q}\underbrace{\left[2\sin\Psi\sin(2\pi Q - \Psi) + \cos 2\pi Q\right]}_{=\cos(2\Psi - 2\pi Q)}\Delta k\ ds. \tag{3.286}$$

Let us label the betatron phase at the point s_0 by Ψ_0. The phase difference at the point s_1 is then given by $\Psi = \Psi(s_1) - \Psi_0$. In a real machine a quadrupole error is always distributed over a finite length, which we will take to run from s_1 to $s_1 + l$. The change in the beta function at the point s_0 due to the error Δk is then

$$\boxed{\Delta\beta(s_0) = -\frac{\beta_0}{2\sin 2\pi Q}\int_{s_1}^{s_1+l} \beta(s)\,\Delta k(s)\,\cos\left[2\Big(\Psi(s) - \Psi_0\Big) - 2\pi Q\right]ds.} \tag{3.287}$$

We again see that $\Delta\beta$ grows without bounds if $\sin 2\pi Q \to 0$. The tune must therefore not have integer or half-integer values. This phenomenon, which we have already seen leads to optical resonances, also leads to unlimited growth of the beta functions and hence the beam size. Stable operation of the accelerator is then impossible.

We must once again note that the formulae derived in this section are only really applicable to very small errors Δk, for which the beta functions remain essentially unchanged. Larger deviations change the entire beam optics and the assumptions used here are no longer valid. In this case it becomes necessary to completely recalculate the machine optics. Similarly, when operating a circular accelerator it is important to bear in mind that a large variation in the individual quadrupole strengths can considerably change the beam optics.

3.16 Chromaticity of beam optics and its compensation

Beam optics are always calculated starting with particles of nominal momentum p_0. A real particle beam, however, always has a certain momentum distribution about this nominal value and so it is important to investigate how the beam optics change for off-momentum particles with a deviation Δp. We may assume that the momentum deviations are very small, $\Delta p \ll p_0$, since the Gaussian distribution function of a particle beam is generally not much broader than $\Delta p/p_0 \approx 10^{-3}$. Instead of the nominal beam optics, particles with a momentum $p = p_0 + \Delta p$ see the modified quadrupole strength

$$k(p) = -\frac{e}{p}g = -\frac{e}{p_0 + \Delta p}g \approx -\frac{e}{p_0}\left(1 - \frac{\Delta p}{p_0}\right)g = k_0 - \Delta k. \tag{3.288}$$

One may thus treat the effect of a momentum deviation as a a quadrupole error Δk, given by

$$\Delta k = \frac{\Delta p}{p}k_0. \tag{3.289}$$

According to (3.273), this quadrupole error induces a tune shift per path element ds of

$$dQ = \frac{\Delta p}{p}\,\frac{1}{4\pi}\;k_0\;\beta(s)\;ds. \tag{3.290}$$

Since the particle maintains the same momentum deviation for many revolutions, it appears to this particle that **all** the quadrupoles in the ring have a quadrupole error proportional to $\Delta p/p$. To calculate the resulting tune shift we must thus integrate over all the quadrupoles. It then follows from (3.290)

$$\boxed{\xi \equiv \frac{\Delta Q}{\Delta p/p} = \frac{1}{4\pi} \oint k(s)\ \beta(s)\ ds.} \tag{3.291}$$

This dimensionless quantity is called the **chromaticity** of the beam optics, by analogy with the chromatic distortion of light in classical optics. It increases with the strength of the beam focusing, with particularly large contributions coming from strong quadrupoles in which the beta functions reach large values.

Storage rings in particular suffer from large chromaticities ξ_x and ξ_z in both planes, as a result of their size and their usually relatively strong focusing. Due to $\Delta Q = \xi\ \frac{\Delta p}{p}$ this leads to considerable tune shifts even for small momentum deviations, with these particles then encountering damaging optical resonances and being lost. A further limiting effect is the so-called **head-tail instability**, which always arises when the value of the chromaticity is negative. This is generally the case in all circular accelerators, as may be seen from the formula (3.291), since the beam focusing means that $k(s)$ has predominantly negative values. This instability restricts the beam current to relatively low values. It is therefore essential to compensate for the chromaticity, particularly in strong focusing machines.

The compensation is performed at points where the particles are naturally sorted according to their momentum, namely anywhere where the dispersion is non-zero. Here the particles travel on average along dispersive trajectories

$$x_{\mathrm{D}}(s) = D(s)\frac{\Delta p}{p}, \tag{3.292}$$

depending on their momenta. At these points additional magnets are installed with a focusing strength which depends on the transverse beam position, i.e. $k \propto x$. **Sextupole magnets** have exactly this property. Their field may be calculated from the potential (3.49), namely

$$\begin{aligned} B_x &= \frac{\partial \Phi}{\partial x} = g'\ x\ z \\ B_z &= \frac{\partial \Phi}{\partial z} = \frac{1}{2}\ g'\ (x^2 - z^2). \end{aligned} \tag{3.293}$$

From this we immediately obtain the gradients along the x- and z-axes

$$\left. \begin{aligned} \frac{\partial B_z}{\partial x} &= g'\ x \\ \frac{\partial B_x}{\partial z} &= g'\ x \end{aligned} \right\} \qquad k_{\mathrm{sext}} = \frac{e}{p}\ g'\ x = m\ x. \tag{3.294}$$

The procedure of chromaticity compensation is illustrated in Fig. 3.41. Particles with nominal momentum $\Delta p/p = 0$ have the correct focusing. They follow

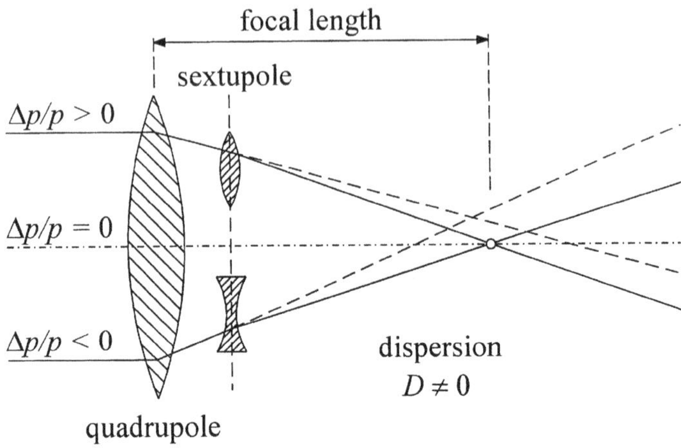

Fig. 3.41 Use of a sextupole to compensate for chromaticity resulting from a quadrupole magnet.

the nominal orbit, with $x = 0$, and so are not influenced by the sextupole situated behind the quadrupole, since its field gradient is zero on the axis. Hence the sextupoles do not affect particles with nominal momentum. When the particle momentum is too large, however, i.e. $\Delta p/p > 0$, the focusing effect of the quadrupole is too weak. The particle travels along a dispersive trajectory $x_D > 0$ and hence passes through the sextupole with a displacement $x = \frac{\Delta p}{p} D > 0$. Because of this non-zero displacement the sextupole field contributes an additional quadrupole strength of

$$k_{\text{sext}} = m\ D\ \frac{\Delta p}{p} \tag{3.295}$$

with the sextupole strength m chosen such that the chromatic effect of the quadrupole vanishes. For particles with lower than nominal momentum the compensation proceeds in exactly the same way, but with the signs reversed.

With sextupoles distributed around the ring, the total chromaticity consists of the sum of the momentum-dependent quadrupole errors $\Delta k = -\frac{\Delta p}{p}$ plus the effect of the sextupoles $k_{\text{sext}} = m\ D\ \frac{\Delta p}{p}\ k_0$. Since all the quadrupoles and sextupoles in the ring contribute, we must once again integrate over a whole revolution. Using (3.291), the effective total chromaticity becomes

$$\boxed{\xi_{\text{tot}} = \frac{1}{4\pi} \oint \Big[m(s)\, D(s) + k(s) \Big]\, \beta(s)\, ds.} \tag{3.296}$$

In principle it is always possible to choose the number, positions and strengths m of the sextupoles so that this integral vanishes. In practice, however, it turns out to be useful to over-compensate for the chromaticity to a certain degree in order to obtain small positive values

$$\xi_{\text{tot}} \approx +1\ \ldots\ +3. \tag{3.297}$$

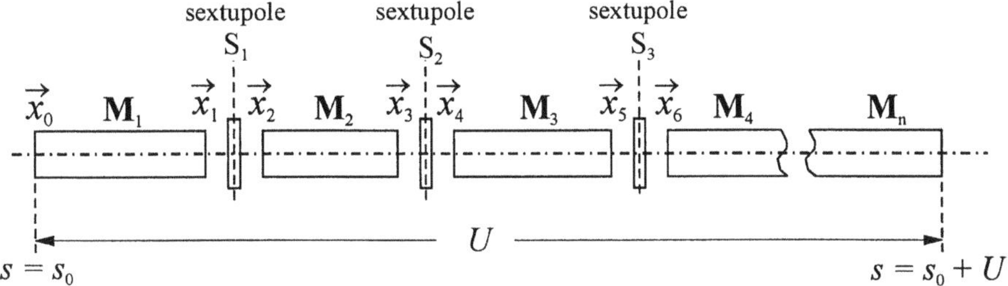

Fig. 3.42 Calculating the trajectory of a particle through a magnet structure with individual sextupoles, assumed to be infinitely thin.

This prevents any small variation in the optics from driving the chromaticity negative again. In addition, a positive value of ξ_{tot} leads to **head-tail damping**.

3.17 Restriction of the dynamic aperture by sextupoles

The sextupole magnets needed to compensate for the chromaticity unfortunately also have negative effects on the dynamic behaviour of the beam. This is essentially due to the fact that the quadratic field distribution in the sextupole means that it exerts non-linear forces on the particle which lead to anharmonic betatron oscillations. Their frequency and hence also the tune are then dependent on the amplitude of the oscillations. Non-linear resonances may thus arise, leading to sudden particle loss above a certain betatron amplitude. Particles with nominal momentum p_0 but large oscillation amplitudes will thus be deflected by the sextupole fields and it will no longer be possible to find a trajectory which is stable over many revolutions. Here one faces the phenomenon of **chaotic particle dynamics**.

It is not possible to treat this non-linear problem analytically, and one is forced to use numerical methods instead. The simplest method can be explained with the help of Fig. 3.42. We make the realistic approximation that the magnet structure of a ring consists of linear sections separated by individual sextupole magnets. These sections are described in the usual way by their transfer matrices $\mathbf{M}_i$. The sextupoles may then be approximated by thin lenses with no longitudinal extent.

At the beginning of the structure we start with a four-dimensional trajectory vector

$$\boldsymbol{X}_0 = \begin{pmatrix} x_0 \\ x_0' \\ z_0 \\ z_0' \end{pmatrix}, \tag{3.298}$$

whose components are chosen at random from within the acceptance ellipse. The trajectory vector immediately before the first sextupole S_1 is obtained by the transformation

$$\boldsymbol{X}_1 = \begin{pmatrix} x_1 \\ x_1' \\ z_1 \\ z_1' \end{pmatrix} = \mathbf{M}_1 \cdot \boldsymbol{X}_0. \tag{3.299}$$

The particle travels through the sextupole with a displacement x_1 and z_1. According to (3.293) the sextupole field at the position of the particle is

$$\boldsymbol{B} = \begin{pmatrix} g'\, x_1\, z_1 \\ \frac{1}{2}\, g'\, (x_1^2 - z_1^2) \end{pmatrix}. \tag{3.300}$$

If l is the effective length of the sextupole then the change in trajectory angle induced by the sextuple field in the two planes is given by

$$\begin{aligned} \Delta x_1' &= \frac{e}{p}\, B_z\, l = \frac{1}{2}\, (ml)\, (x_1^2 - z_1^2) \\ \Delta z_1' &= \frac{e}{p}\, B_x\, l = (ml)\, x_1\, z_1. \end{aligned} \tag{3.301}$$

The evolution of the trajectory through the sextupole cannot be described by the usual matrix formalism because the matrix elements would then themselves depend on the starting values of the trajectory vectors. The changes in angle induced by the sextupole must be explicitly added to the trajectory vector. Beyond the sextupole we thus obtain the new vector

$$\boldsymbol{X}_2 = \begin{pmatrix} x_1 \\ x_1' + \Delta x_1' \\ z_1 \\ z_1' + \Delta z_1' \end{pmatrix}, \tag{3.302}$$

which is then evolved as far as the next sextupole S_2 using the matrix $\mathbf{M}_2$. Here the changes in angle of the particle trajectory are calculated in the same way as before. This procedure, which is termed **particle tracking**, is repeated until sufficiently many revolutions have been calculated, usually several thousand. The result of a computation of this kind is shown in Fig. 3.43 for the horizontal plane. We immediately see from this example that the sextupoles reduce the stable region of phase space. Stable particle motion is only possible for small oscillation amplitudes. If the amplitude exceeds a certain value, which may be very much smaller than the physical acceptance given by the size of the vacuum chamber, then the particle is lost. Hence the space available to the beam is considerably less than that given by the mechanical aperture. The non-linear particle dynamics limit the effective aperture available for the particle motion, and hence this reduced space is termed the **dynamic aperture**.

For stable and reliable operation of the accelerator it is important to have as large a dynamic aperture as possible. This means that the sextupoles must be carefully arranged, but unfortunately the non-linearity means that the optimal arrangement cannot be calculated analytically. In principle one starts with a

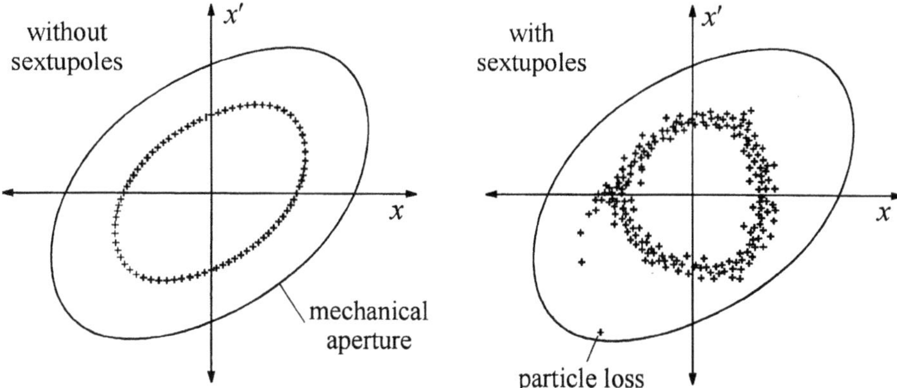

Fig. 3.43 Particle tracking through a circular accelerator (*left* without sextupoles, *right* with sextupoles). In this example only the horizontal phase plane is shown. For each revolution the transverse position of the particle at a particular point s is plotted as a small circle on the phase diagram. The resulting ellipse marks the acceptance of the ring. Without any sextupoles the expected elliptical shape is obtained. With sextupoles present the particle motion is much more complex and in this example leads to a sudden growth in the betatron amplitude and hence to the loss of the beam.

particular distribution of sextupoles and then calculates the size of the dynamic aperture by particle tracking. If this is not sufficient then one must try another arrangement and repeat the particle tracking. In this way the dynamic aperture can be increased step by step. If it proves impossible to achieve the required aperture then the linear optics or even the entire magnet structure must be modified.

Although no analytic solution has yet been found for the optimal distribution of sextupoles in circular accelerators, there are a few general empirical rules to guide us which we will discuss with reference to the model ring accelerator described in Section 3.13.3. The linear beam optics used here have a chromaticity in both planes of $\xi_x = -2.35$ and $\xi_z = -2.36$, which must be reduced to $\xi_x = \xi_z = +1$ by the insertion of sextupoles.

In principle two sextupoles will suffice, one for the horizontal and one for the vertical plane, as shown in the left-hand diagram of Fig. 3.44. The sextupoles are installed immediately after a horizontally (QF) and vertically (QD) focusing quadrupole. Here the beta functions in the respective planes are rather large and so the chromaticity compensation by the sextupoles is very effective, as can be seen from (3.296). In the second case, shown in the right-hand diagram of Fig. 3.44, many sextupoles are equally spaced around the ring. The horizontally acting sextupoles (SF) are again arranged next to the horizontally focusing quadrupoles, and similarly the vertically acting sextupoles (SD) are next to the vertically focusing quadrupoles. Here it is assumed that all the horizontally-acting sextupoles are of equal strength, and likewise that the vertically-acting sextupoles are all equal. If only two sextupoles are used, the integrated sextupole

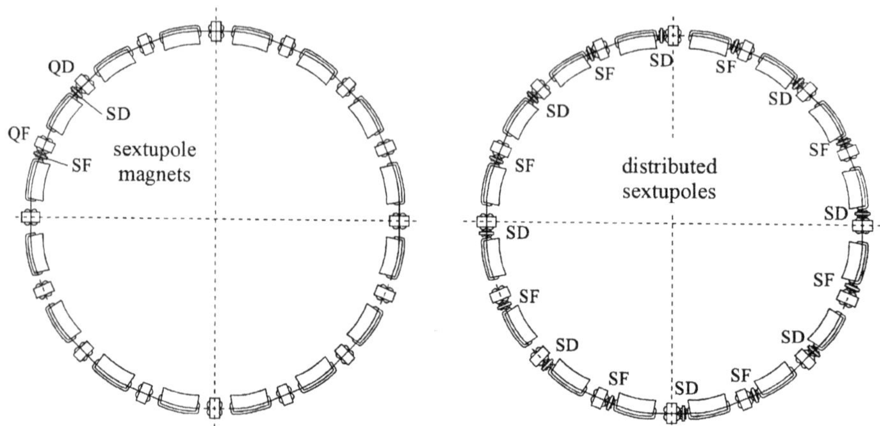

Fig. 3.44 Possible distributions of sextupoles for chromaticity compensation in a model ring. In the case on the left only one sextupole is inserted for each plane, while in the example on the right every space next to a quadrupole around the ring is occupied by a sextupole. In both cases the compensated chromaticity is $\xi_x = \xi_z = +1$.

strengths required are $(m\ l)_{\mathrm{SF}} = -2.34\ \mathrm{m}^{-2}$ and $(m\ l)_{\mathrm{SD}} = 3.81\ \mathrm{m}^{-2}$. The field must be rather stronger in the vertical plane because of the smaller dispersion at the position SD, as may be seen from the optical functions in Fig. 3.32. Particle tracking is performed over some 500 revolutions with this arrangement, in which the emittance of the phase ellipse associated with the trajectory vector $\boldsymbol{X}_0$ is varied in small steps in both the horizontal and vertical planes. This process explores the whole aperture of the beam pipe. After each tracking calculation it is determined whether the particle is already lost or whether its motion is still stable. In this way the range of stable particle motion may be determined. In the case of just two sextupoles this range is extremely narrow, with a horizontal limit of only 15 mm and a vertical range of only 4 mm (Fig. 3.45). It is not possible to operate an accelerator within such narrow limits.

The situation is very different when many evenly distributed sextupoles are employed. Here the required sextupole strengths reduce to $(m\ l)_{\mathrm{SF}} = -0.29\ \mathrm{m}^{-2}$ and $(m\ l)_{\mathrm{SD}} = 0.48\ \mathrm{m}^{-2}$. Furthermore, the perturbations caused by the sextupoles largely cancel out due to the phase differences in the betatron oscillation between the individual sextupoles. As Fig. 3.45 shows, the dynamic aperture is now considerably larger, and indeed is only slightly smaller than the mechanical limit of the vacuum chamber. With this sextupole configuration the accelerator can operate without any problem.

We immediately see from this example that it is always important to distribute as many correspondingly weak sextupoles evenly around the ring as possible. A further advantage comes from connecting the sextupoles acting in one plane into several circuits or **sextupole families** of different strengths. Numerical particle tracking is again required to determine how many families are needed and which sextupole strengths give the largest dynamic aperture.

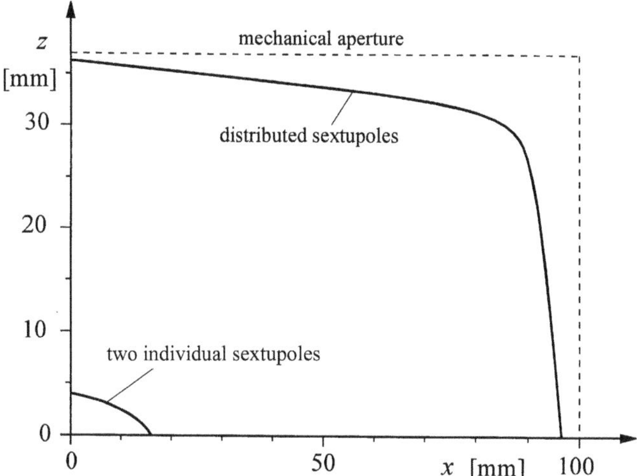

Fig. 3.45 Determination of the dynamic aperture by means of numerical particle tracking for the model ring. The mechanical aperture is shown by the dashed line. The two curves give the limits of stable particle motion for two individual sextupoles and for many evenly distributed sextupoles.

It is worth stressing that this problem of aperture limitation due to non-linear particle dynamics is one of the most difficult and important current problems in particle optics. All higher multiple fields, not just sextupole fields, are involved.

3.18 Local orbit bumps

It is often necessary to deliberately shift the transverse position of the beam within a limited region. This is done using so called **orbit bumps**, which move the beam in a well-defined way without affecting the rest of the ring. The basic principle of the orbit bump is illustrated in Fig. 3.46, which shows the simplest possible case. The beam displacement is always performed using small additional dipole magnets inserted into the ring, known as **steering** or **correcting coils**. These are labelled HK1 and HK2. The beam, which initially lies exactly along the axis, travels from the left through the correcting coil HK1 installed at position s_1 and is deflected through an angle κ_1. As a result of the beam focusing it then begins to perform oscillations about the beam axis, and after an advance in phase of $\Psi = \pi$ it crosses the orbit again at an angle x'_2 at the point s_2. A second steering coil HK2 is installed at exactly this point, which bends the beam through an angle $\kappa_2 = -x'_2$ and so brings it back onto the orbit. The beam then continues exactly along the nominal orbit. Using this principle it is possible to change the position of the circulating particle beam just in the isolated region from s_1 to s_2.

This kind of closed orbit bump requires the strengths of the two coils κ_1 and κ_2 to be in a strict ratio to one another. This is calculated using the transfer matrix $\mathbf{M}$ from s_1 to s_2, which according to (3.164) is uniquely defined by the

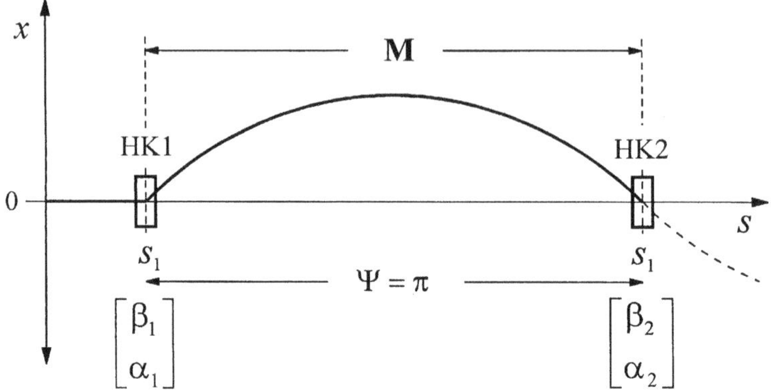

Fig. 3.46 The simplest example of an isolated orbit bump. Two correcting coils HK1 and HK2 are used. The phase difference of the betatron oscillation between the two coils is exactly 180°.

optical functions at these two points, i.e. β_1, α_1 and β_2, α_2. Since the phase is exactly $\Psi = \pi$, the trajectory vector at the point s_2 is

$$\boldsymbol{X}_2 = \begin{pmatrix} -\sqrt{\dfrac{\beta_2}{\beta_1}} & 0 \\ -\dfrac{\alpha_1 - \alpha_2}{\sqrt{\beta_1\,\beta_2}} & -\sqrt{\dfrac{\beta_1}{\beta_2}} \end{pmatrix} \cdot \begin{pmatrix} 0 \\ \kappa_1 \end{pmatrix} = \begin{pmatrix} 0 \\ -\sqrt{\dfrac{\beta_1}{\beta_2}}\,\kappa_1 \end{pmatrix}. \tag{3.303}$$

In order for the correcting coil HK2 to exactly compensate for the angle of the trajectory it must have a bending angle

$$\kappa_2 = +\sqrt{\frac{\beta_1}{\beta_2}}\,\kappa_1. \tag{3.304}$$

This matching condition for the two steering coils gives a closed orbit bump which is often also called a 180° **bump** because of the particular choice of the phase Ψ.

This simplest form of the local orbit bump is not often used in practice, since in general it is not possible to fix the exact phase difference between the steering coils. Instead the principle is extended to three coils HK1, HK2 and HK3, as shown in Fig. 3.47. The first steering coil HK1 bends the beam through the angle κ_1. At the point s_3 the trajectory thus has the vector

$$\boldsymbol{X}_{1\to3} = \mathbf{M}_{1\to3} \cdot \begin{pmatrix} 0 \\ \kappa_1 \end{pmatrix}, \tag{3.305}$$

where $\mathbf{M}_{1\to3}$ gives the evolution from s_1 to s_3. Since the positions of the steering coils may in principle be freely chosen, the resulting trajectory at s_3 in general has both a separation x_2 and an angle x_3' with respect to the ideal orbit. A

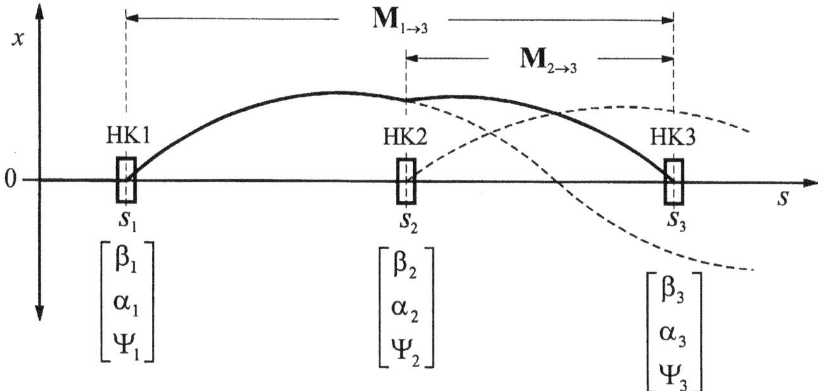

Fig. 3.47 An orbit bump formed from three correcting coils. In this case matching is always possible, regardless of the phase between the individual coils.

second steering coil HK2 is thus used to induce a further shift in the trajectory, which on its own would give a trajectory vector at the point s_3 of

$$\boldsymbol{X}_{2\to 3} = \mathbf{M}_{2\to 3} \cdot \begin{pmatrix} 0 \\ \kappa_2 \end{pmatrix}. \tag{3.306}$$

Here $\mathbf{M}_{2\to 3}$ is the transfer matrix from s_2 to s_3. These two corrections to the trajectory are arranged so that at s_3 the separation goes to zero again and the remaining trajectory angle is compensated for by the bending angle κ_3 of the steering coil HK3. We thus obtain the following relation

$$\begin{aligned} \boldsymbol{X}_3 &= \mathbf{M}_{1\to 3} \cdot \begin{pmatrix} 0 \\ \kappa_1 \end{pmatrix} + \mathbf{M}_{2\to 3} \cdot \begin{pmatrix} 0 \\ \kappa_2 \end{pmatrix} \\ &= \begin{pmatrix} a_{11} & a_{12} \\ a_{21} & a_{22} \end{pmatrix} \cdot \begin{pmatrix} 0 \\ \kappa_1 \end{pmatrix} + \begin{pmatrix} b_{11} & b_{12} \\ b_{21} & b_{22} \end{pmatrix} \cdot \begin{pmatrix} 0 \\ \kappa_2 \end{pmatrix} \\ &= \begin{pmatrix} a_{12}\kappa_1 + b_{12}\kappa_2 \\ a_{22}\kappa_1 + b_{22}\kappa_2 \end{pmatrix}. \end{aligned} \tag{3.307}$$

With the condition

$$\boldsymbol{X}_3 = \begin{pmatrix} 0 \\ -\kappa_3 \end{pmatrix} \tag{3.308}$$

we obtain the equations

$$\begin{aligned} a_{12}\kappa_1 + b_{12}\kappa_2 &= 0 \\ a_{22}\kappa_1 + b_{22}\kappa_2 &= -\kappa_3, \end{aligned} \tag{3.309}$$

which determine the strengths of the correcting coils. For a given value of κ_1, the strengths of the two other coils may be calculated. We again use the matrix

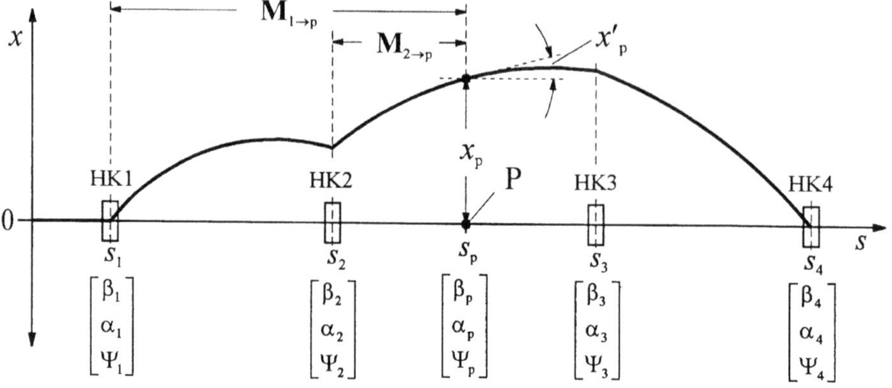

Fig. 3.48 An orbit bump with four correcting coils. This design allows one to simultaneously control both the displacement x_p and the angle x'_p at the point P.

(3.164) to obtain the elements a_{ij} and b_{ij} from the optical functions at the positions s_1, s_2, and s_3. It follows that

$$\kappa_2 = -\frac{a_{12}}{b_{12}}\kappa_1 = -\sqrt{\frac{\beta_1}{\beta_2}}\frac{\sin(\Psi_3-\Psi_1)}{\sin(\Psi_3-\Psi_2)}\kappa_1 \tag{3.310}$$

and

$$\begin{aligned}\kappa_3 &= -a_{22}\kappa_1 - b_{22}\kappa_2 \\ &= \kappa_1\sqrt{\frac{\beta_1}{\beta_3}}\left\{\frac{\sin(\Psi_3-\Psi_1)}{\tan(\Psi_3-\Psi_2)} - \cos(\Psi_3-\Psi_1)\right\}.\end{aligned} \tag{3.311}$$

We see that matching is almost always possible in an orbit bump using three steering coils, regardless of the position of the coils.

The most universal form of orbit bump consists of four steering coils. Here it is possible to choose both the position x_p and the angle x'_p of the shifted beam at the chosen point s_p. Such a system is used, for example, to precisely control the photon beam from an undulator. This type of orbit bump is illustrated in Fig. 3.48. The two steering coils HK1 and HK2 of strengths κ_1 and κ_2 result in a trajectory vector at the point s_p of

$$\begin{aligned}\boldsymbol{X}_\mathrm{p} &= \begin{pmatrix} x_\mathrm{p} \\ x'_\mathrm{p}\end{pmatrix} = \mathbf{M}_{1\to p}\cdot\begin{pmatrix} 0 \\ \kappa_1\end{pmatrix} + \mathbf{M}_{2\to p}\cdot\begin{pmatrix} 0 \\ \kappa_2\end{pmatrix} \\ &= \begin{pmatrix} a_{11} & a_{12} \\ a_{21} & a_{22}\end{pmatrix}\cdot\begin{pmatrix} 0 \\ \kappa_1\end{pmatrix} + \begin{pmatrix} b_{11} & b_{12} \\ b_{21} & b_{22}\end{pmatrix}\cdot\begin{pmatrix} 0 \\ \kappa_2\end{pmatrix} \\ &= \begin{pmatrix} a_{12}\kappa_1 + b_{12}\kappa_2 \\ a_{22}\kappa_1 + b_{22}\kappa_2\end{pmatrix}.\end{aligned} \tag{3.312}$$

The equations determining the field strengths immediately follow

$$\begin{aligned} x_\mathrm{p} &= a_{12}\kappa_1 + b_{12}\kappa_2 \\ x'_\mathrm{p} &= a_{22}\kappa_1 + b_{22}\kappa_2. \end{aligned} \tag{3.313}$$

Since x_p and x'_p are now fixed, these relations immediately give us the required bending angle of the steering coils

$$\begin{aligned} \kappa_1 &= \frac{b_{22}x_\mathrm{p} - b_{12}x'_\mathrm{p}}{a_{12}b_{22} - a_{22}b_{12}} \\ \kappa_2 &= \frac{a_{12}x'_\mathrm{p} - a_{22}x_\mathrm{p}}{a_{12}b_{22} - a_{22}b_{12}}. \end{aligned} \tag{3.314}$$

We now once again express the matrix elements a_{ij} and b_{ij} in terms of the optical functions at the points s_1, s_2, and s_p. This gives

$$\begin{aligned} \kappa_1 = {} & \frac{1}{\sqrt{\beta_1\,\beta_\mathrm{p}}}\,\frac{\cos(\Psi_\mathrm{p}-\Psi_2) - \alpha_\mathrm{p}\sin(\Psi_\mathrm{p}-\Psi_2)}{\sin(\Psi_2-\Psi_1)}\,x_\mathrm{p} \\ & - \sqrt{\frac{\beta_\mathrm{p}}{\beta_1}}\,\frac{\sin(\Psi_\mathrm{p}-\Psi_2)}{\sin(\Psi_2-\Psi_1)}\,x'_\mathrm{p} \end{aligned} \tag{3.315}$$

and

$$\begin{aligned} \kappa_2 = {} & -\frac{1}{\sqrt{\beta_2\,\beta_\mathrm{p}}}\,\frac{\cos(\Psi_\mathrm{p}-\Psi_1) - \alpha_\mathrm{p}\sin(\Psi_\mathrm{p}-\Psi_1)}{\sin(\Psi_2-\Psi_1)}\,x_\mathrm{p} \\ & + \sqrt{\frac{\beta_\mathrm{p}}{\beta_2}}\,\frac{\sin(\Psi_\mathrm{p}-\Psi_1)}{\sin(\Psi_2-\Psi_1)}\,x'_\mathrm{p}. \end{aligned} \tag{3.316}$$

The bending angles required in the steering coils HK3 and HK4 for correct matching of the closed bump may be obtained from (3.315) and (3.316) by simply reflecting the problem about the point P, i.e. by replacing β_1 by β_4, β_2 by β_3, Ψ_1 by Ψ_4 and Ψ_2 by Ψ_3. The signs of the beta function gradients and the orbit displacements are also reversed, i.e. α_p becomes $-\alpha_\mathrm{p}$ and x_p becomes $-x_\mathrm{p}$. We thus obtain

$$\begin{aligned} \kappa_3 = {} & -\frac{1}{\sqrt{\beta_3\,\beta_\mathrm{p}}}\,\frac{\cos(\Psi_4-\Psi_\mathrm{p}) + \alpha_\mathrm{p}\sin(\Psi_4-\Psi_\mathrm{p})}{\sin(\Psi_4-\Psi_3)}\,x_\mathrm{p} \\ & - \sqrt{\frac{\beta_\mathrm{p}}{\beta_3}}\,\frac{\sin(\Psi_4-\Psi_\mathrm{p})}{\sin(\Psi_4-\Psi_3)}\,x'_\mathrm{p} \end{aligned} \tag{3.317}$$

and

$$\kappa_4 = \frac{1}{\sqrt{\beta_4\,\beta_\mathrm{p}}}\,\frac{\cos(\Psi_3-\Psi_\mathrm{p})+\alpha_\mathrm{p}\sin(\Psi_3-\Psi_\mathrm{p})}{\sin(\Psi_4-\Psi_3)}\,x_\mathrm{p} + \sqrt{\frac{\beta_\mathrm{p}}{\beta_4}}\,\frac{\sin(\Psi_3-\Psi_\mathrm{p})}{\sin(\Psi_4-\Psi_3)}\,x'_\mathrm{p}. \tag{3.318}$$

In general a solution always exists for this type of orbit bump too. There are, however, practical limits: if the calculated bending angle exceeds the technical limit of one or more of the steering coils; or if the beam bump causes such a large displacement of the trajectory at a particular point that the beam scrapes against the side of the vacuum chamber. In such cases a different combination of coils must be found.

3.18.1 Examples of local orbit bumps

Since the transverse beam displacement for a given bending angle κ varies in proportion to $\sqrt{\beta(s)}$, the actual shape of an orbit bump is more complicated than is shown in the simplified diagrams of Figs. 3.46 to 3.48. Here we will give two fully worked examples using realistic values.

We start with the model accelerator with its FODO structure, as described in Section 3.13.3, and restrict ourselves exclusively to horizontal orbit shifts. Let the steering coils required to form the orbit bump each be mounted exactly in the middle of the zero-field drift section between the quadrupole and dipole magnets. In choosing this position we must ensure that there are sufficiently large changes in betatron phase between the steering coils, otherwise the effect of the coils will be relatively weak. As a rule of thumb, a phase separation of about $\pi/2$ should be chosen. We start with a bump in which the three correctors, HK1, HK2 and HK3, are installed with a separation of $1\frac{1}{2}$ FODO cells, as shown in Fig. 3.49.

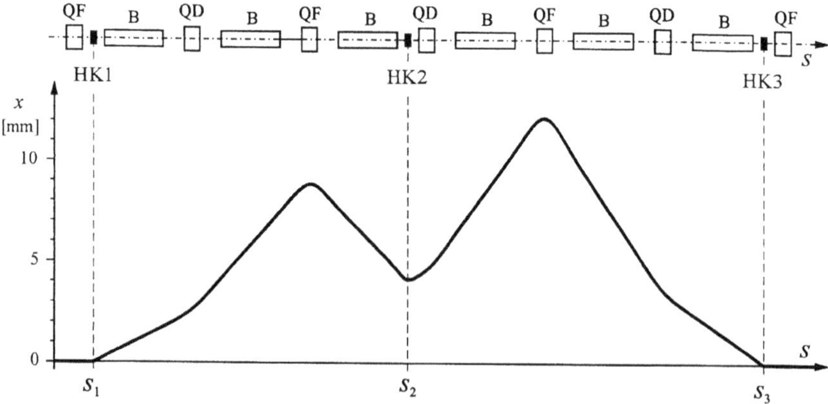

Fig. 3.49 Example of a local orbit bump consisting of three steering coils HK1, HK2, and HK3.

Table 3.5 Optical functions at the positions of the three steering coils in a closed orbit bump in the model ring.

optical function		HK1	HK2	HK3
β_i	[m]	8.16328	2.43100	8.16328
α_i	[rad]	2.08209	0.76732	-2.08209
Ψ_i	[rad]	0.05217	2.01063	4.41756

Table 3.6 Bending angles for the orbit bump steering coils. The bending angle of the first steering coil is arbitrarily set to $\kappa_1 = 1.000$.

steering coil	angle κ_i [mrad]
HK1	1.00000
HK2	2.57075
HK3	1.38109

The values of the optical functions at the three coil positions are summarized in Table 3.5. Inserting these values into equations (3.310) and (3.311), we can calculate the bending angles of the coils HK2 and HK3 required to close the bump. The value of the first steering coil HK1 is arbitrarily chosen to be $\kappa_1 = 1.000$ mrad. The results of this calculation are given in Table 3.6.

The calculation of the full trajectory of the orbit bump proceeds using the usual matrix formalism. The angular changes induced by the steering coils at the points s_1, s_2, and s_3 are simply added to the trajectory vector. The result is shown by the curve in Fig. 3.49.

By introducing a fourth steering coil, HK4, into the ring we may fix the displacement and angle of the beam at a point P, which we are free to choose within the region between the two innermost coils. This is illustrated in Fig. 3.50. Here P is chosen to be the position s_p exactly in the centre of a bending magnet. The required optical functions at all the steering coils and at the point P are listed in Table 3.7.

First of all the beam is only displaced transversely at the point P, without changing its angle with respect to the ideal orbit. This bump thus causes a pure offset in the orbit; it is usually known simply as an **orbit bump**.

The resulting trajectory vector at point P is thus

$$\boldsymbol{X}_\mathrm{p} = \begin{pmatrix} x_\mathrm{p} \\ x'_\mathrm{p} \end{pmatrix} = \begin{pmatrix} 10 \text{ mm} \\ 0 \end{pmatrix} .$$

With these values and the optical functions in Table 3.7 we can use equations (3.315) to (3.318) to calculate the bending angles in the steering coils required for matching. These are listed in Table 3.8.

Table 3.7 Values of the optical functions at the four steering coils HK1 to HK4 and at the freely-chosen point P.

optical function		HK1	HK2	HK3	HK4	P
β_i	[m]	8.16328	2.43100	8.16328	2.43100	4.81931
α_i	[rad]	2.08209	0.76732	-2.08209	0.76732	1.42471
Ψ_i	[rad]	0.05217	2.01063	4.41756	6.48037	3.19624

Table 3.8 Bending angles required in the four steering coils in order to shift the beam by a particular amount $x_{\mathrm{P}} = 10$ mm at the point P, without changing its angle at this point.

steering coil	angle κ_i [mrad]
HK1	-1.62667
HK2	3.14459
HK3	2.15661
HK4	5.57203

Table 3.9 Bending angles required in the four steering coils in order to deflect the beam at the point P through a particular angle $x'_{\mathrm{P}} = -3$ mrad, without changing its transverse position, i.e. $x_{\mathrm{P}} = 0$.

steering coil	angle κ_i [mrad]
HK1	2.30738
HK2	0.01130
HK3	-0.37150
HK4	-4.50274

The resulting shifted orbit near the bump is plotted in Fig. 3.50. The same calculation is now repeated, this time requiring only a change in angle at the point P, without any transverse displacement of the beam. The required trajectory vector is now

$$\boldsymbol{X}_{\mathrm{P}} = \begin{pmatrix} x_{\mathrm{P}} \\ x'_{\mathrm{P}} \end{pmatrix} = \begin{pmatrix} 0 \\ -3 \text{ mrad} \end{pmatrix} .$$

The calculation now yields another orbit bump, made using the same steering coils. The bending angle of the steering coils are found in Table 3.9. This example of a pure **angle bump** is also illustrated in Fig. 3.50.

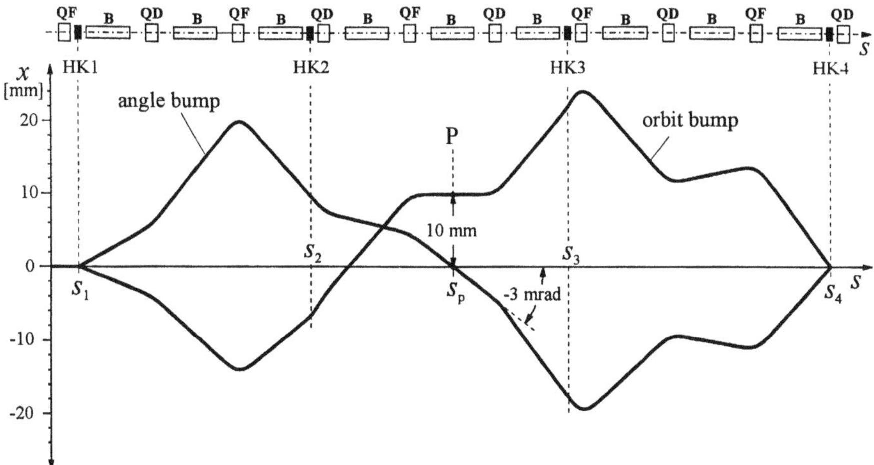

Fig. 3.50 Example of a local orbit bump using four steering coils. Two different bumps are illustrated, both of which use the same coils. In one case the beam is displaced by a distance $x_{\mathrm{p}} = 10$ mm at point P without any change in angle. In the other case the transverse position of the beam is unaltered, but the angle is changed by an amount $x'_{\mathrm{p}} = -3$ mrad.

4

Injection and extraction

4.1 The process of injection and extraction

In circular accelerators, particularly storage rings, the particles follow a rather complicated path from source to experiment. This path may be traced in the diagram of Fig. 4.1. A particle source, depending on its type, produces either electrons or ions with very low initial energies. A pre-accelerator — a linac or a microtron — is then used to bring the particles up to sufficiently high energy to allow injection into the circular main accelerator with as few losses as possible. Here they are raised to the required final energy and then **ejected** out of the ring, at which point they are either sent straight to the experiment or injected into a storage ring. In storage rings this process is repeated many times in order to accumulate large beam currents, as we have seen in Section 1.4.2.

The complicated beam transport system must inject the particle beam into the circular accelerator and then extract it again after a certain time, avoiding

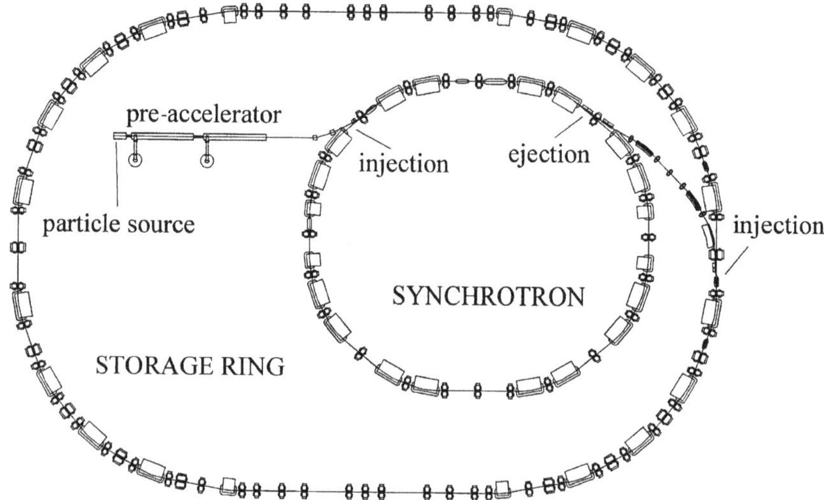

Fig. 4.1 Layout of a storage ring. The particles are emitted from a source and are then brought up to their final energy in a pre-accelerator and a main accelerator (synchrotron) before being injected into the storage ring.

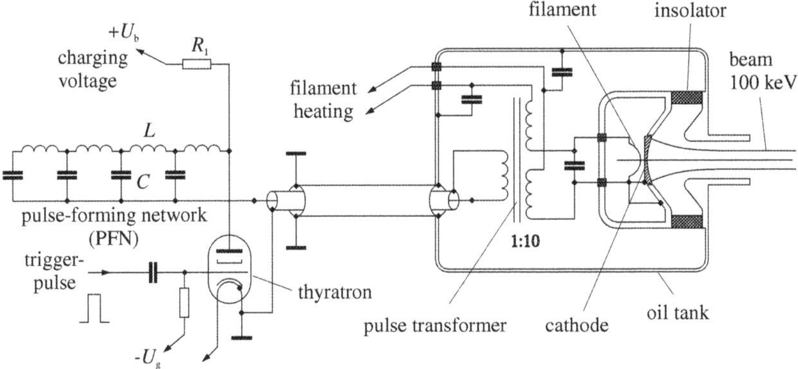

Fig. 4.2 Layout of a diode gun electron source. The voltage between the cathode and anode is produced by discharging the LC network in pulses across the thyratron.

particle losses as far as possible. The process of injection and ejection is non-trivial, since any particle lying outside the vacuum chamber also lies outside the acceptance of the accelerator, as a consequence of Liouville's theorem. Without further assistance it would inevitably hit the wall of the vacuum chamber somewhere and be lost. A procedure must be therefore found to deflect the beam so that the path of the **incoming** beam joins onto (or at least passes close to) the orbit of the circular accelerator, without significantly disturbing the **circulating** high velocity beam. Above all there must be no deflection of the beam during acceleration in a synchrotron or during stable orbiting in a storage ring. In this chapter we will describe the key principles of injection and ejection, along with the fundamental associated problems. We will confine our attention to injection because ejection is essentially only the reverse of injection; it is based on the same principles and uses the same techniques. Before doing so, let us turn briefly to the production of particle beams and consider a few of the most important particle sources.

4.2 Particle sources

With a few exceptions, electron beams are almost always produced using themionic cathodes. In order to achieve the high beam currents required in accelerators, large round cathodes of the type used in high-power tubes such as klystrons are employed. The simplest electron source is the diode, shown in Fig. 4.2. The round cathode is mounted in a vacuum and is heated by an electrical filament until sufficient numbers of electrons are emitted. A short distance from the cathode is the earthed anode, which has a hole bored in its centre through which the beam passes. The source is operated with a voltage of $U = 100$–150 kV between the cathode and anode, which drives a current of a few amperes. To a good approximation the relationship between the current and voltage is given by

$$I \propto U^{\frac{3}{2}}, \tag{4.1}$$

which follows from the space charge law of Langmuir and Schottky. Because of the high currents, the space charge effect must also be taken into account in guiding the beam between the cathode and anode. Specially designed electrodes are used in the so-called **Pierce system** [57], which exactly compensate for the distortions in the potential surface due to the space charge effect. Electron sources, often known as **diode guns**, are usually built according to the Pierce principle.

High currents can only be produced in short pulses in diode guns, due to their very high power requirement. However, this is not a real disadvantage since the linac (for example) which follows the injection stage also operates in pulsed mode. The important components of the pulsed power supply, which produces pulses lasting a few μs, are an LC pulse-forming network with a high-voltage supply, a thyratron which acts as a fast high-current switch, and a pulse transformer. At rest the capacitors in the pulse-forming network are charged to a voltage $U_\mathrm{b} = 10$–20 kV. The thyratron, a gas discharge tube, is then fired with a trigger pulse, causing the pulse-forming network to discharge abruptly through the primary winding of the pulse transformer. The circuit is arranged so that the length of the current pulse is exactly double the time constant of the LC-chain. With a suitable choice of winding ratio in the pulse transformer (e.g. 1:10), the requisite high voltage is produced in the secondary coil, which is directly connected to the cathode. In order to avoid electrical breakdown, the pulse transformer and the parts of the cathode carrying high voltages are placed in an oil-filled tank. The cathode itself is of course in a vacuum.

The diode gun has a relatively simple construction, but has the disadvantage of not being able to produce pulses shorter than $\tau_\mathrm{min} \approx 1\ \mu$s. This is due to its high power requirement, which can only be satisfied by having a high capacitance in the pulse-forming network, and to the limited bandwidth of the pulse transformer. What is more, the oil tank makes it considerably more difficult to replace the cathode, which is necessary from time to time during normal operation.

In order to produce electron beams with pulses lasting in the region of 1 ns, triode guns are nowadays used. These have a **grid** of fine wire mesh mounted between the cathode and anode. The layout of this type of electron source is shown in Fig. 4.3. In this case there is a constant voltage between the anode and the electrode holding the grid, which is usually not higher than around 50 kV to avoid risk of discharge. The grid electrode is housed in a conducting casing which acts as a Faraday cage. The cathode is situated closely behind the grid, which shields out the external fields very effectively. At rest the cathode sits at a potential of about $U_\mathrm{c} \approx +50$ V relative to the grid, so that virtually no electrons are able to escape from the region near the cathode surface. Using a fast transistor switch or an amplifier tube this cathode voltage is briefly switched back to 0 V, at which point the electrons are able to pass through the grid into the high-field region and are then accelerated in the direction of the anode. A short pulse-forming cable ensures that the cathode potential very quickly returns to the rest value and arrests the electron emission. The required pulse amplifier and trigger electronics are well shielded within the cathode housing. The external trigger pulse is carried by an isolating transformer or light guide, in order to safely

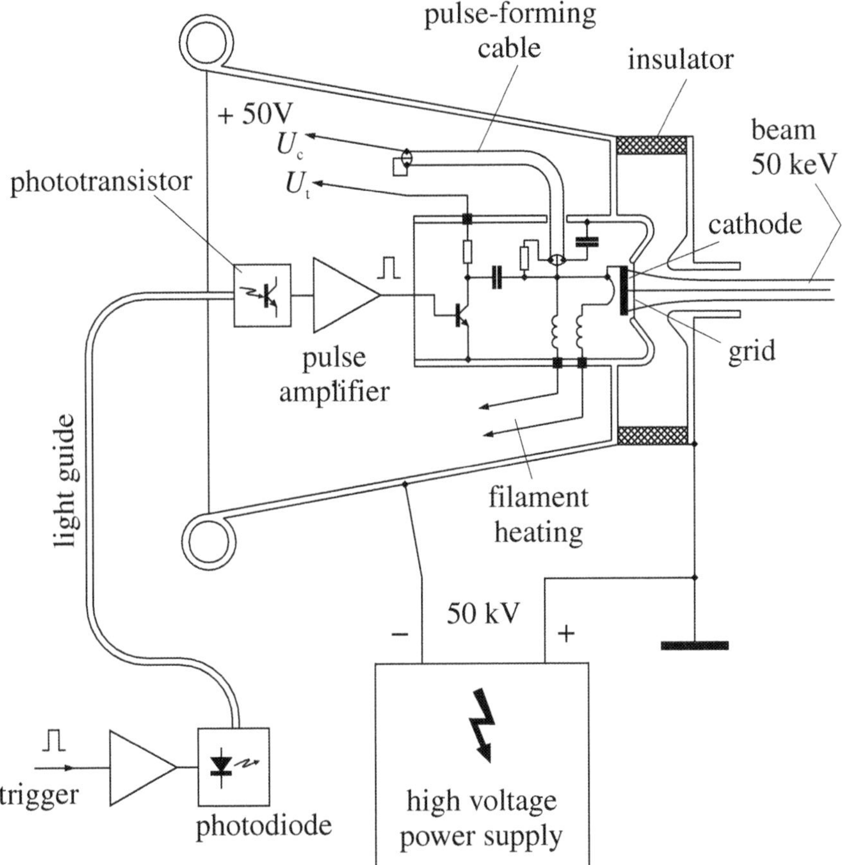

Fig. 4.3 Outline of a triode gun electron source. At rest, a potential of +50 V between the cathode and the grid blocks the flow of electrons. A very short pulse, with a duration determined by the length of the pulse-forming cable, briefly removes this barrier potential so that electrons are able to pass through the grid into the accelerating volume.

bridge the high potential difference of 50 kV.

Since the triode gun only needs to switch through a voltage of around 50 V, compared to around 100 kV in the diode gun, it is clear that this device is able to produce electron pulses that are orders of magnitude shorter. Of course this design also allows the production of much longer pulses if desired: it is merely necessary to use a different pulse amplifier or pulse-forming cable. Thanks to its greater flexibility, the triode gun is nowadays the most commonly used design.

Without going into details, let us also mention that laser guns are sometimes used as electron sources. An intense laser beam is shone on the cathode, which is made of metal or semiconductor (such as gallium arsenide), causing a kind

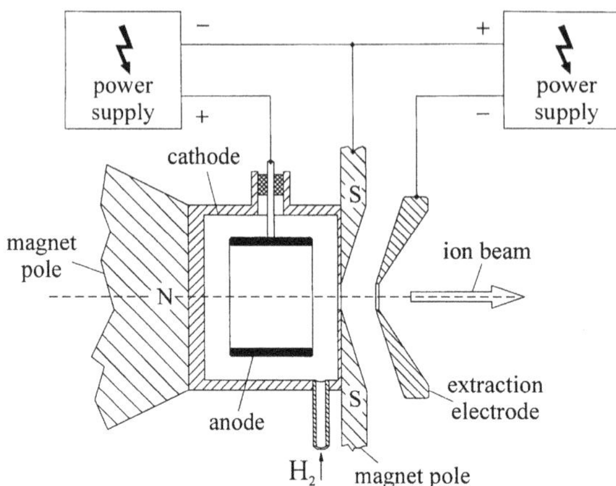

Fig. 4.4 Cross-section through a PIG ion source based on the Penning principle.

of plasma to form on its surface. Very high currents of up to several hundred amperes are achievable, with the pulse length given by the duration of the laser pulse. This can be as low as a few tens to a hundred ps. Another special feature of laser guns is that the well-defined polarization of the electromagnetic waves from the laser leads to the production of polarized electron beams.

Positrons are produced using intense electron beams of sufficiently high energy to allow pair production (Fig. 1.2). The electrons are accelerated up to the optimal energy of $E_e \approx 200$ MeV in a linac and are then fired at a tungsten target. High energy photons are emitted by bremsstrahlung, and are then largely re-absorbed in the same tungsten target via the process of pair production. Significant multiple scattering also occurs within the material of the target. The resulting $e^+ - e^-$ pairs finally exit the target with a broad energy spectrum from 0 to 30 MeV. They are fed into a linac, whose field is arranged so that only the positrons are accelerated.

Proton beams and beams of light and heavy ions are produced in **ion sources**. The simplest such source, shown in Fig. 4.4, is the PIG-source (Philips Ion Gauge) based on the well-known Penning principle [58]. The cathode and the cylindrical anode are situated inside an ionization chamber and have a potential difference of several hundred volts between them. In addition two magnetic poles produce a static magnetic field of around 0.01 T, perpendicular to the electric field. A gas, for example H_2, is pumped into this chamber at a pressure of a few tens to a hundred Pa and is then ionized. The material in the ionization chamber is chosen to ensure that the electrons and positrons recombine as little as possible. Some of the ions exit through an opening on the axis of the chamber, and are then accelerated by an external electric field in the direction of the extraction electrode.

Modern ion sources can be considerably more complicated, especially if specific ions with a well-defined charge or fixed e/m ratio are to be produced. High frequency fields are often used for ionization, usually coupled through coils.

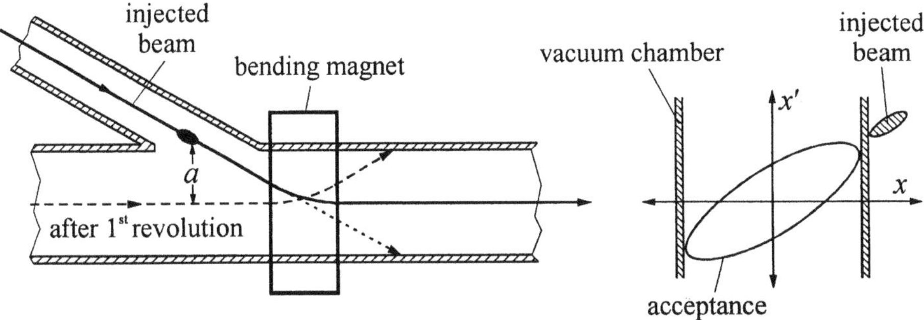

Fig. 4.5 Injection into a circular accelerator. The injected beam lies outside the acceptance ellipse of the accelerator, and must be deflected into the orbit by means of a pulsed magnet.

4.3 The fundamental problem of injection

The task of injection consists of taking a particle beam of well-defined energy from a pre-accelerator and introducing it into a circular accelerator without significant loss of particles. Special measures are required to achieve this, since an external particle beam always lies outside the acceptance of an an accelerator, as shown in Fig. 4.5. Just before it enters the vacuum chamber the beam is at a separation $a > d$ from the orbit, where d is the size of the aperture of the vacuum chamber. The lower limit of the acceptance, given by (3.139), is then

$$A_{\mathrm{inj}} \geq \frac{a^2}{\beta} > \frac{d^2}{\beta}, \tag{4.2}$$

where β denotes the amplitude function of the circular accelerator at this point. The actual value of the acceptance is in general much larger, since the injected beam enters the accelerator at a considerable angle, which is not taken into account at all in (4.2). The injected beam would thus cross the orbit and at a certain distance beyond the injection point would hit the wall of the vacuum chamber and be lost. It is therefore necessary to install a bending magnet immediately behind the injection point. The position and strength of this magnet are carefully chosen so that the incoming beam is deflected as precisely as possible into the orbit. The magnet pushes the acceptance ellipse away from its nominal position so that its centre coincides with that of the injected beam. Unfortunately this is not the end of the problem, for once the beam has completed one full revolution it will encounter this field again and will be deflected through the same angle away from the orbit and into the wall of the vacuum chamber. It is therefore not possible to use a static magnetic field for this task.

If instead a very rapidly pulsed magnet is used, a so-called **kicker magnet** whose field builds up in less than one revolution and then disappears during the revolution following injection, then loss-free injection becomes possible. The incoming beam in deflected only once, as it enters the circular accelerator. By the end of the first revolution the field of the kicker magnet has died away, and so the beam is not deflected a second time. The duration of the kicker magnet pulse must therefore be less that twice the period of revolution of the particles in the ring. In a synchrotron with a circumference of $L = 150$ m the kicker's pulse length is thus $\tau_{\text{kick}} < 1\ \mu$s.

To achieve the high currents needed in storage rings, repeated injection of particles is required. This cannot be achieved with the set-up shown in Fig. 4.5, however, because when the kicker magnet is switched on for the second injection all the particles travelling along the beam orbit from the first injection will be deflected outwards and lost again.

This is an example of a general problem in particle injection, when particles must be injected over several orbits or in several injection stages with arbitrary intervals. The injected beam takes up a finite volume in phase space, which consists of four transverse and two longitudinal dimensions. This phase volume may be expressed in the usual coordinates as follows:

$$\Delta V = \Delta x \cdot \Delta x' \cdot \Delta z \cdot \Delta z' \cdot \frac{\Delta p}{p} \cdot \Delta s \tag{4.3}$$

This brings us to a fundamental rule of particle injection:

Fundamental rule of injection:

It is not possible to reinject particles into an already occupied volume of phase space without losing the particles already present.

4.4 Injection of high proton and ion currents by 'stacking'

A technique which allows the production of high beams currents in circular accelerators by repeated injection is that of 'stacking', most widely used for protons and heavy ions. As we have seen in Chapter 2, these particles produce virtually no synchrotron radiation. They therefore do not lose energy during each revolution, and the betatron oscillations induced during injection continue undamped. Stacking may be employed when the phase volume occupied by the injected beam is less than the total phase space available in the accelerator. The available phase space may be divided up into individual small volumes, which are adjacent but separate, and are filled with particles one after another during each injection. The individually injected beams are literally stacked into the separate sections of phase space until all are filled.

Let us illustrate this process using two examples. In the first case the circumference L of the circular accelerator is divided into n equal sections, each of length $\Delta L = \tau_0\, v$, where τ_0 is the time taken by a particle of velocity v to travel a distance ΔL. This gives n empty phase volumes, shown in Fig. 4.6. Now let us assume that the injected beam is rather shorter than ΔL, allowing one phase volume to be filled with particles during the first injection. In the next injection the timing is shifted by τ_0, so that the neighbouring volume is now filled. Altogether this process is repeated exactly n times, after which the available phase

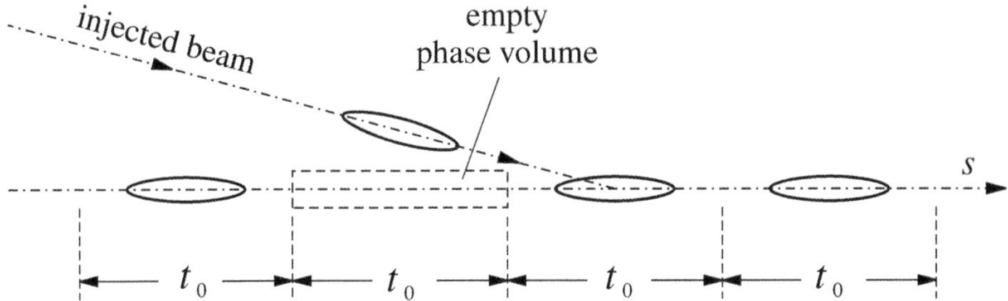

Fig. 4.6 Injection with stacking in longitudinal phase space.

space in the accelerator is completely filled with particles. The total circulating current is now n times as large as that from a single injection.

The difficulty with this process is that the kicker magnet must produce a square pulse not longer than τ_0 whose rising and falling edges must be very short compared to the pulse length. Under no circumstances must the kicker be allowed to disturb the particles circulating in the two neighbouring phase volumes. This problem is generally easier to solve in very large accelerators, where even with a large number of phase volumes the time τ_0 is relatively long. In very small machines, on the other hand, this procedure is not very useful.

Instead of using the longitudinal phase space, stacking can also be performed in the transverse phase volume, as Fig. 4.7 shows. Here a local closed orbit bump is produced for a short time by means of two or three kicker magnets, using the procedure described in Section 3.18, and the strengths of the kickers are varied from one injection to another.

A bending magnet called a **septum** bends the injected beam approximately parallel to the orbit, so that it falls within the acceptance ellipse of the accelerator and follows a path which then exactly crosses the orbit in kicker 2. In the first injection the strength of this kicker is chosen such that the beam is bent exactly onto the orbit, which it then follows. In the second injection the strength of the two kickers is reduced so that the resulting orbit bump brings the beam as close as possible to the septum without losing any particles. With the beam in this position a second beam can now be injected through the septum, very close to the first. Using this bump arrangement the injection procedure may be repeated many times, depending on the choice of working point Q. With each revolution the betatron oscillations cause the centre of mass of the injected beam to move through a particular phase angle about the ideal orbit, so that there is always another free phase volume available to be filled with particles.

Once this 'shell' becomes full a new one can be started and filled by further reducing the kicker strength and hence the bump amplitude. Hence the total available aperture of the accelerator is gradually filled with particles, and very high beam currents can eventually be achieved.

It should be noted that instead of using a variable-amplitude kicker bump,

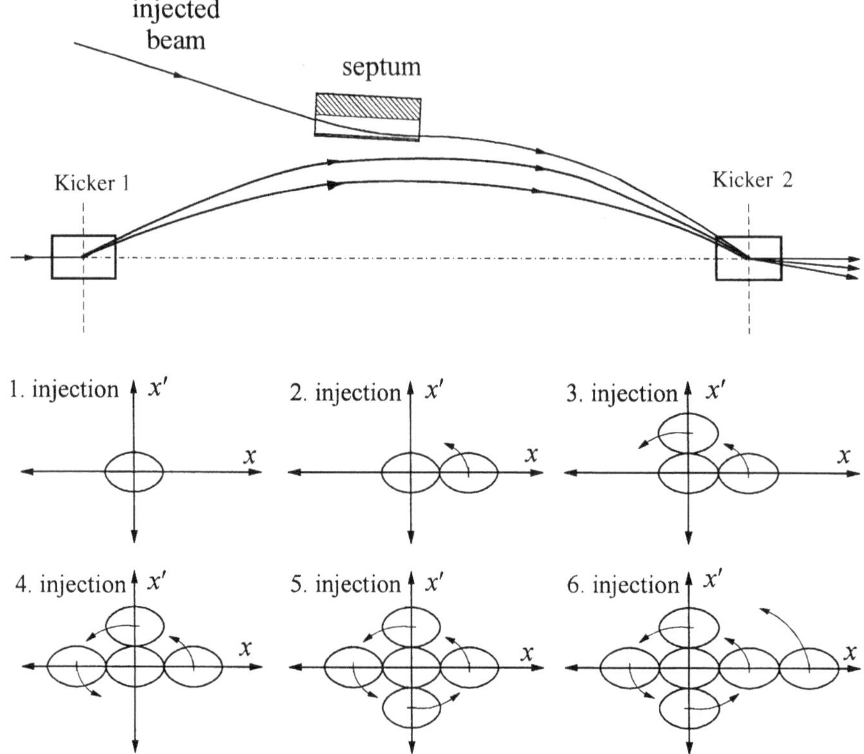

Fig. 4.7 Injection with stacking in transverse phase space. The first injection is performed along the orbit, with all successive ones in close proximity.

transverse stacking can also be performed by varying the particle energy, provided that the dispersion D at the injection point is sufficiently large. By varying their momenta, particles may be injected along different dispersive trajectories with a separation $x_{\mathrm{D}} = D\frac{\Delta p}{p}$ from the orbit. Apart from this the injection process proceeds exactly as in the case of the kicker bump.

4.5 Injection of proton beams using stripping foils

A very elegant method of almost continuous injection into a ring accelerator involves the use of a thin so-called 'stripping' foil. This process, which cannot be used in the case of electron beams, is mostly employed in proton injection. It is based on a very simple principle, illustrated in Fig. 4.8.

H^- ions enriched with electrons are produced in an ion source and then brought up to high energies in a pre-accelerator. After entering the circular accelerator they travel through a bending magnet which bends them onto the orbit. They then encounter a foil, which strips the electrons from the H^- ions as a result of interactions with the material. Positively charged protons thus exit the foil. When, after a full revolution, these particles again reach the first bending

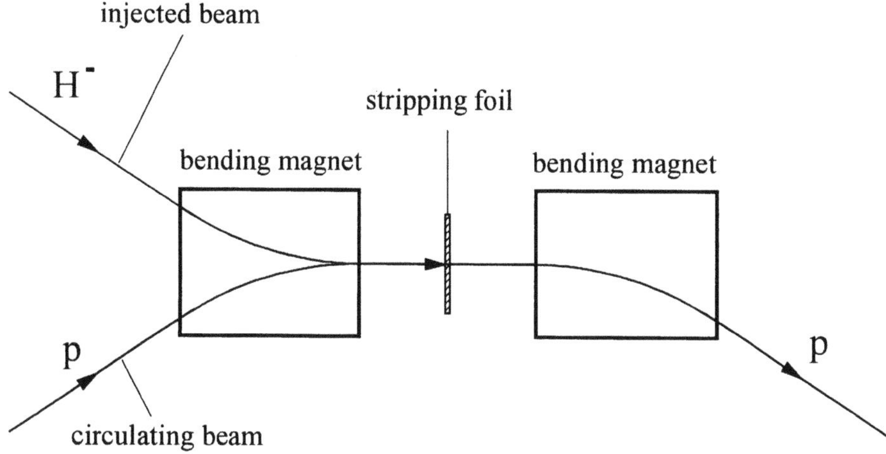

Fig. 4.8 Injection of proton beams by means of a stripping foil.

magnet they are bent in exactly the opposite direction since they now have the opposite charge. The injected and circulating beams are thus automatically kept apart, and so it is not necessary to use pulsed kicker magnets.

This example clearly illustrates how the fundamental rule of injection stated above only applies when charge is conserved. Particles of different charges can indeed be injected into the same phase volume without inducing losses.

4.6 Injection into an electron storage ring

In electron storage rings another technique is used to repeatedly inject particles into a given phase volume of the accelerator without significant particle losses. This relies on the fact that the betatron oscillations of the electrons are damped by the emission of synchrotron radiation. We will consider this phenomenon in detail in Section 6.2. For now we will just remark that all transverse particle oscillations die away with a particular time constant in the focusing field of the electron ring, since energy is removed from the system by the radiation of photons. Obviously this technique is not applicable to protons and ions.

The principle behind this method of building up high particle currents in electron storage rings is illustrated in Fig. 4.9. A local orbit bump is again used, produced for a short time using fast kicker magnets with a pulse length which is generally about equal to one period of revolution. In this case let us assume that a stored beam is already circulating in the ring, requiring a total transverse aperture of at least 7σ in order to guarantee long beam lifetimes. The acceptance ellipse is of course rather larger, extending right to the edge of the vacuum chamber, which here is formed by the so-called **septum sheet**. This is a metal sheet, made as thin as possible, which shields the stored beam from the bending field of the septum. Injection is not possible in this state.

During injection the kicker bump is turned off and for a brief instant the beam is pushed as close to the septum as is possible without inducing significant

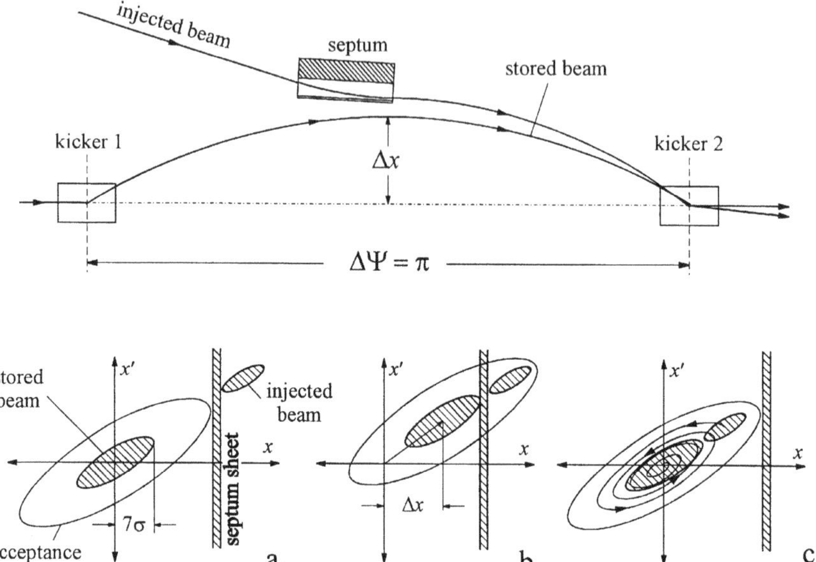

Fig. 4.9 Electron accumulation. The phase volume filled with the injected beam is freed up due to beam damping and becomes available again for the next injection. (a) is the phase diagram in the initial state before injection, (b) shows the state during injection, and (c) the state during the revolutions which follow.

particle loss. The phase ellipse filled with particles is thus pushed a distance Δx in the direction of the septum, which in general also causes a further change in the path angle. The key point is that the whole acceptance ellipse shifts in the phase diagram by the same amount as the orbit. It now juts out beyond the septum sheet and is able to receive the injected beam, capturing it near the stored beam. By the next revolution the kicker pulses have died away again and the stored beam is returned to its nominal position on the orbit.

The newly injected beam performs large betatron oscillations around the stored beam, but these oscillations are stable because they lie within the acceptance of the storage ring. As a result of the damping mentioned above, the amplitude of these oscillations decays away with time and the new particles move in a spiral inside the phase ellipse towards the orbit. After a few damping periods, lasting from 1 ms to a few tens of ms, the injected particles are incorporated into the stored beam, increasing its intensity. The phase volume occupied during the injection is now free again. This procedure may in principle be repeated as often as desired, yielding very high beam currents, until technical or physical limits are reached.

This injection technique circumvents the fundamental rule of injection stated above because energy is lost by radiation, and so Liouville's theorem no longer holds.

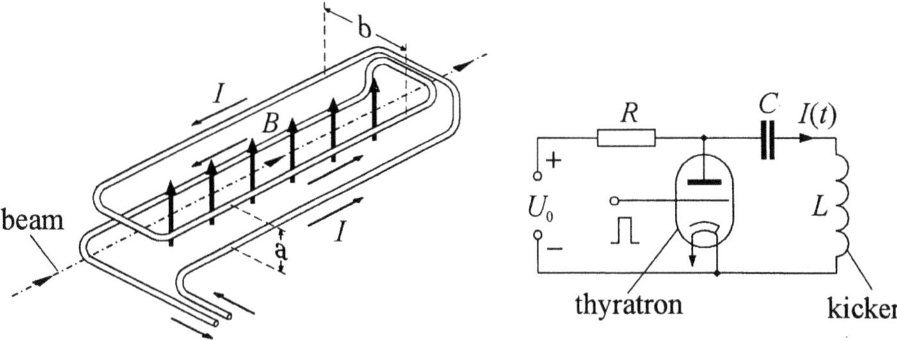

Fig. 4.10 A pulsed kicker magnet consisting of four parallel conductors. The current pulse is produced by discharging a capacitor across a thyratron.

4.7 Kicker and septum magnets

Apart from the special case of the stripping foil, the injection methods described here all make use of very fast pulsed magnets, the kickers. These are the most important technical components of the injection stage, and so we will describe the fundamentals of their construction and operation here.

The duration of the kicker pulse depends upon the circumference of the accelerator and on the layout of the injection system, but typically lies in the region of 1 μs. Such short pulses can only be achieved using magnets with a very small inductance, since otherwise higher voltages are required than are technically possible. In the simplest case a coil with just a few turns and without an iron core is used, as illustrated in Fig. 4.10. In the example shown, the kicker magnet consists of four parallel conductors with a horizontal separation b and a vertical separation a. They are arranged symmetrically about the orbit and so are separated from it by a distance

$$r = \frac{1}{2}\sqrt{a^2 + b^2}. \tag{4.4}$$

The field due to a wire at this separation is simply calculated by

$$\mu_0 I = \oint \boldsymbol{B} d\boldsymbol{r} = 2\pi r |\boldsymbol{B}| \qquad \longrightarrow \qquad |\boldsymbol{B}| = \frac{\mu_0 I}{2\pi r}. \tag{4.5}$$

The symmetrical arrangement means that the x-components of the fields due to the four conductors all cancel out along the orbit. The resulting net field acts in the z-direction with a strength

$$B_z = \frac{4\mu_0 b}{\pi(a^2 + b^2)} I. \tag{4.6}$$

Here we assume that the kicker length is much greater than its width, so that the end effects of the longitudinal fields may be neglected. The inductance of the kicker may be determined simply by calculating the voltage generated by the time-varying magnetic field. Using (4.6) this is given by

$$U = n \iint_A \dot{\boldsymbol{B}} d\boldsymbol{f} = n \frac{4\mu_0 b^2 l}{\pi(a^2 + b^2)} \dot{I}. \tag{4.7}$$

Here l is the length of the kicker magnet, $A = l \times b$ is its cross-sectional area, and $n = 2$ is the number of current turns. The inductance is immediately obtained from (4.7) as

$$L = \frac{U}{\dot{I}} = \frac{8\mu_0 b^2 l}{\pi(a^2 + b^2)}. \tag{4.8}$$

The expressions given here for the kicker field and inductance are of course only rather rough approximations, due to the simplifying assumptions we have made about the geometry. However, they are generally sufficient for a rough order-of-magnitude calculation. In many cases it must also be borne in mind that the conductors which make up the kicker magnet are installed in a vacuum, and so are surrounded by a vacuum chamber, which has a certain shielding effect on the field. The field is therefore weaker than that given by (4.6). This effect is very well described using an empirical correction factor, where it is assumed that the surrounding vacuum chamber is a round cylinder of radius R_0. The effective kicker field at the orbit is then

$$B_{\text{eff}} = k_c B_z \qquad \text{with} \qquad k_c = 1 - \frac{a\, b}{4 \arctan \dfrac{a}{b}\, R_0^2}. \tag{4.9}$$

As an example let us calculate the dimensions of a typical kicker magnet, ignoring the vacuum chamber. Here the bending angle κ to be produced by the kicker is fixed, and may be calculated from the magnetic field and the particle energy with the help of equation (3.4), giving

$$\kappa = \frac{l}{R} = \frac{e}{p} B_z l = 0.2998 \frac{B_z l}{E} \qquad \text{with } E \text{ in [GeV]}. \tag{4.10}$$

The other values follow immediately from (4.6) and (4.8):

kicker properties:		calculated values :
$\kappa = 3$ mrad		
$a = 0.04$ m	$\longrightarrow$	$B_z = \dfrac{\kappa E}{0.2998 l} = 0.05$ T
$b = 0.08$ m		
$l = 1.0$ m		$I = 3127$ A
$E = 5.0$ GeV		$L = 2.56\ \mu$H

Since there are only a few conductors, a current pulse of several thousand amperes is required. This is most easily obtained by charging a capacitor to a high voltage and then discharging it through the kicker magnet. The basic circuit diagram is shown in Fig. 4.10. The capacitor C is charged through a resistor R up to the voltage U_0 and then at the moment of injection is discharged using a thyratron. Thyratrons are gas discharge tubes which act as very high current switches at high voltages. After the thyratron fires we have an LC-oscillator circuit in which a time varying current $I(t)$ flows. This current obeys the well-known relation

$$\ddot{I}(t) + \omega I(t) = 0 \qquad \text{with} \qquad \omega = \frac{1}{\sqrt{LC}}, \tag{4.11}$$

which has the general solution

$$I(t) = I_1 \cos\omega t + I_2 \sin\omega t. \tag{4.12}$$

At time $t = 0$ we have $I(0) = 0$, so it immediately follows that $I_1 = 0$. We thus have $I(t) = I_2 \sin\omega t$. Hence the current varies sinusoidally, with a maximum value given by I_2. After half a period of oscillation the process stops, because the thyratron cannot carry negative current. The result is a half-wave pulse of duration

$$\tau_{\text{kick}} = \frac{\pi}{\omega} = \pi\sqrt{LC} \qquad \longrightarrow \qquad C = \left(\frac{\tau_{\text{kick}}}{\pi}\right)^2 \frac{1}{L} \tag{4.13}$$

In order to reach a given maximum current $I_{\text{max}} = I_2$, the capacitor must be charged to a certain voltage U_0 at time $t = 0$. The voltage varies according to (4.12) as

$$U(t) = L\dot{I} = \omega L I_{\text{max}} \cos\omega t. \tag{4.14}$$

The required initial voltage is thus

$$U_0 = \omega L I_{\text{max}} = \sqrt{\frac{L}{C}} I_{\text{max}}. \tag{4.15}$$

If we require the kicker magnet in this example to have a pulse length $\tau_{\text{kick}} = 1\ \mu\text{s}$ then we may use (4.13) and (4.15) to calculate the required value of the capacitance and the initial voltage. In this example we obtain

$$\begin{aligned} C &= \left(\frac{\tau_{\text{kick}}}{\pi}\right)^2 \frac{1}{L} = 39.6\ \text{nF} \\ U_0 &= \sqrt{\frac{L}{C}} I_{\text{max}} = 25.1\ \text{kV}. \end{aligned} \tag{4.16}$$

If very short pulses are required, then the kicker magnets must be operated at high voltages, despite their relatively small inductance.

In addition to the kicker, another special type of magnet used in injection is the septum. As we have seen, it has the task of bending the injected beam immediately before it enters the circular accelerator so that it comes as close as possible to the already circulating beam and also at a very small angle to it. The septum is thus a bending magnet whose field only acts within the gap between its poles, and so only deflects the injected beam. The circulating beam stored in the ring, which passes by just outside the septum, is not affected.

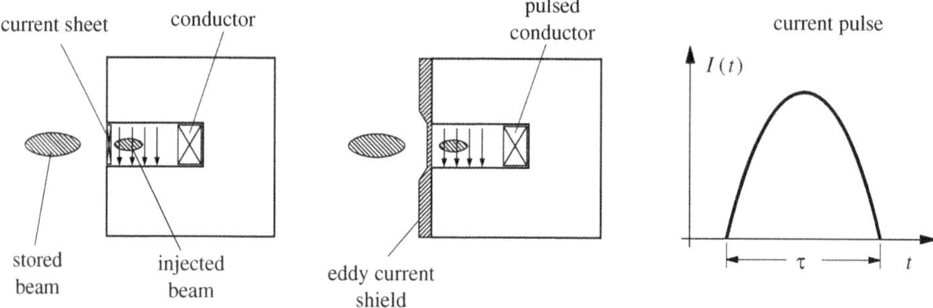

Fig. 4.11 Septum magnet with current sheet and eddy current field. On the *left* the field inside the gap in the septum is shielded by a thin current sheet. This magnet may be operated in pulsed or continuous mode. The septum on the *right* uses an eddy-current shield made from a good conductor. This type of septum can only be operated in pulsed mode.

A simple magnet with an open gap, such as the one shown in Fig. 3.8, is not suitable for this purpose because its stray field extends far beyond the gap and so would of course also deflect the circulating beam. In some cases this would cause considerable distortion of the orbit. One way to shield this field is to cover the open side of the gap with a current sheet carrying the same current as in the conductors which produce the field, but in the opposite direction (Fig. 4.11). Since the separation between the injected and circulating beams must be very small during injection, the current sheet is very thin and generally has a thickness of only a few mm. Owing to the very high currents, this leads to very high current densities, which in continuous operation would require expensive cooling. However, the septum is only ever required during the very brief moment of injection. It therefore makes sense to operate the septum using short current pulses, in the same way as the kicker magnets. The time-integrated average power is thus very much smaller, and a cooling system is not required. The current pulses in the septum do not need to be quite as short as in the kickers, and the pulse length typically ranges from a few tens of μs to 1 ms.

Another very interesting way to shield out the stray field from the septum is the eddy-current shield, which closes off the open side of the septum (right-hand example in Fig. 4.11). This consists of a sheet of a good-conducting material, such as copper. When the magnet is operated with a half-wave pulse of duration τ, eddy currents arise in the sheet which oppose and so weaken the magnetic field which is trying to penetrate it. If the pulse length τ is short, the stray field can be very effectively suppressed by a thick enough shield.

To make a coarse estimate of the shielding effect, let us start with a half-wave pulse of length τ, for which the lowest significant Fourier component has the frequency

$$\omega = \frac{\pi}{\tau}. \tag{4.17}$$

The current density and hence the field distribution in the copper sheet are determined by the **skin effect**

$$i(x) = i_0 \exp\left(-\frac{x}{d_s}\right), \tag{4.18}$$

where

$$d_s = \sqrt{\frac{2}{\omega\sigma\mu_r\mu_0}} \tag{4.19}$$

gives the penetration depth. It is immediately evident that a high frequency ω and good conductivity in the eddy-current shield will give the best shielding effect. If we take a pulse length of $\tau = 50\ \mu$s, then from (4.17) the lowest frequency component is $\omega = 6.3 \times 10^4\ \mathrm{s}^{-1}$. Using the conductivity of copper $\sigma_{\mathrm{Cu}} = 5.9 \times 10^7\ \Omega^{-1}\ \mathrm{m}^{-1}$ and the relative permeability $\mu_r = 1$, equation (4.19) gives $d_s = 0.66$ mm. For a 2 mm thick eddy-current shield, equation (4.18) predicts a reduction of the stray field to 5% of the unshielded value. In reality, measurements have shown the shielding to be considerably better than this, mainly because the geometrical distribution of the stray field is far more complicated than we have assumed in the calculation of the skin depth. As a first approximation, however, and staying a factor 2 to 3 on the safe side, this calculation is perfectly valid.

5

RF systems for particle acceleration

In Chapter 1 we showed that the fundamental limitations to the final beam energy achievable in static accelerators may be overcome by the use of high-frequency voltages. Virtually all modern accelerators use powerful radiofrequency systems to produce the requisite strong electric fields, with frequencies ranging from a few hundred MHz to several GHz. In such systems waveguides are preferred as resonators and to conduct the beam, since they have low losses and can deliver very high power. We will begin by considering the physics of waveguides and resonant cavities, and will discuss their most important properties. Further details may be found, for example, in references [59], [110] and [109].

5.1 Waveguides and their properties

The propagation of an electromagnetic wave in a waveguide is described by the general wave equation

$$\Delta \boldsymbol{E} - \frac{1}{c^2}\ddot{\boldsymbol{E}} = 0. \tag{5.1}$$

In the discussion which follows we are only interested in the spatial distribution of the wave. We separate out the periodic time dependence which has a frequency ω, and write

$$\boldsymbol{E}(\boldsymbol{r},t) = \boldsymbol{E}(\boldsymbol{r})e^{i\omega t} \qquad \text{with} \qquad \boldsymbol{r} = (x,y,z). \tag{5.2}$$

Here the coordinate system has been chosen so that x and y represent the horizontal and vertical coordinates and z the propagation direction along the waveguide (Fig. 5.1). Substituting into (5.1) yields the time-independent wave equation

$$\Delta \boldsymbol{E}(\boldsymbol{r}) + k^2 \boldsymbol{E}(\boldsymbol{r}) = 0 \tag{5.3}$$

with the wavenumber

$$k = \frac{\omega}{c} = \frac{2\pi}{\lambda}. \tag{5.4}$$

If, to begin with, we only consider the z-component of the field then (5.3) becomes

$$\frac{\partial^2 E_z}{\partial x^2} + \frac{\partial^2 E_z}{\partial y^2} + \frac{\partial^2 E_z}{\partial z^2} = -k^2 E_z, \tag{5.5}$$

which we may solve using the trial solution

$$E_z(x, y, z) = f_x(x)\ f_y(y)\ f_z(z). \tag{5.6}$$

From (5.5) it then follows that

$$\frac{f_x''}{f_x} + \frac{f_y''}{f_y} + \frac{f_z''}{f_z} = -k^2. \tag{5.7}$$

Defining

$$k_x^2 \equiv -\frac{f_x''}{f_x} \qquad k_y^2 \equiv -\frac{f_y''}{f_y} \qquad k_z^2 \equiv -\frac{f_z''}{f_z} \tag{5.8}$$

equation (5.7) yields the relation

$$k_x^2 + k_y^2 + k_z^2 = k^2, \tag{5.9}$$

which, if we set

$$k_x^2 + k_y^2 = k_\mathrm{c}^2 \tag{5.10}$$

then becomes

$$k_z = \sqrt{k^2 - k_\mathrm{c}^2}. \tag{5.11}$$

Using (5.8), the wave propagation along the waveguide is described by the equation

$$f_z'' + k_z^2\ f_z = 0. \tag{5.12}$$

Multiplying this by $f_x \times f_y$ and using (5.6) results in a differential equation describing the electrical field component along the axis of the waveguide, namely

$$\frac{\partial^2 E_z}{\partial z^2} + k_z^2\ E_z = 0. \tag{5.13}$$

The solution to this equation is the function

$$E_z = E_0\ e^{ik_z\ z}, \tag{5.14}$$

as may easily be seen by substitution. If the wavenumber k_z is complex, then the amplitude of the wave travelling through the waveguide falls off exponentially, i.e. loss-free wave propagation is not possible. Loss-free propagation occurs only when k_z is real. Using (5.11), it is thus possible to define two regimes for waveguide operation, namely

$$k_z = \begin{cases} \text{complex} & \text{if} \quad k_\mathrm{c}^2 \geq k^2 \quad \text{(damping)} \\ \text{real} & \text{if} \quad k_\mathrm{c}^2 < k^2 \quad \text{(propagation)} \end{cases} \tag{5.15}$$

The special value of the wavenumber k_c is termed the **cut-off wavenumber** and divides waveguide operation into the regimes of free propagation and damping. For practical purposes it is therefore crucial whether the wavenumber k of the electromagnetic wave propagating in free space is larger or smaller than this cut-off value. Using

$$k_c = \frac{2\pi}{\lambda_c} \tag{5.16}$$

we may also define the associated **cut-off wavelength** λ_c. From equation (5.11) it then follows that

$$\frac{1}{\lambda^2} = \frac{1}{\lambda_c^2} + \frac{1}{\lambda_z^2} \tag{5.17}$$

or, solving for λ_z,

$$\lambda_z = \frac{\lambda}{\sqrt{1 - \left(\frac{\lambda}{\lambda_c}\right)^2}}. \tag{5.18}$$

It is worth noting that in the regime of loss-free wave propagation in the waveguide, the wavelength λ_z is always larger than the wavelength in free space. This means that the **phase velocity** of the wave within the waveguide is greater than the speed of light:

$$v_\varphi = \frac{\omega\, \lambda_z}{2\pi} > c. \tag{5.19}$$

Re-expressing (5.17) using $\omega = 2\pi c/\lambda$ and solving for the frequency, we obtain the important **dispersion relation** for waveguides

$$\boxed{\omega = c\sqrt{k_z^2 + \left(\frac{2\pi}{\lambda_c}\right)^2}.} \tag{5.20}$$

5.1.1 Rectangular waveguides

Having established the general features of wave propagation in a waveguide, let us now take a closer look at the two most important designs used in accelerators. To transport the wave from the transmitter to the accelerator, **rectangular waveguides** (Fig. 5.1) used in communications systems are employed. As we have seen, a knowledge of the cut-off wavelength is crucial in choosing the dimensions of a waveguide. To calculate this wavelength for a rectangular waveguide, let us start with the following equations which result from (5.8):

$$\begin{aligned} f_x'' + k_x^2\, f_x &= 0 \\ f_y'' + k_y^2\, f_y &= 0. \end{aligned} \tag{5.21}$$

The general solution has the form

$$\begin{aligned} f_x(x) &= A\sin(k_x\, x) + B\cos(k_x\, x) \\ f_y(y) &= C\sin(k_y\, y) + D\cos(k_y\, y). \end{aligned} \tag{5.22}$$

The constants of integration A, B, C and D are fixed by the boundary conditions of wave propagation in the waveguide. These state that all electric fields

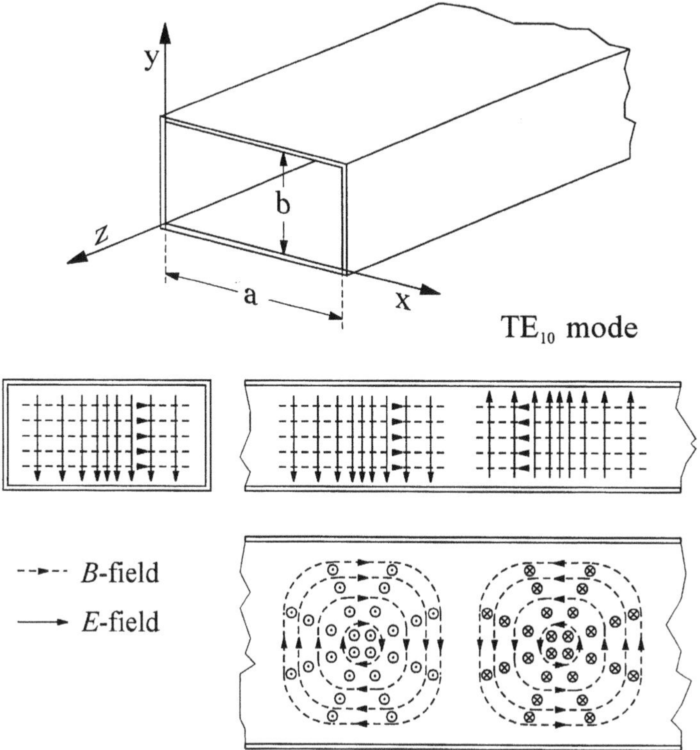

Fig. 5.1 Rectangular waveguide with TE_{10} wave.

perpendicular to the conducting walls of the waveguide vanish at the surface of the walls, i.e. $f_x(0) = 0$ and $f_y(0) = 0$. From this it immediately follows that $B = D = 0$. If a is the width and b the height of the waveguide, then we also have $f_x(a) = 0$ and $f_y(b) = 0$. This condition is satisfied when

$$\begin{aligned} k_x\, a &= m\, \pi \\ & \qquad\qquad \text{with} \quad m, n = \text{integers}. \\ k_y\, b &= n\, \pi \end{aligned} \tag{5.23}$$

Inserting this into the definition of the cut-off wavenumber (5.10) gives

$$k_c^2 = \left(\frac{m\,\pi}{a}\right)^2 + \left(\frac{n\,\pi}{b}\right)^2 \tag{5.24}$$

and the cut-off wavelength

$$\lambda_c = \frac{2}{\sqrt{\left(\frac{m}{a}\right)^2 + \left(\frac{n}{b}\right)^2}}. \tag{5.25}$$

Applying the boundary conditions on the electric field plus the additional condition that all magnetic field lines perpendicular to the conducting walls must vanish at the surface due to the production of eddy currents, we can calculate the possible field configurations for waves in the cavity. As is clear from (5.23), the are an unlimited number of configurations, which we call **waveguide modes**. Only a few, however, are of practical use. The field line diagram for the most important mode in the rectangular waveguide is shown in Fig. 5.1. Since for this mode the electric field lines only run perpendicular to the direction of wave motion, it is termed a TE_{10} mode (transverse electric) or H_{10} wave, since the magnetic field lines run in the waveguide direction. The indices in TE_{10} give the number of nodes in the waveguide cross-section in the horizontal and vertical direction. The individual electric and magnetic field components of this mode are:

$$\begin{aligned}
E_x &= 0 \\
E_y &= \hat{E}\sin\left(\frac{\pi\,x}{a}\right) e^{-i\,k_z\,z} \\
E_z &= 0 \\
H_x &= \frac{\hat{E}}{Z_0}\frac{\lambda}{\lambda_z}\sin\left(\frac{\pi\,x}{a}\right) e^{-i\,k_z\,z} \\
H_y &= 0 \\
H_z &= -\,i\,\frac{\hat{E}}{Z_0}\frac{\lambda}{2a}\cos\left(\frac{\pi\,x}{a}\right) e^{-i\,k_z\,z}.
\end{aligned} \tag{5.26}$$

Here $\hat{E}$ is an arbitrary amplitude and Z_0 is the impedance of the waveguide. Further details of various types of waveguides and their characteristics can be found, for example, in [59], [110] and [109].

5.1.2 Cylindrical waveguides

Electromagnetic waves may of course also be carried by cylindrical waveguides. To calculate the form of the wave we move to a cylindrical coordinate system (Θ, r, z), as shown in Fig. 5.2. In these coordinates the solution of the wave equation is given in terms of Bessel functions $J_n(x)$ instead of trigonometric functions. Apart from this, the same boundary conditions apply at the surface of the conducting cylinder wall as in rectangular waveguides. In cylindrical waveguides the most important mode for accelerator physics applications is the TM_{01} mode, in which only transverse magnetic field lines are present. This means that the electrical field runs parallel to the cylinder axis and can accelerate charged particles as they travel through the waveguide. The field configuration for this TM_{01} wave, which is sometimes also called the E_{01} wave because of the longitudinal electric field, is illustrated in Fig. 5.2. The calculation of the field components yields the analogous result to (5.26):

$$\begin{aligned}
E_r &= -i\,\hat{E}\frac{k_z}{k_c}\,J_0'(k_c r)e^{-i\,k_z\,z} \\
E_\Theta &= 0 \\
E_z &= \hat{E}J_0(k_c r)e^{-i\,k_z\,z}
\end{aligned}$$

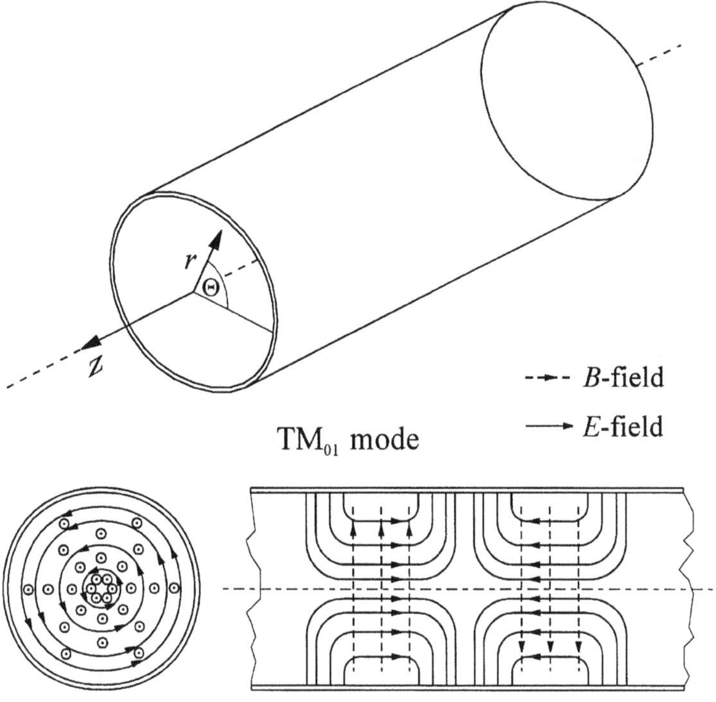

Fig. 5.2 Cylindrical waveguide with TM_{01} wave.

$$\begin{aligned} H_r &= 0 \\ H_\Theta &= -i\,\frac{\hat{E}}{Z_0}\frac{k}{k_c}\;J_0'(k_c r)e^{-i\,k_z\,z} \\ H_z &= 0 \end{aligned} \tag{5.27}$$

Using these relations we may now obtain the important cut-off wavenumber k_c. As already mentioned, the electrical field components running parallel to the conducting cylinder must vanish at its surface. If D is the diameter of the cylinder it then follows that

$$E_z\left(\frac{D}{2}\right) = 0. \tag{5.28}$$

From (5.27) we see that this condition can only be satisfied if the Bessel function completely vanishes, i.e.

$$J_0\left(k_c\frac{D}{2}\right) = 0. \tag{5.29}$$

If x_1 is the first zero of the Bessel function, then

$$k_c = \frac{2\,x_1}{D} \qquad \text{with} \qquad x_1 = 2.40483 \tag{5.30}$$

The corresponding cut-off wavelength is then

$$\lambda_\mathrm{c} = \frac{\pi\, D}{x_1}. \tag{5.31}$$

5.2 Resonant cavities

The general solution of the wave equation (5.1) can always be written in the form

$$W(\boldsymbol{r},t) = Ae^{i(\omega t + \boldsymbol{k}\cdot\boldsymbol{r})} + Be^{i(\omega t - \boldsymbol{k}\cdot\boldsymbol{r})}. \tag{5.32}$$

This describes a sum of two waves, one moving in one direction and another in the opposite direction, with arbitrary amplitudes A and B. If a wave is totally reflected at a surface then both amplitudes are the same, i.e. $A = B$ and (5.32) yields

$$\begin{aligned} W(\boldsymbol{r},t) &= Ae^{i\omega t}\left(e^{i\,\boldsymbol{k}\cdot\boldsymbol{r}} + e^{-i\,\boldsymbol{k}\cdot\boldsymbol{r}}\right) \\ &= 2A\cos(\boldsymbol{k}\cdot\boldsymbol{r})\; e^{i\omega t}. \end{aligned} \tag{5.33}$$

The field configuration resulting from the superposition of a forwards- and a backwards-travelling wave has a **static** amplitude $2A\cos\boldsymbol{k}\cdot\boldsymbol{r}$, i.e. it is a **standing wave**. Hence there are positions at which the amplitude is zero, namely when $\boldsymbol{k}\cdot\boldsymbol{r} = (n+\frac{1}{2})\pi$. If metal walls are introduced at these positions the field configuration does not change. The same is true in a waveguide if its entrance and exit are closed off by two perpendicular conducting sheets a distance l apart. A stable standing wave can always form in this fully closed cavity if the condition

$$l = q\frac{\lambda_z}{2} \qquad \text{with} \qquad q = 0, 1, 2, \ldots \tag{5.34}$$

is satisfied. Hence only certain well-defined wavelengths λ_r are present in the cavity. These are termed **resonant wavelengths**. Inserting (5.34) into (5.17) gives the general resonance condition for a **resonant cavity**

$$\boxed{\frac{1}{\lambda_\mathrm{r}^2} = \frac{1}{\lambda_\mathrm{c}^2} + \frac{1}{4}\left(\frac{q}{l}\right)^2.} \tag{5.35}$$

Near to the resonant wavelength a resonant cavity behaves like an electrical oscillator, but with a very much higher quality factor (Q-value) and correspondingly lower losses than resonators made of individual coils and capacitors. It is this key advantage which is exploited to generate high accelerating voltages.

5.2.1 Rectangular waveguides as resonant cavities

We take the cut-off wavelength in a rectangular waveguide directly from (5.25), namely

$$\left(\frac{2}{\lambda_c}\right)^2 = \left(\frac{m}{a}\right)^2 + \left(\frac{n}{b}\right)^2. \tag{5.36}$$

By simply inserting this expression into the resonance condition (5.35) we obtain the resonant wavelength

$$\lambda_r = \frac{2}{\sqrt{\left(\frac{m}{a}\right)^2 + \left(\frac{n}{b}\right)^2 + \left(\frac{q}{l}\right)^2}} \quad \text{with} \quad m, n, q = \text{integers}. \tag{5.37}$$

The integers m, n and q again define the various modes in the resonant cavity. The number of modes is once again virtually unlimited, but only a few of them are used in practical applications, namely those with m, n and q between 0 and 2.

5.2.2 Cylindrical resonant cavities

In the same way, resonators can also be built from cylindrical waveguides, and indeed this is the preferred design for producing accelerating voltages. In the following discussion we will again only consider the TM_{01} wave. The resonant wavelength may be immediately obtained by inserting the expression for the cut-off wavelength (5.31) into the general resonance condition (5.35), yielding

$$\frac{1}{\lambda_r^2} = \left(\frac{x_1}{\pi D}\right)^2 + \frac{1}{4}\left(\frac{q}{l}\right)^2 \quad \text{with} \quad q = 0, 1, 2, \ldots \tag{5.38}$$

or alternatively

$$\lambda_r = \frac{1}{\sqrt{\left(\frac{x_1}{\pi D}\right)^2 + \frac{1}{4}\left(\frac{q}{l}\right)^2}}. \tag{5.39}$$

$x_1 = 2.40483$ is again the first zero of the Bessel function. In the cylindrical resonators used in accelerators the mode with $q = 0$ is used, termed the TM_{010} mode. The calculation of the resonant wavelength thus reduces to the simple form

$$\lambda_r = \frac{\pi D}{x_1}. \tag{5.40}$$

For this TM_{010} mode the length l does not affect the resonant wavelength. l may thus be chosen relatively freely. As an example of this type of cavity let us consider the single-cell accelerating structure developed for the storage ring DORIS at the German Electron Synchrotron laboratory, DESY. This cavity is illustrated in Fig. 5.3.

The inner diameter measures $D = 462$ mm and the length $l = 276$ mm. Inserting these values into (5.40), we obtain the resonant wavelength and corresponding frequency

$$\begin{aligned} \lambda_r &= 0.60354 \text{ m} \\ f_r &= \frac{c}{\lambda_r} = 496.7 \text{ MHz}. \end{aligned}$$

This frequency is somewhat lower than the operating frequency of $f = 500$ MHz.

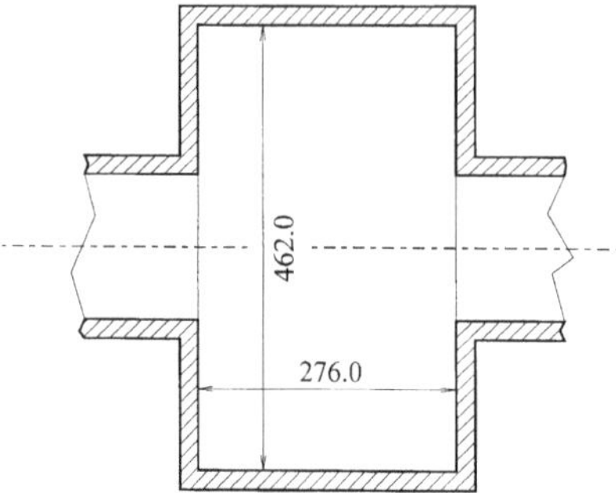

Fig. 5.3 Example of a single-cell cavity. It is chosen to have the dimensions $D = 462$ mm and $l = 276$ mm used in the accelerating structure developed for the storage ring DORIS, designed for a resonant frequency of 500 MHz.

This difference is quite intentional, however, since the exact frequency is adjusted using a tuning plunger which is inserted radially through a round opening in the resonator (Fig. 5.4). This reduces the volume in the cavity and so increases the resonant frequency. In a cylindrical cavity, operated in the TM_{010} mode, the magnetic field lines run in concentric circles around the beam axis, with the field strength increasing with radius and reaching a maximum at the wall. This means that an inductive loop can couple very effectively to the field at this point. The loop is excited via a coaxial cable. Using this arrangement the accelerating TM_{010} wave can be generated in the cavity.

When there is a large distance between the power generator and the cavity, or when very high power over 100 kW is transmitted, it no longer makes sense to use coaxial cables to transport the RF power. The non-negligible ohmic resistance of the relatively thin inner conductor leads to large losses, which for extremely high loads can lead to critical heating. Instead rectangular cavities are used, operated in the TE_{10} mode. Because of their large surface area these waveguides have considerably lower losses, and if necessary they can easily be externally cooled. Only a very short section of coaxial cable, with a specially cooled inner conductor, is then used to connect directly to the cavity,

The connection between the waveguide and the coaxial cable is performed using a standard technique in communications engineering. The end of the cavity is closed off by a conducting wall, so that a standing wave forms with an antinode (electric field maximum) at a distance $\lambda/4$ from the wall. The coaxial cable is connected to the waveguide at this point, as Fig. 5.4 shows. Specially shaped connectors ensure that no reflections are produced and that the wave passes from the waveguide to the coaxial cable with virtually no losses. A ceramic window inside the coaxial cable separates the waveguide section, which is at normal pressure, from the ultra-high vacuum inside the cavity, without impeding the passage of the RF wave. This **cavity window** is a critical component whose importance should not be overlooked. Very high power and high voltages are produced in an extremely small volume, and if spontaneous gas emission from the surface of the cavity should occur at high RF power, then there is the risk

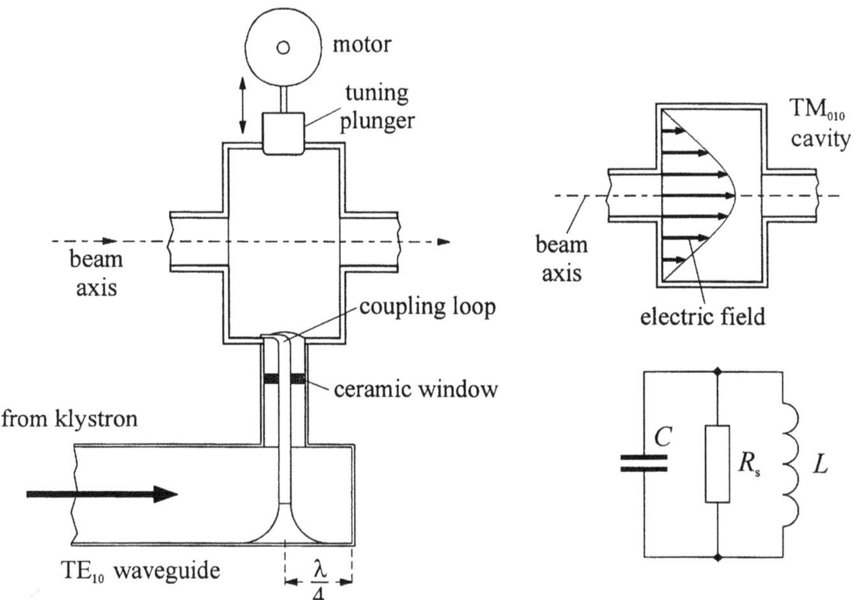

Fig. 5.4 Design of a single-cell accelerating structure using the TM_{010} mode. The exact resonant frequency is adjusted using a tuning plunger. The resonator is excited by an inductive coupling loop.

of sudden glow discharge resulting in overheating of this window.

A stable standing wave can only exist in the cavity if the resonance condition (5.39) or (5.40) is strictly satisfied. If the coupled wavelength deviates even very slightly from this precise value there is a marked reduction in amplitude. The cavity thus behaves like an electrical oscillator, but with a very high quality factor

$$Q = \frac{\omega_r}{\Delta\omega} = \frac{R_s}{Z}. \tag{5.41}$$

Here ω_r is the resonant frequency and $\Delta\omega$ is the frequency shift at which the amplitude is reduced by –3 dB relative to the resonance peak. The electrical response of a cavity may thus be described by a parallel circuit containing C, L and R_s, as illustrated in Fig. 5.4. On resonance the impedance has the value

$$Z = \omega\, L = \frac{1}{\omega\, C} \tag{5.42}$$

and R_s is the so-called **shunt impedance**, which describes the ohmic loss in the cavity. On resonance the entire average coupled RF power P_{RF} is converted to heat in the shunt impedance. Thus the peak voltage produced in the cavity is

$$U_{cav} = \sqrt{2P_{RF}R_s}. \tag{5.43}$$

As an example of the cavity voltages routinely achieved today, let us again consider the single-cell DORIS cavities (Fig. 5.3), constructed entirely from copper:

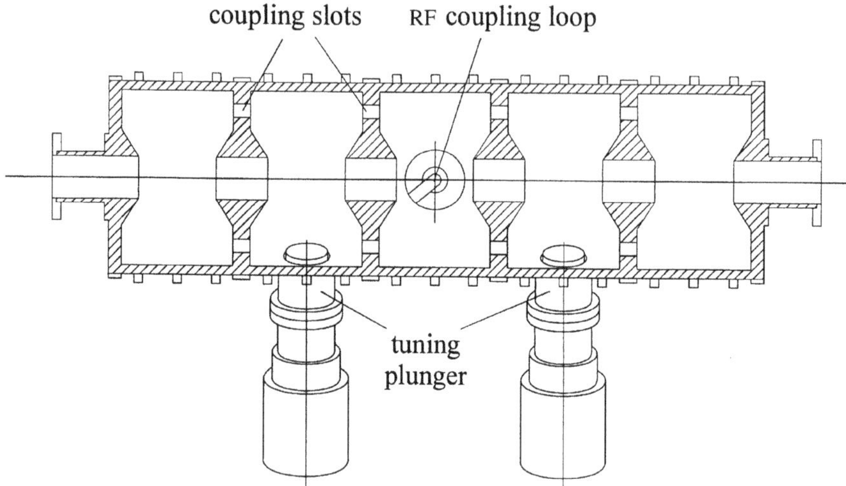

Fig. 5.5 Layout of a five-cell accelerating structure. The power feed is coupled to the middle cell and two tuning plungers are sufficient for the entire structure.

$$f_{\mathrm{RF}} = 500 \text{ MHz}, \quad R_{\mathrm{s}} = 3.0 \times 10^6\ \Omega, \quad Z = 80\ \Omega, \quad Q = 38\,000$$

$$P_{\mathrm{RF}} = 50 \text{ kW}, \quad U_{\mathrm{cav}} = 548 \text{ kV}$$

Much higher voltages of up to several hundred MV are required in today's electron accelerators, which have final beam energies well above 10 GeV, as may be seen for example in Table 2.1 in Chapter 2. Several hundred single-cell accelerating sections are then required, and sufficient free space must be provided for these in the accelerator. To reduce the resources and space required, compact units have been designed in which three or five individual cells are combined. As an example, Fig. 5.5 shows the five-cell cavity [60] developed for the storage ring PETRA at the German Electron Synchrotron laboratory DESY. The high frequency power is supplied to the middle cell via an inductive loop, just as in a single cell. The neighbouring cells are connected by suitably sized coupling slots in the dividing walls between the cells. The drift sections inside the dividing walls are longer around the beam axis, so that RF waves cannot pass through. It is thus possible to design the channel through which the beam passes to suit the requirements of the beam optics, without having to worry about the RF coupling between the cells. The RF coupling is achieved completely independently by means of the coupling slots.

Only two tuning plungers, installed in the second and fourth cells, are required to tune all five cells. The other cells are tuned at the same time because of the coupling between them. It turns out that in this case the tuning is more difficult than in a single-cell cavity, since it is first necessary to set the resonant frequency and then to ensure that the RF power is evenly distributed over the five cells.

Overall, however, this design is considerably less complex technically than five individual structures.

As an example of a five-cell accelerating section, let us look at the key parameters of the PETRA cavities (Fig. 5.5), which were again constructed from copper:

$$f_{\mathrm{RF}} = 500 \text{ MHz} \qquad P_{\mathrm{RF}} = 125 \text{ kW}$$

$$R_{\mathrm{s}} = 18.0 \times 10^6\ \Omega \qquad U_{\mathrm{cav}} = 2.12 \text{ MV}$$

It should be noted here that at full power the shunt impedance is lower than that measured at low power in the laboratory. This is due to the strong heating of the cavity walls, which reduces the conductivity of the copper. In continuous operation at 125 kW a value of $R_{\mathrm{s}} \approx 14.5 \times 10^6\ \Omega$ results and the cavity voltage reduces to $U_{\mathrm{cav}} = 1.90$ MV.

5.3 Accelerating structures for linacs

In linacs, cylindrical waveguides may be used. A TM_{01} wave is generated, which has a longitudinal electric field which reaches a maximumalong the beam axis. Particle acceleration is not yet possible with this arrangement, however, since according to (5.19) the phase velocity of the wave exceeds that of light. The particles, which are travelling more slowly, thus undergo acceleration from the passing wave for half a period but then experience an equal deceleration. Averaged over a long time interval, there is no net transfer of energy to the particles. The phase velocity of the wave must therefore be matched to the velocity of the particle in order to continuously supply power to the beam over long sections.

It is thus necessary to modify the waveguide to reduce the phase velocity of the wave to a value $v_\varphi \leq c$. This is achieved using iris-shaped screens, which in normal linac structures are installed at a constant separation in the waveguide. The cross-section of this structure, sometimes called a 'disc-loaded structure' is shown in Fig. 5.6. The effect of the irises can be seen in Fig. 5.7, which shows

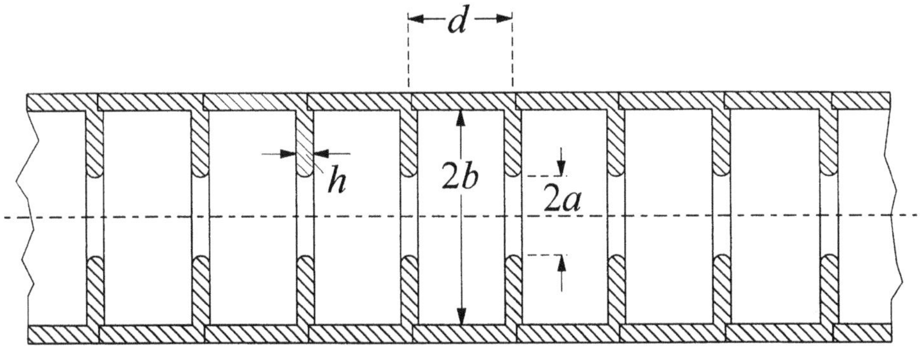

Fig. 5.6 Cross-section through a typical linac structure. The phase velocity of the RF wave is reduced to the particle velocity by the insertion of irises.

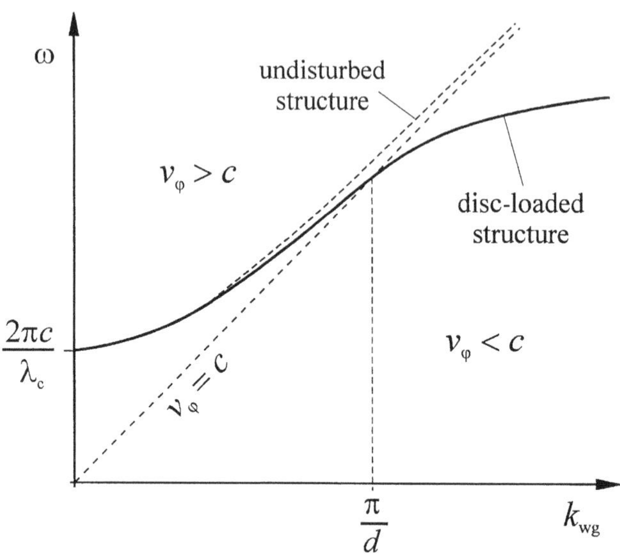

Fig. 5.7 Dispersion curve for a cylindrical waveguide, with and without irises. The frequency ω is plotted as a function of the wavenumber k_z of the waveguide.

the relationship between the wavenumber k_z and the frequency ω. From (5.20), the dispersion relation in a waveguide is

$$\omega = c\sqrt{k_z^2 + \left(\frac{2\pi}{\lambda_c}\right)^2}$$

and the curve always lies in the region $v_\varphi > c$. With the installation of irises the curve flattens off and crosses the boundary line $v_\varphi = c$ at the point $k_z = \pi/d$, where d is the separation of the irises. Above this value the phase velocity of the wave is below the velocity of light. This is the domain in which particle acceleration is possible. With a suitable choice of iris separation d the phase velocity can in principle be set to any value. Hence the irises are arranged very close together at the very beginning of all linacs, since here the particles still have non-relativistic velocities ($\beta = v/c \ll 1$). The phase velocity v_φ is then matched to the velocity of the accelerated particle ('beta matching') by continuously increasing the separation d, in the same way as we saw in the Wideröe linac in Section 1.3.5. Once the particles are travelling at close to the speed of light, waveguide structures with a constant iris separation are used.

The standard operation of a linac structure is in the S-band, namely at a wavelength of exactly $\lambda = 0.100$ m, which corresponds to a frequency of $f_{\mathrm{RF}} = 2.998$ GHz. As in radar technology, for example, the RF power is usually provided by pulsed power tubes such as klystrons and is then fed into the linac structure by means of a TE_{10} wave in a rectangular waveguide. At the interface the rectangular TE_{10} waveguide is connected perpendicular to the cylindrical TM_{01} cavity, as shown in Fig. 5.8.

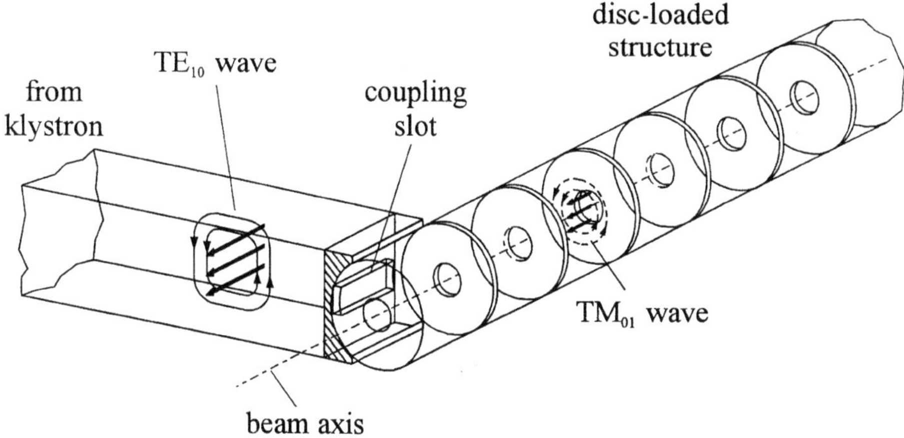

Fig. 5.8 Coupling of the TE_{10} waveguide to the linac structure. The transfer of the wave is achieved without reflections via an appropriately sized coupling slot.

It is evident that in this arrangement both waves have essentially the same field configuration at the interface, and so the wave is able to pass easily from one mode to the other. As the geometry is not completely identical, however, reflections arise at the interface which must be compensated for by a coupling slot of suitable size.

Linac structures can be operated using both travelling waves and standing waves. The choice depends only on whether the structure is closed by a reflection-free boundary or not. The two modes of operation are illustrated in Fig. 5.9. In the more common case there is another coupler at the end of the structure which leads the wave out into an absorber. With correct matching there is no reflection and the solution to the wave equation yields only a travelling wave. In the other case the wave is reflected with virtually no losses at the end of the structure and overlaps with the incoming wave to produce a standing wave.

In smooth waveguides, waves with a wide range of frequencies can propagate without losses, provided that the wavelength lies below the cut-off wavelength. This is no longer the case in a linac structure. The irises form a periodic structure within the cavity, reflecting the wave as it passes through and causing interference. This process is analogous to the interference of light in a diffraction grating. Loss-free wave propagation can only occur if the wavelength is an integer multiple of the iris separation d, namely

$$\lambda_z = p\, d \qquad \text{with} \qquad p = 1, 2, 3, \ldots . \tag{5.44}$$

It immediately follows that

$$\frac{2\pi}{p} = \frac{2\pi}{\lambda_z}\, d = k_z\, d \qquad \text{with} \qquad p = 1, 2, 3, \ldots . \tag{5.45}$$

Hence the irises only allow certain wavelengths, characterized by the number p, to travel in the longitudinal direction. These fixed wave configurations are also

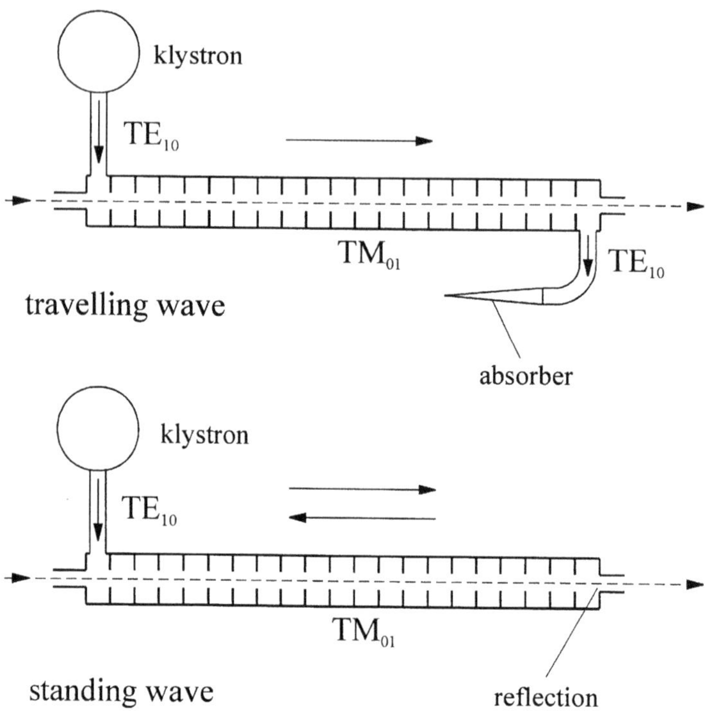

Fig. 5.9 The two modes of operation of the linac structure. The upper diagram shows the more commonly used travelling wave mode in which an absorber is installed at the end of the structure to prevent reflections. In the second case the wave is reflected virtually without losses, resulting in a standing wave.

termed modes. In principle there are arbitrarily many such modes, but only the following three are actually used for acceleration:

$$k_z\, d = \begin{cases} \pi & (\pi \text{ mode} \quad \text{i.e.} \quad \lambda_z = 2d) \qquad \text{if} \quad p = 2 \\[2ex] \dfrac{2\pi}{3} & (2\pi/3 \text{ mode} \quad \text{i.e.} \quad \lambda_z = 3d) \qquad \text{if} \quad p = 3 \\[2ex] \dfrac{\pi}{2} & (\pi/2 \text{ mode} \quad \text{i.e.} \quad \lambda_z = 4d) \qquad \text{if} \quad p = 4 \end{cases} \tag{5.46}$$

The π mode requires relatively long settling time, i.e. it takes a relatively long time for the transient oscillations to die away and a stationary state to be reached. This mode is therefore not suitable for fast pulsed operation. The $\pi/2$ wave has a relatively low shunt impedance per unit length, and so for fixed RF power the energy gain per structure is rather small. The best compromise between short settling time and high shunt impedance is the $2\pi/3$ mode, which is preferred in modern linacs. The field configurations for these three modes are sketched in Fig. 5.10. An important quantity in a linac structure is the maximum possible

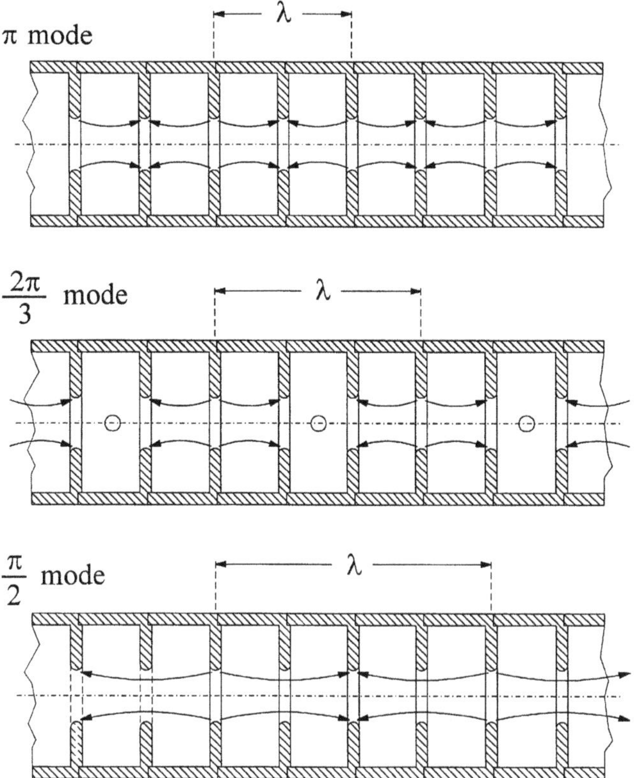

Fig. 5.10 The field configurations of the three most important modes in linac structures.

energy gain of a particle as it travels a distance l through the structure. This value depends only on the potential difference crossed by the particle, which is calculated from

$$U = K\,\sqrt{P_{\mathrm{RF}}\;l\;r_0}. \tag{5.47}$$

Here P_{RF} is the supplied RF power, l the length of the structure, r_0 the shunt impedance per metre and K a correction factor, which generally has a value of around $K \approx 0.8$. The shunt impedance per metre can be estimated to reasonable accuracy using the following empirical formula:

$$r_0 = 5.12 \times 10^8 \frac{\beta_z(1-\eta)^2}{p + 2.61\beta_z(1-\eta)} \left(\frac{\sin D/2}{D/2}\right)^2 \tag{5.48}$$

with (see Fig. 5.6)

$$\begin{aligned}
\beta_z &= \frac{v_\varphi}{c} \qquad \text{phase velocity} \\
\eta &= \frac{h}{d} \qquad (h = \text{thickness},\ d = \text{iris separation}) \\
p &= \text{number of irises per wavelength (equal to the mode number)} \\
D &= \frac{2\pi}{p}(1-\eta)
\end{aligned}$$

As an example let us calculate the value for the SLAC structure, a now standard structure developed at the Stanford Linear Accelerator Center in California and employed in the 3 km long linac there. It has the design shown in Fig. 5.6 with the following dimensions

$$\begin{aligned}
2b &= 82.474 \text{ mm} \\
2a &= 22.606 \text{ mm} \\
h &= 5.842 \text{ mm} \\
d &= 35.001 \text{ mm}
\end{aligned}$$

We set the phase velocity $\beta_z = 1$ and use the $2\pi/3$ mode (i.e. $p = 3$). Putting these values into equation (5.48), we find that the shunt impedance per metre of the SLAC structure is

$$r_0 = 53 \times 10^6 \ \frac{\Omega}{\text{m}}.$$

The total accelerating voltage for a structure of length $l = 3$ m is given by (5.47)

$$U = K\ \sqrt{P_{\text{RF}}\ l\ r_0} = 59.7 \text{ MV},$$

if the supplied power is $P_{\text{RF}} = 35$ MW. Such high power can of course only be maintained in pulsed operation for a duration of a few μs, otherwise it is no longer possible to deal with the amount of heat which is generated. In any case the linac structure must be operated at a well-defined temperature, and so a control system is required to keep the temperature constant to within the narrow range $\Delta T = \pm 0.1°$. This is necessary because it is not technically feasible to mechanically align so many individual cells during operation.

The gradient, i.e. the energy gain per metre, in the SLAC structure is in this case

$$\frac{dE}{ds} = 19.9 \text{ MeV/m}.$$

Today gradients of between 15 MeV/m and 20 MeV/m are routinely achieved, and values of over 100 MeV/m have been measured in laboratory tests using very short structures. It will be necessary to develop extremely high-gradient linac structures if we wish to build linacs with energies of several hundred GeV or even a few TeV. For this reason, intensive work in this field is underway in several major research institutes. A thorough treatment of the physics and technology of linacs may be found for example in [61].

5.4 Klystrons as power generators for accelerators

Cavities installed in circular accelerators and linac structures require RF power of at least a few tens of kW and as much as several MW in large high-energy accelerators or linacs. The klystron has proven to be the most effective power generator for accelerator applications, and we will very briefly describe the principle behind this device here. A comprehensive treatment of the fundamental principles and the technology of klystrons can be found, for example, in [62].

The principle of this classical power tube is outlined in Fig. 5.11. Electrons are emitted from a round cathode with large surface area, and are accelerated by a voltage of a few tens of kV. This yields a round beam, with a current ranging from a few amperes up to more than 10 amperes, depending on the power of the tube. Suitably shaped electrodes near the cathode are used to focus the beam in the same way as we saw for electron sources in Section 4.2. Sometimes several solenoids are added along the tube to ensure good collimation of the beam.

The beam comes out of the cathode with a very well-defined particle velocity and passes through a first cylindrical cavity, which is operated not in the familiar TM_{010} mode but instead in the TM_{011} or even TM_{012} mode, in order to achieve a better coupling to the beam. A wave is excited in this resonator by an external preamplifier with a power output of a few tens of watts, which, depending on the phase, will accelerate, brake, or simply not influence the particles in the beam. The velocity of the particles through the cavity is thus modulated, with a frequency exactly equal to the resonant frequency. In the zero-field drift section which follows, the faster particles move ahead, while the slower ones lag behind. This leads to a change in the hitherto uniform particle density distribu-

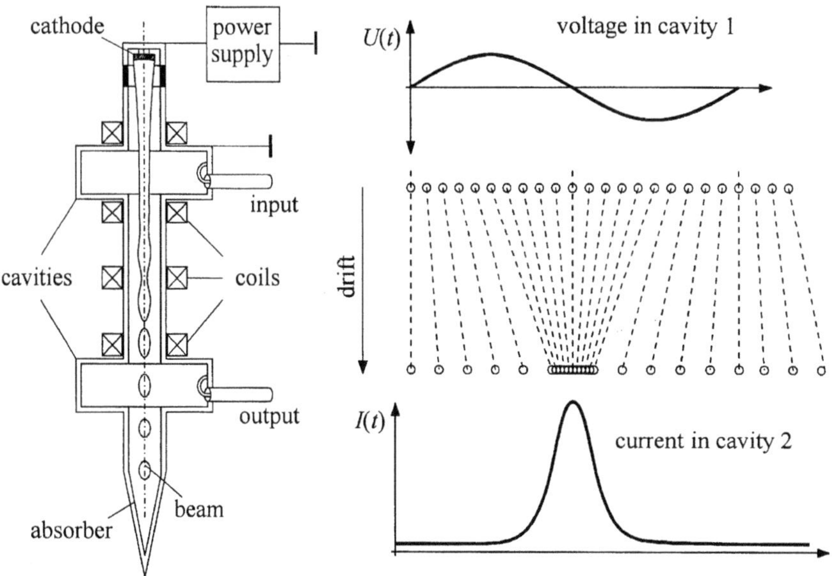

Fig. 5.11 The classical microwave klystron, operating in the ten centimetre region.

tion within the beam. After a certain distance bunches of particles are formed, with a separation given by the wavelength of the driving wave.

The continuous current from the cathode thus becomes a pulsed current, with a frequency equal to the frequency of the coupled driving supply. A second cavity mounted at this point is resonantly excited by the pulsed current, and the RF wave generated in this cavity can then be coupled out. In an optimal design the beam is almost completely stopped by the RF field it excites in the second cavity, leaving just a small residual energy, i.e. the kinetic energy stored in the beam is transformed into RF energy. The output power of the klystron can be written in the simple generalized form

$$P_{\text{klystron}} = \eta \, U_0 \, I_{\text{beam}}. \tag{5.49}$$

Here U_0 is the supply voltage of the klystron, I_{beam} the beam current and η the efficiency of the klystron, which nowadays ranges from 45% and 65%. A typical accelerator klystron of average power operates at a voltage of $U_0 = 45$ kV and has a beam current of $I_{\text{beam}} = 12.5$ A and an efficiency of $\eta = 0.45$. From (5.49) the resulting high-frequency power output is $P_{\text{klystron}} = 253$ kW. Peak values of up to 1.2 MW per tube are now achieved in continuous wave mode. These tubes mostly operate in the frequency range between 350 and 500 MHz.

A better coupling of the beam to the output cavity can be achieved by inserting two or three further resonators between those shown in Fig. 5.11, each tuned to frequencies close to the operating frequency. All modern high-power klystrons use an arrangement of several resonators.

As we have already seen, higher power output can only be achieved by generating a higher voltage in the output cavity. Since this can never significantly exceed the accelerating voltage at the cathode, this accelerating voltage must be increased. In particular, klystrons used to drive linac structures operating in the S-band at the commonly-used frequency of 2.998 GHz require voltages between 250 and 300 kV, traversed by beam currents of around 250 A. For an efficiency of $\eta = 45\%$ this yields a power output of $P_{\text{kl}} = 30 - 35$ MW. Naturally such high power can only be handled in pulsed operation. The pulse length in linacs is a few μs with a repetition rate of a few hundred Hz. This yields an average power output of a few tens of kW, which is relatively easy to manage.

A sensitive klystron utilizing considerably higher beam energies of a few MeV is currently under development, but such energies can no longer be produced by a simple high voltage. Because of their high energy the electrons travel at relativistic velocities, and so this device is also called the **relativistic klystron**. The basic principle is outlined in Fig. 5.12. The electrons emitted from the high-current cathode are accelerated in several steps up to the required energy of a few MeV. Due to the high beam current of several thousand amperes this acceleration can no longer be achieved using resonators, and inductive accelerating sections are used instead. These have a very low impedance and in fact just consist of a single winding around a ferrite core. Using very strong pulsed currents a voltage of a few hundred kV is generated for short periods, which accelerates the particles

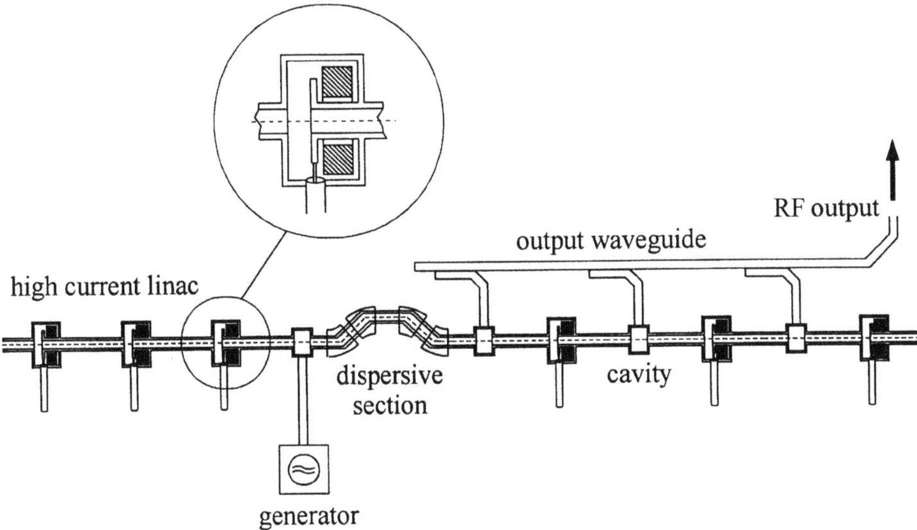

Fig. 5.12 The relativistic klystron. An inductive high-current linac accelerates the electrons up to an energy of a few MeV, with currents as high as a few thousand amperes. The energy modulation is performed in the usual way using a modulating cavity, which is followed by a dispersive section to bunch the beams. The beam now has a high-frequency structure and traverses several output cavities, with re-acceleration performed between each one to replace the lost energy.

as they cross. A whole series of these inductive accelerating sections, arranged one after another, are needed to reach the required final energy.

Modulator cavities are then used to modulate the energy of the electrons in the same way as before, although their velocity hardly changes. The beam must therefore be passed through a dispersive section consisting of a series of bending magnets, in which the particles of different energies follow different paths. This results in a bunching of the beam, as we saw in the drift region in the classical klystron. This section is followed by the output cavities, with further accelerating sections between the cavities to replace the particle energy lost as RF power. This method can achieve extremely high output power, and a few laboratories have successfully performed initial trials of such klystrons.

5.5 The klystron modulator

The electrical power used by high-power klystrons in storage rings comes from power supplies regulated by thyristors which provide a constant high voltage. This technology is also employed in electron synchrotrons, in which the klystron power must be increased with every accelerating cycle to match the particle energy. A completely different approach is needed for linac klystrons operating in pulsed mode, however. Here a voltage of up to 300 kV and a current of around 250 A must be supplied for a period of a few μs, corresponding to a pulsed power

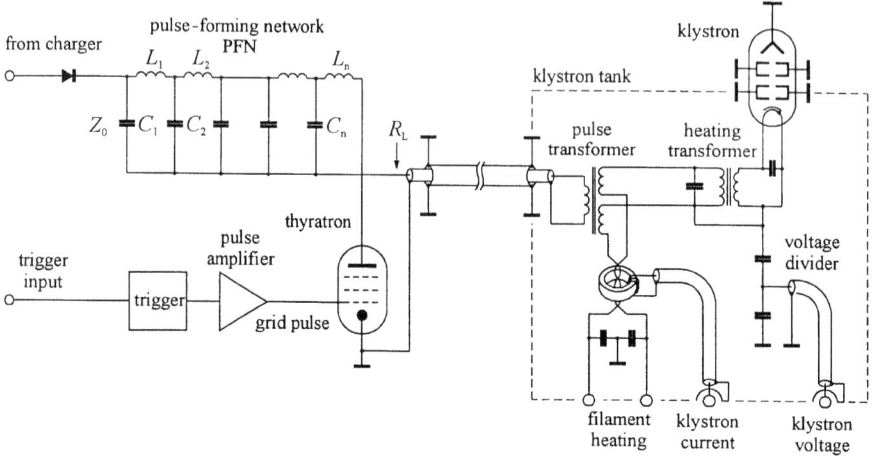

Fig. 5.13 The pulsed power generator used in the klystron modulator. The energy required for a single pulse is stored in the capacitors of the pulse-forming network (PFN). The pulse generated when the network is discharged is raised to the high voltage required for the klystron by means of a transformer.

of up to 80 MW. S-band klystrons with this level of pulsed electrical power use power supplies called **klystron modulators**, which we will now describe.

Short pulses of extremely high power are produced, usually by charging capacitors up to high voltages and then using a suitable switch to discharge them through a very low-resistance load. This process is also used in the klystron modulator, whose basic layout is shown in Fig. 5.13. If P_k denotes the input power needed by the klystron then for a pulse of of duration τ_tot an energy $W = P_\mathrm{k}\tau_\mathrm{tot}$ is required. Before the pulse starts, this energy is stored in the capacitors C_1, C_2 to C_n of the pulse-forming network, which are charged to a voltage U_L. A pulse-forming network built from inductors and capacitors has the advantage over a simple capacitor that when it is discharged a relatively good rectangular pulse with steep rising and falling edges is formed. It behaves just like an unterminated cable which is charged up and then suddenly discharged at one end, ideally through a load resistor. The wavefront produced during the discharge travels to the end of the cable where it is reflected and eventually returns to the start, at which point the discharge current is interrupted. With this form of discharge the pulse lasts exactly twice the time it takes the wave to travel from the start of the cable to the end. In the *LC* pulse-forming network it does not in principle matter over how many elements the total capacitance and inductance are distributed. In general the sharpness of the pulse edges increases with more *LC* elements in the chain, i.e. if the individual values of L and C are smaller. The energy stored in a pulse-forming network of this kind is therefore

$$E = P_\mathrm{k}\tau_\mathrm{tot} = \frac{1}{2}C_\mathrm{tot}U_\mathrm{L}^2 \qquad \text{with} \qquad C_\mathrm{tot} = \sum_{i=1}^{n} C_i. \tag{5.50}$$

The total capacitance immediately follows

$$C_{\text{tot}} = \frac{2P_{\text{k}}\tau_{\text{tot}}}{U_{\text{L}}^2}. \tag{5.51}$$

To choose the other parameters of the pulse-forming network we need to calculate the inductances L_i, which we will here set to be all equal, and likewise the capacitances C_i. When the PFN is discharged the current pulse flows through a load R_{L} which we will assume to be real. To achieve optimal energy matching the load which terminates the chain must exactly match its impedance Z_0, namely

$$Z_0 = \sqrt{\frac{L_i}{C_i}} = \sqrt{\frac{L_{\text{tot}}}{C_{\text{tot}}}} = R_{\text{L}}. \tag{5.52}$$

R_{L} thus has exactly half the load voltage $U_{\text{L}}/2$ across it. The power delivered to the load is then

$$P_{\text{L}} = P_{\text{k}} = \left(\frac{U_{\text{L}}}{2}\right)^2 \frac{1}{R_{\text{L}}} \qquad \longrightarrow \qquad R_{\text{L}} = \frac{U_{\text{L}}^2}{4P_{\text{k}}}. \tag{5.53}$$

Thus we have fixed the size of the load. Since $Z_0 = R_{\text{L}}$, the total inductance required follows immediately from (5.52)

$$L_{\text{tot}} = Z_0^2 C_{\text{tot}}. \tag{5.54}$$

The transit time along the network is thus

$$\tau_{\text{PFN}} = \sqrt{L_{\text{tot}} C_{\text{tot}}} = Z_0 C_{\text{tot}} = \frac{\tau_{\text{tot}}}{2}, \tag{5.55}$$

as mentioned above. To discharge the network a thyratron is used, a gas discharge tube which can switch voltages of several tens of kV and currents of several kA. This tube is fired by sending a trigger pulse to its control grid. This switching process lasts less than a μs. The current discharged from the PFN then flows through the load R_{L}, which in reality is the primary coil of a pulse transformer with between four and six windings. The transformer has two identical bifilar coils as secondaries, each with about 70 windings. The high voltage for the klystron heating is supplied via these two windings, which are isolated from each other. An additional filament transformer is usually installed at this point so that the secondary winding of the pulse transformer does not have to carry too high a filament current.

The secondary current and secondary voltage are monitored by a ring-core transformer or a capacitive voltage divider. Since the secondary side of the pulse transformer and everything connected to it are at voltages of up to 300 kV relative to earth, this section is placed in an oil-filled tank, as is usual in high-voltage transformers. As an example, the parameters of a typical klystron modulator are given in Table 5.1.

The PFN can in principle be charged up to U_{L} using a suitable power supply with the requisite supply voltage. However, a much more elegant and also more

Table 5.1 Parameters of a typical klystron modulator.

parameter	value
pulse power	$P_{\mathrm{k}} = 80$ MW
charging voltage	$U_{\mathrm{L}} = 40$ kV
pulse length	$\tau_{\mathrm{tot}} = 3\ \mu$s
impedance of PFN	$Z_0 = 5\ \Omega$
total capacitance	$C_{\mathrm{tot}} = 300$ nF
total inductance	$L_{\mathrm{tot}} = 7.5\ \mu$H
pulse transformer	
number of primary windings	$n_{\mathrm{p}} = 5$
number of secondary windings	$n_{\mathrm{s}} = 72$
primary voltage	$U_{\mathrm{p}} = 20$ kV
primary current	$I_{\mathrm{p}} = 4000$ A
secondary voltage	$U_{\mathrm{s}} = 288$ kV
secondary current	$I_{\mathrm{s}} = 278$ A

economical method in terms of energy usage is to use the principle of **resonant charging**, illustrated in Fig. 5.14. Here a power supply is used which only generates half the charging voltage, i.e. $U_0 = U_{\mathrm{L}}/2$. An inductance L is connected between the power supply and the PFN. This is very much larger than the inductances L_i in the network, and so to a good approximation we can assume the capacitances C_i to be simply connected in parallel. The inductance L at the end of the network has the total capacitance C_{tot} of the PFN as its load. The voltage in this capacitance depends on the charging current $I(t)$ and has the general form

$$U_{\mathrm{c}}(t) = \frac{1}{C_{\mathrm{tot}}} \int I(t)\, dt = \frac{1}{C_{\mathrm{tot}}}\, \xi(t). \tag{5.56}$$

The supply current flows through the inductance L, with ohmic losses characterized by the resistors R_{v} connected in series with the inductor. Using $\xi(t) = \int I(t)dt$, the equation describing the charging process then follows

$$\ddot{\xi}(t) + \frac{R_{\mathrm{v}}}{L}\dot{\xi}(t) + \frac{1}{LC_{\mathrm{tot}}}\xi(t) = \frac{U_0}{L} = \text{const.} \tag{5.57}$$

This inhomogeneous equation has the solution

$$U_{\mathrm{c}}(t) = (A\cos\omega t + B\sin\omega t)e^{-\zeta t} + U_0 \qquad \text{with} \qquad \left\{ \begin{array}{rcl} \omega & = & \sqrt{\omega_0^2 - \zeta^2} \\ \omega_0 & = & \dfrac{1}{\sqrt{LC_{\mathrm{tot}}}} \\ \zeta & = & \dfrac{R_{\mathrm{v}}}{2L} \end{array} \right. \tag{5.58}$$

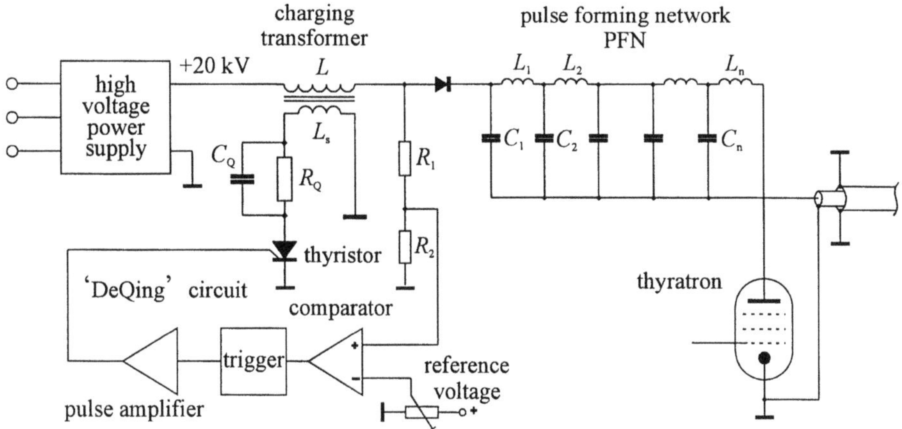

Fig. 5.14 Resonant charging of a pulse-forming network. A precisely defined voltage is maintained by the 'deQing' circuit.

If we assume very slight damping, i.e. $\zeta \ll \omega_0$, the initial conditions then yield

$$\begin{array}{rcl} U_c(0) & = & 0 \\ \dot{U}_c(0) & = & 0 \end{array} \quad \longrightarrow \quad \begin{array}{rcl} A & = & -U_0 \\ B & = & 0 \end{array} .$$

For this virtually undamped oscillator the voltage on the capacitors has the form

$$U_c(t) = U_0(1 - \cos \omega t). \tag{5.59}$$

Half an oscillation period after the discharge of the PFN, the voltage on the capacitors reaches the maximum value $U_{max} = 2\,U_0$. We thus obtain a charging voltage which is twice the value delivered by the power supply. The diode connected in series with the inductance L prevents the oscillation from continuing once the peak value has been reached and discharging the pulse-forming network again.

The maximum voltage achieved by resonant charging is inherently rather variable. This is because fluctuations in the power supply, which is usually not regulated, propagate through to the output. Precise regulation of this supply would be rather costly, due to the high voltage, and in any case would not remove fluctuations caused by temperature drifts in the other components involved in the resonant charging process. A principle known as 'deQing' has therefore been developed, which is illustrated in Fig. 5.14. It is based on the idea of first of all of raising the input voltage U_0 and so increasing the maximum charging voltage to create a voltage surplus. The voltage in the pulse-forming network is constantly monitored during the resonant charging procedure. After a certain time the charging voltage reaches the nominal level, at which point the charging process is immediately stopped and the remaining energy stored in the inductance L is rapidly converted into heat. The charging process is thus halted at a well-defined voltage.

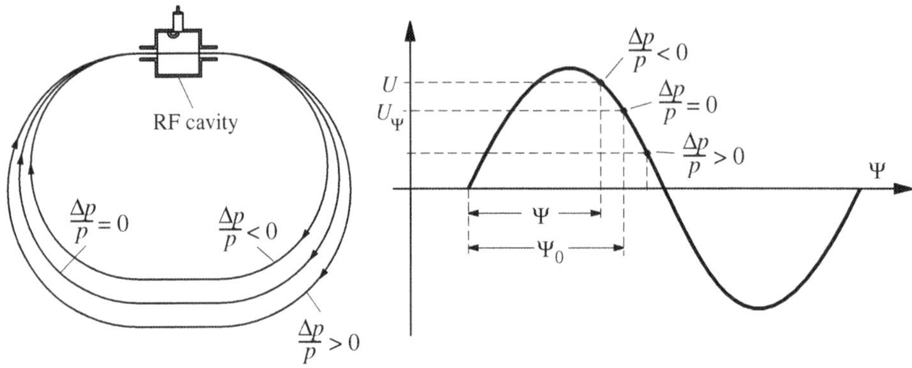

Fig. 5.15 Phase focusing of relativistic particles in circular accelerators.

In the simplest case a resistor and a thyratron are connected in parallel with the inductance L. When the nominal voltage is reached, the initially open thyratron switch closes and the relatively low resistance induces large losses from the oscillating system, resulting in virtually adiabatic damping. The quality factor Q of the oscillator is drastically reduced to a value of around $Q \approx 1$, hence the name 'deQing' switch. Because of the poor quality factor, the resonant charging stops immediately.

As is shown in Fig. 5.14, the same effect can be achieved by fitting the inductance L with a secondary coil L_s and adding a capacitor C_Q and resistance R_Q connected in parallel, thus creating a strongly damped oscillator. The advantage of this arrangement lies in the low voltage, which allows semiconductor switches (thyristors) to be used. The monitoring of the charging voltage is performed by the voltage dividers R_1 and R_2 at the low voltage potential. The voltage is compared with a threshold reference voltage, which when reached fires the trigger and switches the thyristor. Using this system the charging voltage in the pulse-forming network can be stabilized to within an accuracy of about 0.1%.

5.6 Phase focusing and synchrotron frequency

In a circular accelerator, just as in a linac, it is important to ensure that the circulating particles maintain a well-defined fixed average phase relative to the accelerating high-frequency voltage. It is therefore necessary to perform **phase focusing**, the principle of which we will now outline with the help of Fig. 5.15. For the sake of simplicity we assume that the particles have relativistic velocities ($v = c$). An ideal on-momentum particle ($\Delta p/p = 0$) travels along the nominal trajectory fixed by the construction of the ring and has the nominal phase Ψ_0 relative to the RF voltage when it passes through the RF cavity. Here the energy required by the particle, for acceleration and to compensate for losses from synchrotron radiation, is supplied exactly. A particle whose momentum is too low ($\Delta p/p < 0$) will travel along an inner dispersive trajectory, shorter than the nominal trajectory. Hence it will arrive at the cavity somewhat earlier, with a smaller phase $\Psi = \Psi_0 - \Delta\Psi$, and so will see a correspondingly larger voltage.

As a result it is more strongly accelerated than an ideal particle and so catches up with the nominal phase. The opposite happens to a particle with too much energy. In reality the particles oscillate about the ideal phase and so are kept stable within the accelerator. This periodic longitudinal particle motion about the nominal phase is called **synchrotron oscillation**.

In order to study this motion quantitatively, let us calculate the energy sum for a particle over a full revolution. For a nominal particle ($\Delta p/p = 0$) we have

$$E_0 = eU_0 \sin \Psi_0 - W_0. \tag{5.60}$$

Here U_0 is the peak voltage in the cavity and W_0 is the energy requirement per revolution of the particle. We choose an arbitrary particle which has a slight momentum deviation and a phase deviation of $\Delta\Psi$. In this case the energy balance is

$$E = eU_0 \sin(\Psi_0 + \Delta\Psi) - W, \tag{5.61}$$

where we write the energy requirement to linear approximation in the form

$$W = W_0 + \frac{dW}{dE}\Delta E. \tag{5.62}$$

Taking the difference between (5.60) and (5.61) gives

$$\Delta E = E - E_0 = eU_0 \Big[\sin(\Psi_0 + \Delta\Psi) - \sin \Psi_0\Big] - \frac{dW}{dE}\Delta E. \tag{5.63}$$

The duration of the synchrotron oscillation is very long compared to the period of revolution, with one oscillation period generally lasting for a few hundred revolutions. This means that instead of differentiating with respect to time we can simply divide by the revolution time T_0, giving

$$\Delta\dot{E} = \frac{\Delta E}{T_0} = \frac{eU_0}{T_0}\Big[\sin(\Psi_0 + \Delta\Psi) - \sin \Psi_0\Big] - \frac{dW}{dE}\frac{\Delta E}{T_0}. \tag{5.64}$$

The phase deviation $\Delta\Psi$ is due to the different revolution time of the off-momentum particle. The revolution time of an ideal particle is

$$T_0 = \frac{L_0}{v_0}, \tag{5.65}$$

where L_0 is the length of the ideal path and v_0 is the nominal particle velocity. Of course the particles usually have very small deviations ΔL and Δv from these nominal values, so the general revolution time for an arbitrary particle is

$$T = \frac{L_0 + \Delta L}{v_0 + \Delta v} \approx (L_0 + \Delta L)\frac{v_0 - \Delta v}{v_0^2} \approx \frac{1}{v_0^2}(L_0 v_0 + v_0 \Delta L - L_0 \Delta v). \tag{5.66}$$

The time difference is thus

$$\Delta T = T - T_0 = \frac{v_0 \Delta L - L_0 \Delta v}{v_0^2} \tag{5.67}$$

or, rearranging,

$$\frac{\Delta T}{T_0} = \frac{\Delta L}{L_0} - \frac{\Delta v}{v_0}. \tag{5.68}$$

Using the momentum compaction factor α defined in (3.110) we have

$$\frac{\Delta L}{L_0} = \alpha \frac{\Delta p}{p} \tag{5.69}$$

and from (B.11) (see appendix) we obtain the relativistic relation

$$\frac{\Delta v}{v_0} = \frac{1}{\gamma^2} \frac{\Delta p}{p}. \tag{5.70}$$

This leads to the relative time deviation for an arbitrary relativistic particle from (5.68)

$$\frac{\Delta T}{T_0} = \left(\alpha - \frac{1}{\gamma^2} \right) \frac{\Delta p}{p}. \tag{5.71}$$

Since the accelerating frequency has the period T_{RF}, we can immediately calculate the resulting phase shift, namely

$$\Delta \Psi = 2\pi \frac{\Delta T}{T_{\mathrm{RF}}} = \omega_{\mathrm{RF}} \Delta T. \tag{5.72}$$

As the ideal particle exactly encounters the RF voltage at exactly the nominal phase on each revolution, the RF frequency ω_{RF} must be an integer multiple of the revolution frequency ω_{rev}, namely

$$q = \frac{\omega_{\mathrm{RF}}}{\omega_{\mathrm{rev}}} \qquad \text{with} \qquad q = \text{integer}. \tag{5.73}$$

This ratio q is called the **harmonic number** of the ring. It thus follows from (5.71) and (5.72) that

$$\Delta \Psi = q \omega_{\mathrm{rev}} \Delta T = 2\pi q \frac{\Delta T}{T_0} = 2\pi q \left(\alpha - \frac{1}{\gamma^2} \right) \frac{\Delta p}{p}. \tag{5.74}$$

Now let us again use (B.16) to substitute for the momentum

$$\frac{\Delta p}{p} = \frac{1}{\beta^2} \frac{\Delta E}{E} \tag{5.75}$$

to obtain the following equation for the phase variation of an arbitrary particle as it circulates around the ring

$$\Delta \Psi = \frac{2\pi q}{\beta^2} \left(\alpha - \frac{1}{\gamma^2} \right) \frac{\Delta E}{E}, \tag{5.76}$$

which when differentiated with respect to time gives the expression

$$\Delta\dot{\Psi} = \frac{\Delta\Psi}{T_0} = \frac{2\pi q}{\beta^2 T_0}\left(\alpha - \frac{1}{\gamma^2}\right)\frac{\Delta E}{E}. \tag{5.77}$$

We first calculate the particle motion for very small deviations from the nominal phase, i.e. $\Delta\Psi \ll \Psi_0$. In this case the trigonometric functions in (5.64) may be simplified using

$$\sin(\Psi_0 + \Delta\Psi) - \sin\Psi_0 = \sin\Psi_0 \cos\Delta\Psi + \cos\Psi_0 \sin\Delta\Psi - \sin\Psi_0 \approx \Delta\Psi \cos\Psi_0 \tag{5.78}$$

and we obtain

$$\Delta\dot{E} = \frac{eU_0}{T_0}\,\Delta\Psi\,\cos\Psi_0 - \frac{dW}{dE}\frac{\Delta E}{T_0}. \tag{5.79}$$

To obtain the equation of motion we differentiate this expression again with respect to time

$$\Delta\ddot{E} = \frac{eU_0}{T_0}\,\Delta\dot{\Psi}\,\cos\Psi_0 - \frac{dW}{dE}\frac{\Delta\dot{E}}{T_0}. \tag{5.80}$$

Inserting (5.77) back into the above, we immediately obtain

$$\boxed{\Delta\ddot{E} + 2a_{\mathrm{s}}\,\Delta\dot{E} + \Omega^2\,\Delta E = 0} \tag{5.81}$$

with

$$a_{\mathrm{s}} = \frac{1}{2T_0}\frac{dW}{dE} \tag{5.82}$$

and

$$\Omega = \omega_{\mathrm{rev}}\sqrt{-\frac{eU_0 q\cos\Psi_0}{2\pi\beta^2 E}\left(\alpha - \frac{1}{\gamma^2}\right)}. \tag{5.83}$$

This equation describes a harmonic oscillation, and may be solved using the general expression

$$\Delta E(t) = \Delta E_0\;e^{\omega t}. \tag{5.84}$$

Inserting this gives the characteristic equation $\omega^2 + 2a_{\mathrm{s}}\omega + \Omega^2 = 0$, from which the unknown $\omega = -a_{\mathrm{s}} \pm \sqrt{a_{\mathrm{s}}^2 - \Omega^2}$ is obtained. The damping is in general very small, i.e. $a_{\mathrm{s}} \ll \Omega$, so as a solution of (5.81) we obtain the weakly damped oscillation

$$\Delta E(t) = \Delta E_0\;e^{-a_{\mathrm{s}}t}\;e^{i\,\Omega t}. \tag{5.85}$$

The expression (5.83) for Ω gives the frequency of this synchrotron oscillation. It is immediately apparent that stable oscillations can only exist if

$$\left(\alpha - \frac{1}{\gamma^2}\right)\cos\Psi_0 < 0. \tag{5.86}$$

Depending on the particle energy there are thus two distinct regions for the nominal phase, namely

$$\begin{aligned} \frac{\pi}{2} < \Psi_0 < \frac{3\pi}{2} \qquad &\text{if} \qquad \alpha > \frac{1}{\gamma^2} \\ -\frac{\pi}{2} < \Psi_0 < \frac{\pi}{2} \qquad &\text{if} \qquad \alpha < \frac{1}{\gamma^2}. \end{aligned} \tag{5.87}$$

The **transition energy** involved in crossing from one region to the other is thus given by

$$\gamma_t = \frac{1}{\sqrt{\alpha}}. \tag{5.88}$$

Here the synchrotron frequency goes to zero ($\Omega \to 0$) and the phase focusing effect vanishes with it, and so stable machine operation can no longer be maintained. Where possible it is important to avoid crossing the transition energy during accelerator operation, or to take special measures to ensure that the particles only spend a very short time in this region. This is not difficult in electron accelerators, since the electrons almost always have relativistic energies far above γ_t. Proton synchrotrons are a different matter however, and here it is often impossible to avoid crossing the transition energy.

5.7 Region of phase stability (separatrix)

In the previous section the phase oscillation or, equivalently, the energy oscillation was only calculated for very small amplitudes for which the motion is linear to a good approximation. In circular accelerators, however, quite a number of particles in the beam always have such large phase amplitudes that they behave non-linearly, due to the shape of the potential produced by the sinusoidal RF voltage. For very large amplitudes they may even escape from the stable region, leading to particle losses. The boundary between the stable and unstable regions in the ΔE-$\Delta\Psi$ phase plane is called the **separatrix**. It allows the phase and energy acceptance of the accelerator to be directly determined. In this section we will calculate the non-linear oscillator potential and the separatrix of the synchrotron oscillation. We start from equations (5.64) and (5.77) but neglect the damping, which is in any case very weak:

$$\begin{aligned} \Delta\dot{E} &= \frac{eU_0}{T_0}\Big[\sin(\Psi_0 + \Delta\Psi) - \sin\Psi_0\Big] \\ \Delta\dot{\Psi} &= \frac{2\pi q}{\beta^2 T_0 E}\left(\alpha - \frac{1}{\gamma^2}\right)\Delta E. \end{aligned} \tag{5.89}$$

It can be seen immediately from the second equation in (5.89) that the angular velocity is proportional to the energy deviation. Differentiating this equation once again with respect to time and inserting the result into the first equation yields

$$\begin{aligned}\Delta\ddot{\Psi} &= \frac{2\pi q}{\beta^2 T_0 E}\left(\alpha - \frac{1}{\gamma^2}\right)\Delta\dot{E}\\ &= \frac{2\pi q e U_0}{\beta^2 T_0^2 E}\left(\alpha - \frac{1}{\gamma^2}\right)\left[\sin(\Psi_0 + \Delta\Psi) - \sin\Psi_0\right].\end{aligned} \tag{5.90}$$

The non-linear equation of motion thus has the form

$$\Delta\ddot{\Psi}(t) + \chi\left[\sin(\Psi_0 + \Delta\Psi(t)) - \sin\Psi_0\right] = 0 \tag{5.91}$$

with

$$\chi = -\frac{2\pi q e U_0}{\beta^2 T_0^2 E}\left(\alpha - \frac{1}{\gamma^2}\right). \tag{5.92}$$

This cannot be solved by analytical means, and instead numerical methods must be used to calculate the particle motion in the accelerating potential. It is generally useful to take the values at time $t = 0$ for the angular deviation $\Delta\Psi(0)$ and the energy offset $\Delta E(0)$ as initial conditions for the calculation. The angular velocity $\Delta\dot{\Psi}(0)$ may then be calculated directly from these values using (5.89). Figure 5.17 shows a few particle trajectories with different initial conditions in the $\Delta\Psi$-ΔE phase plane, calculated using (5.91).

Despite the non-linearity of the particle motion in the accelerating RF field, the oscillator potential and the equation of the separatrix can be calculated analytically. Here we again start with the equations (5.89) and cross-multiply them. This gives

$$\underbrace{2\Delta E\Delta\dot{E}}_{=\frac{d}{dt}(\Delta E)^2} = \frac{\beta^2 e U_0 E}{\pi q\left(\alpha - \frac{1}{\gamma^2}\right)}\left[\underbrace{\Delta\dot{\Psi}\sin(\Psi_0 + \Delta\Psi)}_{=-\frac{d}{dt}\cos(\Psi_0+\Delta\Psi)} - \Delta\dot{\Psi}\sin\Psi_0\right]. \tag{5.93}$$

In this form the equation can be integrated analytically with respect to time, yielding

$$(\Delta E)^2 = -\frac{\beta^2 e U_0 E}{\pi q\left(\alpha - \frac{1}{\gamma^2}\right)}\left[\cos(\Psi_0 + \Delta\Psi) + \Delta\Psi\ \sin\Psi_0\right] + H. \tag{5.94}$$

The constant of integration H is the **Hamiltonian function** of the oscillation, $(\Delta E)^2$ is the kinetic energy and

$$V(\Delta\Psi) = \frac{\beta^2 e U_0 E}{\pi q\left(\alpha - \frac{1}{\gamma^2}\right)}\left[\cos(\Psi_0 + \Delta\Psi) + \Delta\Psi\ \sin\Psi_0\right] \tag{5.95}$$

is the potential energy. Figure 5.16 shows the form of this potential. For small amplitudes $\Delta\Psi \ll \pi$ it is almost identical to the potential of a simple harmonic

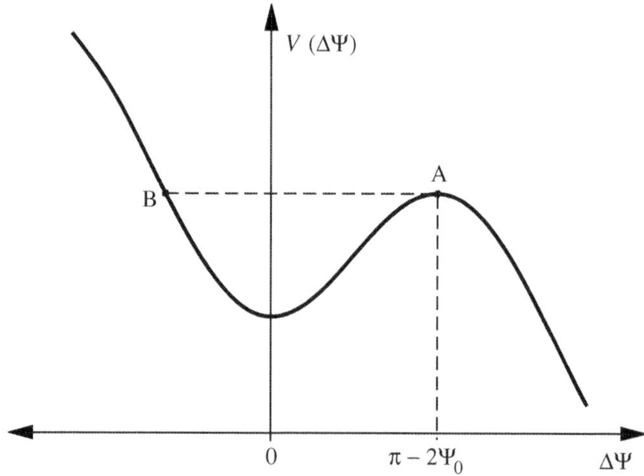

Fig. 5.16 Form of the potential produced by the accelerating RF voltage. The maximum region of phase stability is marked by the points A and B.

oscillator. If the amplitudes grow larger and larger, however, one moves more and more into the non-linear region, in which the potential also becomes asymmetric. The maximum region of phase stability is marked by the points A and B. Outside this region the particles can no longer be contained and travel out of time with the RF phase of the accelerating field. Point A marks one turning point of the potential function. Its position can be easily found using the condition

$$\frac{dV}{d(\Delta\Psi)} = \frac{\beta^2 e U_0 E}{\pi q \left(\alpha - \dfrac{1}{\gamma^2}\right)} \Big[-\sin(\Psi_0 + \Delta\Psi) + \sin\Psi_0 \Big] = 0. \tag{5.96}$$

The trivial solution $\Delta\Psi = 0$ is of no interest here. A further solution may be found by rewriting the sine function in the form

$$\sin\Psi_0 = \sin(\pi - \Psi_0) = \sin(\Psi_0 + \underbrace{\pi - 2\Psi_0}_{=\Delta\Psi}) \tag{5.97}$$

i.e. there is a second solution

$$\Delta\Psi_{\mathrm{max}} = \pi - 2\Psi_0. \tag{5.98}$$

For $\Delta E = 0$ this phase fixes the phase limits of the oscillation. Oscillations with $\Delta\Psi < \Delta\Psi_{\mathrm{max}}$ are stable, all others are unstable. The boundary between the stable and unstable regions is marked by the separatrix mentioned above. This is calculated by setting the energy deviation $\Delta E = 0$ for the maximum phase $\Delta\Psi = \Delta\Psi_{\mathrm{max}} = \pi - 2\Psi_0$ in equation (5.94). This yields the Hamiltonian

$$H = \frac{\beta^2 e U_0 E}{\pi q \left(\alpha - \dfrac{1}{\gamma^2}\right)} \Big[\cos(\pi - \Psi_0) + (\pi - 2\Psi_0)\ \sin\Psi_0 \Big], \tag{5.99}$$

which determines the particle motion at the boundary of the stable region. Inserting this into (5.94) gives the equation of the separatrix

$$(\Delta E)^2 + \frac{\beta^2 e U_0 E}{\pi q \left(\alpha - \dfrac{1}{\gamma^2}\right)} \Big[\cos(\Psi_0 + \Delta\Psi) + \cos\Psi_0 + (2\Psi_0 + \Delta\Psi - \pi)\sin\Psi_0\Big] = 0, \tag{5.100}$$

or

$$\Delta E = \pm\sqrt{-\frac{\beta^2 e U_0 E}{\pi q \left(\alpha - \dfrac{1}{\gamma^2}\right)} \Big[\cos(\Psi_0 + \Delta\Psi) + \cos\Psi_0 + (2\Psi_0 + \Delta\Psi - \pi)\sin\Psi_0\Big]}. \tag{5.101}$$

We again see from this relation that a real solution is only possible if Ψ_0 lies in a particular phase range, namely

$$\begin{aligned} \frac{\pi}{2} < \Psi_0 < \frac{3\pi}{2} \qquad &\text{if} \qquad \alpha > \frac{1}{\gamma^2} \\ -\frac{\pi}{2} < \Psi_0 < \frac{\pi}{2} \qquad &\text{if} \qquad \alpha < \frac{1}{\gamma^2}. \end{aligned} \tag{5.102}$$

The separatrix calculated from (5.101) is plotted in a $\Delta\Psi$-ΔE diagram in Fig. 5.17. Notice that the periodicity of the accelerating voltage results in a series of stable regions, with a phase separation of 2π. Altogether an accelerator has a total of q such regions around the circumference, within which the particles

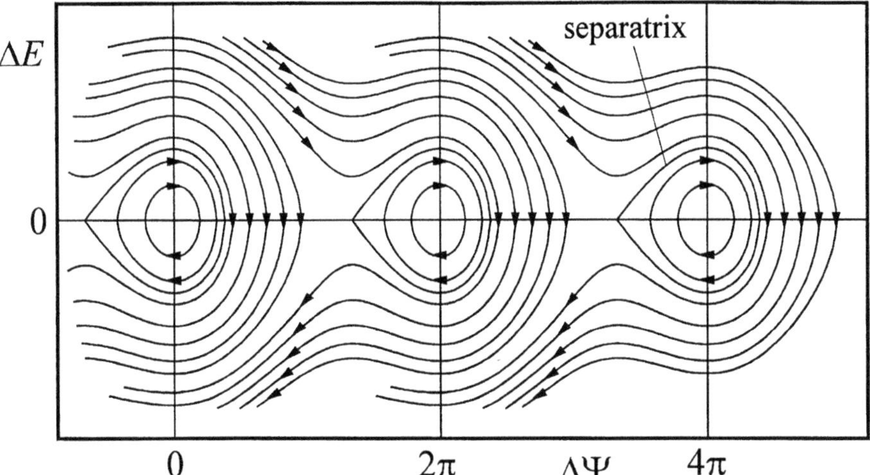

Fig. 5.17 Particle motion in an accelerating RF field. The separatrix marks the boundary between stable synchrotron oscillations and unstable ones, for which acceleration is not possible.

undergo synchrotron oscillations and from which they cannot escape. They are trapped, rather like drops of water in a bucket. For this reason these regions of phase stability within a separatrix are also termed **buckets**. In Fig. 5.17 several particle trajectories calculated using (5.91) are plotted, both inside and outside the separatrix. Using the equation of the separatrix (5.101) we can directly calculate the energy acceptance of the accelerator, in so far as it is limited by the RF acceleration.[1] This is the maximum energy deviation a stably circulating particle with exactly the nominal phase (i.e. $\Delta\Psi = 0$) may have, namely

$$\Delta E_{\mathrm{max}} = \pm \sqrt{\frac{2\beta^2 e U_0 E}{\pi q \left(\alpha - \dfrac{1}{\gamma^2}\right)} \left[\cos\Psi_0 + \left(\Psi_0 - \frac{\pi}{2}\right)\sin\Psi_0\right]}. \tag{5.103}$$

One immediately sees that the energy acceptance increases with increasing RF voltage, and that this dependence goes as

$$\Delta E_{\mathrm{max}} \propto \sqrt{U_0}. \tag{5.104}$$

This relation does not hold strictly, since when the RF voltage U_0 increases, the nominal phase Ψ_0 also changes. Since this effect is rather weaker, however, (5.104) may be reliably used to make estimates. In any case it makes sense to use a sufficiently large overvoltage to provide a large energy acceptance. Alternatively, if the maximum energy variation in the beam is known, (5.103) can be used to calculate the minimum total voltage that the accelerating sections must produce.

[1] The energy acceptance can of course also be limited by the magnet structure if, for example, large dispersion or large chromaticity arise.

6
Radiative effects

In Chapter 2 we showed that electron beams of sufficiently high energy emit synchrotron radiation according to (2.15) with a power

$$P_s = \frac{e^2 c}{6\pi\varepsilon_0} \frac{1}{(m_0 c^2)^4} \frac{E^4}{R^2}.$$

Because of their high rest mass, this effect is negligible for protons until they reach extremely high energies of at least 1 TeV. In this chapter we will consider how the emission of synchrotron radiation determines many of the properties of electron beams, in particular by damping synchrotron and betatron oscillations. In what follows it is helpful to replace the bending radius R by the magnetic field in (2.15), as follows

$$\frac{1}{R} = \frac{e}{p} B = \frac{ec}{E} B \qquad \Longrightarrow \qquad \frac{E^2}{R^2} = e^2 c^2 B^2. \tag{6.1}$$

Following this substitution we obtain

$$P_s = \mathcal{C} E^2 B^2 \qquad \text{with} \qquad \mathcal{C} = \frac{e^4 c^3}{6\pi\varepsilon_0 (m_0 c^2)^4}. \tag{6.2}$$

We will use this radiation formula in the treatment of radiative effects which follows.

6.1 Damping of synchrotron oscillations

The loss of energy in the form of synchrotron radiation results in a damping of the particle oscillations. Let us first of all consider synchrotron oscillations. We begin with the equation of motion (5.81)

$$\Delta\ddot{E} + 2a_s \Delta\dot{E} + \Omega^2 \Delta E = 0$$

in which the damping term has the form

$$a_s = \frac{1}{2T_0} \frac{dW}{dE}. \tag{6.3}$$

To evaluate this term we need to determine the ratio dW/dE. We consider the radiation from a particle with an energy deviation and use the fact that this particle follows a dispersive trajectory. The path element ds' through a bending magnet of bending radius R, along a dispersive trajectory with a displacement Δx, may be found using similar triangles and is

$$ds' = \left(1 + \frac{\Delta x}{R}\right) ds, \tag{6.4}$$

where ds is the corresponding path element along the ideal orbit. Using $ds'/dt = c$, the energy loss per revolution may be written as

$$W = \int_0^{T_0} P_s dt = \oint P_s \frac{ds'}{c} = \frac{1}{c} \oint P_s \left(1 + \frac{\Delta x}{R}\right) ds. \tag{6.5}$$

The transverse beam displacement due to the energy deviation at a point of dispersion D is

$$\Delta x = D \frac{\Delta E}{E}, \tag{6.6}$$

which gives

$$W = \frac{1}{c} \oint P_s \left(1 + \frac{D}{R} \frac{\Delta E}{E}\right) ds. \tag{6.7}$$

Differentiating with respect to energy finally yields the required expression

$$\frac{dW}{dE} = \frac{1}{c} \oint \left[\frac{dP_s}{dE} + \frac{D}{R}\left(\frac{dP_s}{dE}\frac{\Delta E}{E} + P_s \frac{1}{E}\right)\right] ds. \tag{6.8}$$

Since the energy deviation ΔE varies periodically about the nominal value, on average we have $\langle \Delta E/E \rangle = 0$. Hence this term does not contribute to the total and we may write

$$\frac{dW}{dE} = \frac{1}{c} \oint \left[\frac{dP_s}{dE} + \frac{DP_s}{RE}\right] ds. \tag{6.9}$$

In this expression, dP_s/dE is still unknown. It may be determined by differentiating (6.2), yielding

$$\begin{aligned} \frac{dP_s}{dE} &= 2CEB^2 + 2CE^2 B \frac{dB}{dE} \\ &= 2\left(\frac{CE^2B^2}{E} + \frac{CE^2B^2}{B}\frac{dB}{dE}\right) \\ &= 2P_s\left(\frac{1}{E} + \frac{1}{B}\frac{dB}{dE}\right). \end{aligned} \tag{6.10}$$

In quadrupole magnets with non-zero dispersion the magnetic field varies with energy according to

$$\frac{dB}{dE} = \frac{dB}{dx}\frac{dx}{dE}. \tag{6.11}$$

Since $dx = DdE/E$, it follows that

$$\frac{dB}{dE} = \frac{dB}{dx}\frac{D}{E}. \tag{6.12}$$

Inserting this expression into (6.10), we obtain

$$\begin{aligned}\frac{dW}{dE} &= \frac{1}{c}\oint\left[2P_s\left(\frac{1}{E}+\frac{D}{BE}\frac{dB}{dx}\right)+P_s\frac{D}{RE}\right]ds \\ &= \underbrace{\frac{2}{cE}\oint P_s\,ds}_{=\frac{2W_0}{E}}+\frac{1}{cE}\oint DP_s\left(\frac{2}{B}\frac{dB}{dx}+\frac{1}{R}\right)ds. \end{aligned}\tag{6.13}$$

This relation, along with the relation (6.3), yields the required damping constant

$$a_s = \frac{1}{2T_0}\frac{dW}{dE} = \frac{W_0}{2T_0E}\left[2+\frac{1}{cW_0}\oint DP_s\left(\frac{2}{B}\frac{dB}{dx}+\frac{1}{R}\right)ds\right], \tag{6.14}$$

or alternatively

$$a_s = \frac{W_0}{2T_0E}(2+\mathcal{D}) \tag{6.15}$$

with

$$\mathcal{D} = \frac{1}{cW_0}\oint DP_s\left(\frac{2}{B}\frac{dB}{dx}+\frac{1}{R}\right)ds.$$

The integral may be further simplified by replacing dB/dx by the quadrupole strength k and the magnetic field B by the associated bending radius R:

$$\left.\begin{aligned} k &= \frac{ec}{E}\frac{dB}{dx} \quad\rightarrow\quad \frac{dB}{dx} = \frac{kE}{ec} \\ \frac{1}{R} &= \frac{ec}{E}B \quad\rightarrow\quad \frac{1}{B} = \frac{ec}{E}R \end{aligned}\right\}\Longrightarrow \frac{1}{B}\frac{dB}{dx} = kR. \tag{6.16}$$

Furthermore, the radiated power P_s can be written, according to (2.15) and (6.2), in the form

$$P_s = \frac{\mathcal{C}}{e^2c^2}\frac{E^4}{R^2}. \tag{6.17}$$

The integral in (6.15) then becomes

$$\begin{aligned}\oint DP_s\left(\frac{2}{B}\frac{dB}{dx}+\frac{1}{R}\right)ds &= \frac{\mathcal{C}E^4}{e^2c^2}\oint\frac{D}{R^2}\left(2kR+\frac{1}{R}\right)ds \\ &= \frac{\mathcal{C}E^4}{e^2c^2}\oint\left[\frac{D}{R}\left(2k+\frac{1}{R^2}\right)\right]ds. \end{aligned}\tag{6.18}$$

The energy radiated by a particle travelling along the ideal orbit is

$$W_0 = \int_0^{T_0} P_s\,dt = \frac{1}{c}\oint P_s\,ds = \frac{\mathcal{C}E^4}{e^2c^3}\oint\frac{ds}{R^2}. \tag{6.19}$$

Inserting this into (6.15) yields the expression for the damping of synchrotron oscillations in the form:

$$\boxed{\begin{aligned} a_s &= \frac{W_0}{2T_0 E}(2+\mathcal{D}) \\ \text{with} \qquad \mathcal{D} &= \frac{\oint \left[\frac{D}{R}\left(2k+\frac{1}{R^2}\right)\right] ds}{\oint \frac{ds}{R^2}} \end{aligned}} \tag{6.20}$$

Notice that the damping of synchrotron oscillations at a given beam energy E is entirely determined by the magnet structure, since apart from the bending radius R and the quadrupole strength k, (6.20) only contains the dispersion, which is itself completely determined by the magnet structure.

In circular accelerators with separate bending magnets and quadrupoles (separated function magnets) we never have both $k \neq 0$ and $1/R \neq 0$ at the same time. This means that in general $|\mathcal{D}| \ll 1$ in machines of this type. In this case the damping constant a_s is a positive quantity and the synchrotron oscillations are damped. The situation can be very different when dipole and quadrupole fields are combined in one magnet and $\mathcal{D}$ reaches very large values, sometimes with a negative sign. For $\mathcal{D} < -2$ the damping constant becomes negative and the synchrotron oscillations are enhanced and grow with time. Naturally it is impossible to operate an electron storage ring under such conditions.

6.2 Damping of betatron oscillations

Synchrotron radiation also has a damping effect on transverse particle oscillations. We will now study this effect, considering for simplicity the vertical oscillations in a planar accelerator with no vertical dispersion. We further assume that the beta function is relatively constant, i.e. $\alpha = -\beta'(s)/2 \approx 0$. The displacement z of the particle may then be written in the form

$$\left.\begin{aligned} z &= b\sqrt{\beta(s)}\cos\phi \\ z' &= -\frac{b}{\sqrt{\beta(s)}}\sin\phi \end{aligned}\right\} \quad \underset{\Longrightarrow}{A \equiv b\sqrt{\beta(s)}} \quad \left\{\begin{aligned} z &= A\cos\phi \\ z' &= -\frac{A}{\beta(s)}\sin\phi \end{aligned}\right. \tag{6.21}$$

It immediately follows that

$$A^2 = A^2\cos^2\phi + A^2\sin^2\phi = z^2 + \left[\beta(s)\ z'\right]^2. \tag{6.22}$$

This expression allows us to calculate the amplitude A from z and z'. The synchrotron radiation is emitted in the direction of motion of the electron and so the particle's momentum $\boldsymbol{p}$ changes by $\delta\boldsymbol{p}$ but its direction is unaltered (Fig. 6.1). After emitting a photon the electron has the momentum

$$\boldsymbol{p}^* = \boldsymbol{p} - \delta\boldsymbol{p}. \tag{6.23}$$

The longitudinal component p_s is restored to its nominal value by acceleration in the cavities, while the transverse component remains uncorrected. This means

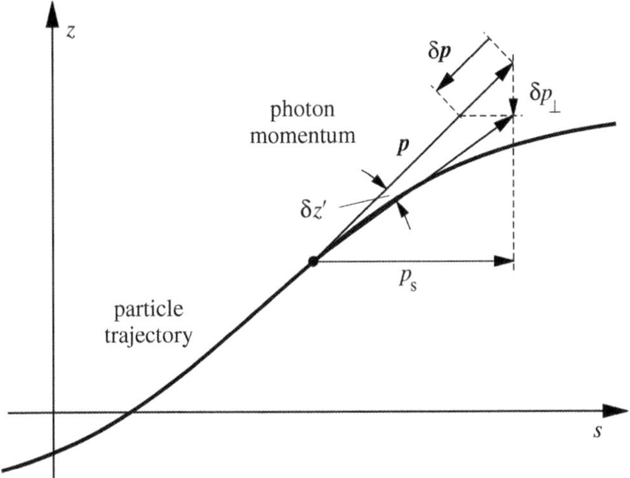

Fig. 6.1 Damping of betatron oscillations by the emission of a photon and restoration of the longitudinal momentum p_s by re-acceleration.

that the angle z' between the particle momentum and the s-axis, which is generally very small, is reduced by the amount

$$\delta z' = -\frac{\delta p_\perp}{|\boldsymbol{p}|}. \tag{6.24}$$

The change in energy of the particle, which is extremely relativistic ($\beta = 1$), is

$$\delta E = \frac{c^2}{v}\delta p_\perp \tag{6.25}$$

or with $v = z'c$

$$\delta E = \frac{c}{z'}\delta p_\perp. \tag{6.26}$$

With $E = c|\boldsymbol{p}|$ we finally obtain

$$\delta z' = -\frac{\delta E}{E}z'. \tag{6.27}$$

The momentum of the emitted photon changes the amplitude A of the betatron oscillation. This change may be immediately obtained from (6.22):

$$\delta(A^2) = \underbrace{\delta(z^2)}_{=0} + \delta(z'^2\beta^2(s)) = \beta^2(s)\delta(z'^2). \tag{6.28}$$

Here $\delta(z^2) = 0$, since z' changes but z does not. If follows that

$$2A\delta A = 2\beta^2(s)z'\delta z' \qquad \Longrightarrow \qquad A\delta A = \beta^2(s)z'\delta z'. \tag{6.29}$$

Inserting (6.27) into this expression gives

$$A\delta A = -\beta^2(s)z'^2\frac{\delta E}{E}. \tag{6.30}$$

Since z' is oscillating we must average over z'^2, with z' given in (6.21). This average is calculated as follows

$$\langle z'^2 \rangle = \frac{A^2}{2\pi\beta^2(s)}\int_0^{2\pi}\sin^2\phi\; d\phi = \frac{A^2}{2\beta^2(s)}. \tag{6.31}$$

From (6.30) we thus obtain

$$A\langle\delta A\rangle = -\frac{A^2}{2\beta^2(s)}\beta^2(s)\frac{\delta E}{E} = -\frac{A^2}{2}\frac{\delta E}{E}. \tag{6.32}$$

Over a full revolution the averaged individual energy losses sum together to give the total loss W_0. The average change in amplitude per revolution is then $\sum\langle\delta A\rangle = \Delta A$. Using this expression it follows from (6.32) that

$$\frac{\Delta A}{A} = -\frac{W_0}{2E}. \tag{6.33}$$

The amplitude thus decreases, i.e. the vertical betatron oscillations are damped. The vertical damping constant a_z may be immediately obtained from the general relation

$$\frac{dA}{A} = -a_z\; dt. \tag{6.34}$$

Over a revolution period $\Delta t = T_0$ the damping constant is then

$$a_z = -\frac{\Delta A}{A\;\Delta t} = \frac{W_0}{2ET_0}. \tag{6.35}$$

For a given particle energy E the damping constant depends only on the amount of energy W_0 radiated per revolution in the form of photons.

The damping constant a_x of horizontal betatron oscillations may be calculated in the same way, but here the effect of the dispersion must also be taken into account. For further details see, for example, the treatment by M. Sands [51]. The result of this calculation yields the value

$$a_x = \frac{W_0}{2ET_0}(1 - \mathcal{D}), \tag{6.36}$$

where $\mathcal{D}$ is defined in (6.20).

6.3 The Robinson theorem

Collecting together the damping constants calculated in (6.20), (6.35) and (6.36) we have:

$$\begin{aligned} a_s &= \frac{W_0}{2ET_0}(2+\mathcal{D}) = \frac{W_0}{2ET_0}J_s \qquad \text{with} \qquad J_s = 2+\mathcal{D} \\ a_z &= \frac{W_0}{2ET_0} \qquad\qquad\;\; = \frac{W_0}{2ET_0}J_z \qquad \text{with} \qquad J_s = 1 \\ a_x &= \frac{W_0}{2ET_0}(1-\mathcal{D}) = \frac{W_0}{2ET_0}J_x \qquad \text{with} \qquad J_x = 1-\mathcal{D}. \end{aligned} \tag{6.37}$$

The important theorem discovered by K.W. Robinson[63] immediately follows:

$$\boxed{J_x + J_z + J_s = 4} \tag{6.38}$$

The sum of the damping constants is an invariant quantity. In a planar machine J_x and J_s (or equivalently a_x and a_s) may vary according to the value of $\mathcal{D}$, but the condition (6.38) is always strictly respected.

Machines with separated function magnets generally have $\mathcal{D} \ll 1$ and one talks about a **natural damping distribution**, i.e. $J_x = 1$, $J_z = 1$ and $J_s = 2$. Such machines always have oscillation damping in all three planes. Traditional synchrotrons with a magnet structure consisting of combined bending magnets and alternating-gradient quadrupoles, on the other hand, have $\mathcal{D} \approx 1$. In this situation Robinson's theorem implies that the damping constants have values of $J_s \approx 3$ and $J_x \approx 0$. The horizontal betatron oscillation is thus no longer damped, and may even be amplified. As a result, machines of this type are not suitable for use as electron storage rings.

It is possible to vary the damping distribution in a circular electron machine by deflecting the beam onto a dispersive trajectory of slightly higher energy. When the beam passes through the quadrupoles its centre of mass is then off axis by a distance $\Delta x_{\rm D} = D\,\Delta p/p$, causing each dipole to act like a superposition of a dipole and a quadrupole. As the energy increases, the machine behaves more and more like a combined function accelerator, changing the value of $\mathcal{D}$ and hence also the damping distribution.

The beam can be deliberately pushed onto a dispersive trajectory by varying the frequency of the accelerator RF system. This changes the wavelength and hence also the circumference of the path. The circumference must always be an integer multiple of the RF wavelength, which means that the harmonic number q remains fixed as the frequency is varied, giving

$$L = q\,\lambda = q\frac{c}{f_{\rm RF}} \qquad \Longrightarrow \qquad dL = -qc\frac{df}{f^2}. \tag{6.39}$$

It immediately follows that

$$\frac{\Delta L}{L} = -\frac{qc}{L}\frac{\Delta f}{f^2} = -\frac{\Delta f}{f}. \tag{6.40}$$

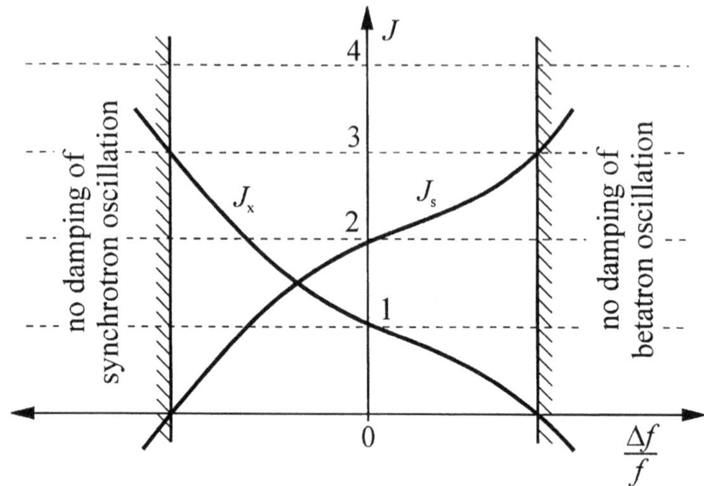

Fig. 6.2 Changing the damping distribution by altering the frequency of the accelerating system.

The relation between the change in path length and the energy shift is given by the momentum compaction factor(3.110):

$$\frac{\Delta L}{L} = \alpha \frac{\Delta E}{E} \qquad \Longrightarrow \qquad \frac{\Delta E}{E} = \frac{1}{\alpha}\frac{\Delta L}{L} = -\frac{1}{\alpha}\frac{\Delta f}{f}. \tag{6.41}$$

The relative change in the frequency thus pushes the orbit onto the dispersive trajectory

$$x_{\mathrm{D}}(s) = -D(s)\frac{1}{\alpha}\frac{\Delta f}{f}. \tag{6.42}$$

The relationship between the frequency shift $\Delta f/f$ and the damping distribution between J_s and J_x is shown in the curves in Fig. 6.2.

6.4 The beam emittance

The beam cross-section is given by the expression $\sigma = \sqrt{\varepsilon\,\beta}$, as was shown in Chapter 3. Here β is a position-dependent quantity which describes the beam focusing at that point. It can be varied over a wide range according to the choice of magnet structure and the strength of the quadrupoles. We have already met the **emittance**, ε, as the area of the phase ellipse which, as a consequence of Liouville's theorem, has the same value at every point in a beam transport system or circular accelerator. At the time we treated the emittance as a predefined constant and did not discuss how it arises or what determines its value. Let us now consider this question.

First of all we will briefly outline the implications of the emittance for experiments at storage rings, beginning with synchrotron radiation sources. For most experiments which use this type of radiation, an important quantity is the intensity, namely the number of photons emitted within a particular energy range per second. It is useful to define the energy interval to be a range of 0.1% of the useful photon energies (0.1% BW). The **flux** F of photons in this energy range is usually normalized to a beam current of 1 A, leading to the definition

$$F = \frac{\text{photons}}{\text{s } 0.1\% \text{ BW A}}. \tag{6.43}$$

This definition is not always sufficient, however, for example in experiments requiring high position resolution. Such experiments rely on the fact that the radiation source is virtually point-like, i.e. the transverse extent and divergence of the radiation should be extremely small. The quality of the beam is described by the **brightness** and the **brilliance**. The brightness only describes the angular divergence of the beam, which is given by $\sigma'_{x,z} = \sqrt{\varepsilon_{x,z}/\beta_{x,z}}$. It is defined as

$$\begin{aligned} S &= \frac{F}{2\pi\ \sigma'_x\ \sigma'_z} = \frac{F\sqrt{\beta_x\ \beta_z}}{2\pi\ \sqrt{\varepsilon_x\ \varepsilon_z}} \\ &= \frac{\text{photons}}{\text{s } 0.1\% \text{ BW mrad}^2 \text{ A}}. \end{aligned} \tag{6.44}$$

The brilliance also relates to the transverse beam dimensions

$$\sigma_{x,z} = \sqrt{\varepsilon_{x,z}\ \beta_{x,z}}.$$

The definition of the brilliance, based on that of the brightness, is

$$\begin{aligned} B &= \frac{F}{4\pi^2\ \sigma_x\ \sigma_z\ \sigma'_x\ \sigma'_z} = \frac{F}{4\pi^2\ \varepsilon_x\ \varepsilon_z} \\ &= \frac{\text{photons}}{\text{s } 0.1\% \text{ BW mm}^2 \text{ mrad}^2 \text{ A}}. \end{aligned} \tag{6.45}$$

It should be pointed out that these definitions refer only to Gaussian-shaped electron beams. Furthermore, the definitions found in the literature do not always agree. The important point, however, is that both the brilliance and the brightness are essentially determined by the beam emittance. As a result, storage rings designed for the production of synchrotron radiation usually have as small an emittance $\varepsilon_{x,z}$ as possible.

The situation is very different in storage rings used for high energy physics experiments, in which two particle beam circulate in opposite directions and collide at an interaction point. The maximum current is limited by the space charge forces which the beams exert on each other as they cross. This effect will be studied quantitatively in Chapter 7. One consequence is that in order to achieve the maximum event rate in the detectors the emittances must be very large, practically filling the available aperture of the ring.

It is thus clear that high-energy experiments and experiments using synchrotron radiation cannot be conducted simultaneously at the same storage ring,

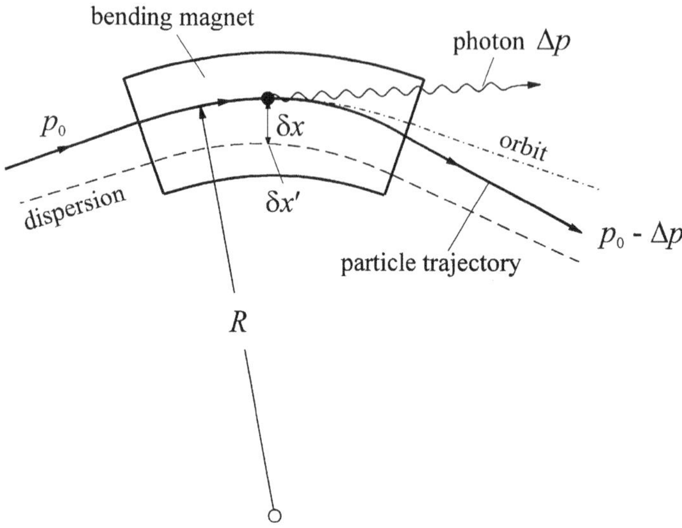

Fig. 6.3 The origin of natural emittance due to photon emission in a bending magnet.

since the two disciplines place conflicting demands on the beam.[1] We will therefore not consider the emittance of proton and heavy ion beams any further.

In electron beams, on the other hand, the amplitude of the betatron oscillations is damped by the emission of synchrotron radiation, as we have already seen in this chapter. If this damping were the only effect, then after a while the oscillation amplitude would reduce to zero and the beam would have a negligibly small emittance. This is not the case, however, because synchrotron radiation is not emitted continuously in arbitrarily small packets of energy. Instead it is radiated stochastically, as can be seen from the very broad spectrum, sometimes in rather large energy bursts. This has the effect of constantly exciting betatron oscillations. The emittance is determined by the balance between excitation and damping of the betatron oscillations. As we will see, this ultimately depends upon the magnet structure. The properties of proton and heavy-ion beams are fixed by their history, but electrons on the other hand will 'forget' everything in their past if they undergo a period of damping. This is a great advantage and means that the emittance in electron storage rings can be controlled without worrying about the conditions in the pre-accelerator.

In order to understand the origins of the intrinsic emittance, i.e. the emittance due only to the magnet structure, it is important to clarify the mechanism by which photon emission excites transverse particle oscillations. This is illustrated in Fig. 6.3. Since synchrotron radiation is only emitted in bending magnets, we may restrict ourselves to this region of the accelerator. For simplicity we assume

[1]This statement is not strictly correct, as the emittance may slowly increase due to effects such as collisions of the particles with residual gas atoms in the vacuum chamber. In addition, there are techniques which dramatically reduce the emittance of proton or heavy-ion beams in a storage ring by so-called **stochastic cooling** or by **electron cooling**.

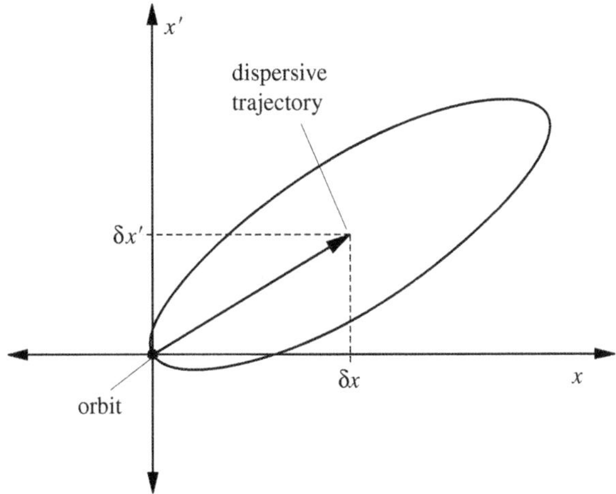

Fig. 6.4 The phase ellipse around the dispersive trajectory, resulting from photon emission.

that the electron is travelling exactly along the orbit with nominal momentum p_0 before it enters the magnet. It thus has zero initial emittance $\varepsilon_i = 0$. After travelling a certain distance through the magnet it spontaneously emits a photon of momentum Δp. The particle thus continues with the reduced momentum $p_0 - \Delta p$ and is no longer able to remain in the ideal orbit. The new equilibrium trajectory is displaced from the ideal orbit by a distance

$$\delta x = D\,\frac{\delta p}{p} \qquad \text{and angle} \qquad \delta x' = D'\,\frac{\delta p}{p}, \tag{6.46}$$

where D is the dispersion at the point at which the photon is emitted and D' is its gradient. During the revolutions which follow, the focusing and bending magnets cause the particle to perform oscillations about this new path, and clearly it now has a non-zero emittance. The corresponding phase ellipse is shown in Fig. 6.4 and can be calculated by simply inserting (6.46) into (3.135). The emittance of an individual particle is given by

$$\begin{aligned} \varepsilon_i &= \gamma\,\delta x^2 + 2\,\alpha\,\delta x\,\delta x' + \beta\,\delta x'^2 \\ &= \left(\frac{\delta p}{p}\right)^2 \left(\gamma\,D^2 + 2\,\alpha\,D\,D' + \beta\,D'^2\right) \\ &= \left(\frac{\delta p}{p}\right)^2 \mathcal{H}(s). \end{aligned} \tag{6.47}$$

The optical functions β, α and γ are evaluated at the point of photon emission. The function $\mathcal{H}(s)$, which has the form of the ellipse equation, is clearly important in determining the emittance.

We now know the emittance of an individual particle after it has emitted a photon of well-defined energy. To calculate the emittance of an entire beam we

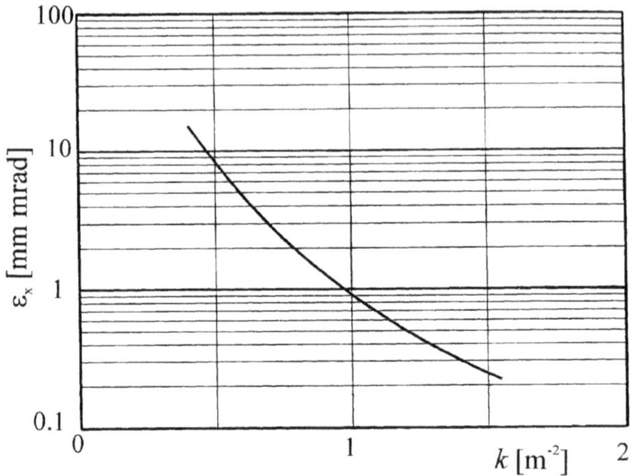

Fig. 6.5 Dependence of the horizontal emittance on the quadrupole strength k in a model FODO ring.

must average over all possible photon energies and emission probabilities. This is a rather lengthy calculation which we will omit here, but a good derivation can be found in M. Sands [51], for example. This calculation ultimately yields the general expression for the horizontal emittance of an electron beam, assuming a ring with bending only in the horizontal plane:

$$\boxed{\varepsilon_x = \frac{55}{32\sqrt{3}} \frac{\hbar}{mc} \gamma^2 \frac{\left\langle \frac{1}{R^3} \mathcal{H}(s) \right\rangle}{J_x \left\langle \frac{1}{R^2} \right\rangle}}. \tag{6.48}$$

Here $J_x = 1 - \mathcal{D}$. The average $\langle \ldots \rangle$ only needs to be calculated along the length of the bending magnet, since the bending radius R is non-zero only in this region. Since most storage rings use bending magnets of identical radius, and strongly focusing machines generally have $J_x \approx 1$, we can usually rewrite the expression (6.48) in the more convenient form

$$\varepsilon_x = 1.47 \times 10^{-6} \frac{E^2}{R} \frac{1}{l} \int_0^l \mathcal{H}(s)\, ds, \tag{6.49}$$

where the energy E is given in GeV, the bending radius R in m, and the emittance ε_x in mrad. We immediately see that a large bending radius and small value of the function $\mathcal{H}(s)$ lead to a small emittance. Looking more closely at $\mathcal{H}(s)$ in (6.47), we see that the dispersion and the beta function in the magnet should be kept as small as possible.

Varying the quadrupole strengths in a ring will change the beta function and dispersion, and with them the emittance. The effect is relatively strong, since

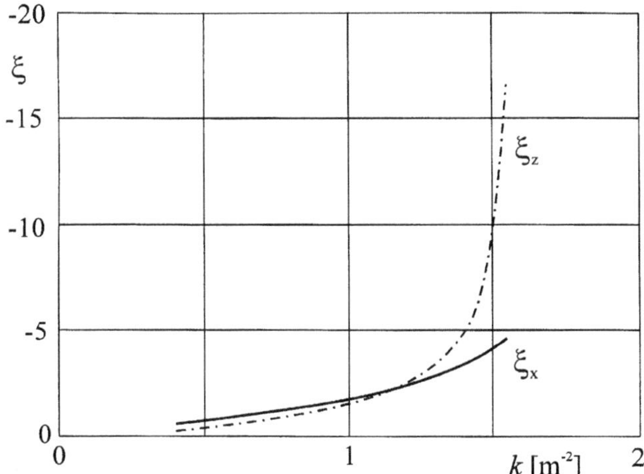

Fig. 6.6 Dependence of the horizontal chromaticities $\xi_{x,z}$ on the quadrupole strength k in a model FODO ring.

the dispersion enters quadratically into the expression for the emittance. We can show this effect quantitatively, using our model ring with the simple FODO structure (Section 3.13.3) as an example. If both quadrupole sections QF and QD are chosen to have the same strength (apart from the sign) then the range $0.4\ \mathrm{m}^{-2} < k < 1.55\ \mathrm{m}^{-2}$ gives stable beam-optical conditions, i.e. symmetric solutions exist. If we use (6.48) or (6.49) to calculate the emittance ε_x for various values of k, we obtain the dependence shown in Fig. 6.5. It is noticeable that ε_x is reduced by almost two orders of magnitude as the quadrupole strength is increased from the lowest stable value up to the maximum. Relatively strong quadrupoles are thus required in order to achieve small emittances.

The minimization of the emittance by the use of strong quadrupoles reaches a hard limit, as may be seen from Fig. 6.6, which shows the behaviour of the horizontal and vertical chromaticities. Above $k = 1.3\ \mathrm{m}^{-2}$ the vertical chromaticity increases sharply. To compensate for this effect, strong sextupoles are then required which drastically reduce the dynamic aperture. It clearly makes no sense to continue to reduce the emittance by increasing the quadrupole strength still further.

Modern storage rings with extremely small emittances, specially designed as sources of synchrotron radiation, all operate at this emittance limit imposed by the chromaticity. Very expensive arrangements of sextupole magnets are sometimes used in order to extend this limit as far as possible.

6.4.1 The lower limit of the beam emittance: the low emittance lattice

Although increasing the quadrupole strength generally results in smaller emittances, it is far from clear that this approach leads to the lowest achievable value. In the following section we will calculate the theoretical minimum emittance, us-

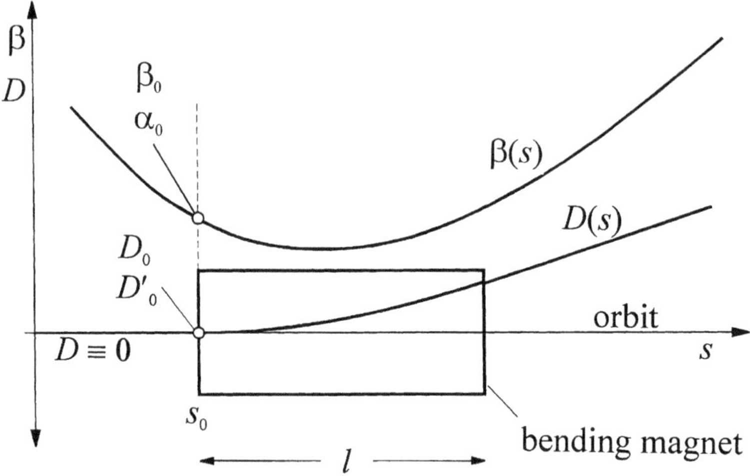

Fig. 6.7 Form of the optical functions in a bending magnet for given initial values β_0, α_0, D_0 and D'_0.

ing a few necessary boundary conditions. It is sufficient to consider the beam optics in only one bending magnet, assuming for the sake of simplicity that the optical functions have the same form in every other bending magnet. This assumption is justified by the symmetrical design of most extremely low emittance storage rings. We will also consider only planar machines, which restricts the problem to the horizontal plane.

The bending magnet is fully described optically by its bending radius R and its length l. If at the point $s_0 = 0$ at the beginning of the magnet the amplitude function is given by the values β_0 and α_0 and the dispersion by D_0 and D'_0, then the evolution of these functions through the rest of the dipole is uniquely defined (Fig. 6.7). According to (6.48), the emittance is then also uniquely fixed. The minimum beam emittance can be determined simply by varying the starting values of the optical functions until the minimum emittance value is found. In principle it is possible to vary all four initial values β_0, α_0, D_0, and D'_0. It is, however, useful to insert a straight section on one side of the magnet, in which the dispersion and its gradient go to zero. The undulator or wiggler magnet is then installed on this side. As these magnets produce intense synchrotron radiation, a non-zero dispersion in this region would significantly increase the emittance. Requiring a zero-dispersion section imposes the initial values $D_0 = 0$ and $D'_0 = 0$. The equations (3.102) then give the evolution of these functions in the dipole magnet:

$$
\begin{aligned}
D(s) &= R\left(1 - \cos\frac{s}{R}\right) \approx \frac{s^2}{2R} \\
D'(s) &= \sin\frac{s}{R} \approx \frac{s}{R}
\end{aligned}
\tag{6.50}
$$

The above approximations are generally justified, since the length of the magnet is small compared to the bending radius and so the condition $s/R \ll 1$ virtually always holds. Only the initial values of the beta function may now be varied in order to minimize the emittance. For simplicity we will neglect weak focusing, which in general only has a very small effect on the emittance. From the point of view of beam focusing, the dipole magnet effectively behaves like a drift region, the transfer matrix $\mathbf{M}$ of which is given in in (3.72). Using the initial values β_0 and α_0, (3.149) yields the transformation

$$\begin{pmatrix} \beta(s) & -\alpha(s) \\ -\alpha(s) & \gamma(s) \end{pmatrix} = \begin{pmatrix} 1 & s \\ 0 & 1 \end{pmatrix} \cdot \begin{pmatrix} \beta_0 & -\alpha_0 \\ -\alpha_0 & \gamma_0 \end{pmatrix} \cdot \begin{pmatrix} 1 & 0 \\ s & 1 \end{pmatrix}, \tag{6.51}$$

from which we directly obtain the form of the optical functions:

$$\begin{aligned} \beta(s) &= \beta_0 - 2\alpha_0 s + \gamma_0 s^2 \\ \alpha(s) &= \alpha_0 - \gamma_0 s \\ \gamma(s) &= \gamma_0 = \text{const.} \end{aligned} \tag{6.52}$$

Using the dispersion function (6.50) and the beta function (6.52), equation (6.48) allows us to specify the function $\mathcal{H}(s)$, important for calculating the emittance:

$$\begin{aligned} \mathcal{H}(s) &= \gamma(s)D^2(s) + 2\alpha(s)D(s)D'(s) + \beta(s)D'^2(s) \\ &= \frac{1}{R^2}\left(\frac{\gamma_0}{4}s^4 - \alpha_0 s^3 + \beta_0 s^2\right). \end{aligned} \tag{6.53}$$

If we again assume that all the bending magnets in the ring are identical and that the damping number $J_x \approx 1$, it then follows from (6.48) and (6.53) that

$$\varepsilon_x = C_\gamma \frac{\gamma^2}{R}\frac{1}{l}\int_0^l \mathcal{H}(s)\, ds = C_\gamma \gamma^2 \left(\frac{l}{R}\right)^3 \left(\frac{\gamma_0 l}{20} - \frac{\alpha_0}{4} + \frac{\beta_0}{3l}\right) \tag{6.54}$$

with

$$C_\gamma = \frac{55}{32\sqrt{3}}\frac{\hbar}{mc} = 3.832 \times 10^{-13}\ \text{m}.$$

Recognising that the ratio $l/R = \Theta$ corresponds to the bending angle per dipole magnet, we may write the emittance in the form

$$\boxed{\varepsilon_x = C_\gamma \gamma^2 \Theta^3 \left(\frac{\gamma_0 l}{20} - \frac{\alpha_0}{4} + \frac{\beta_0}{3l}\right).} \tag{6.55}$$

Since Θ enters this expression raised to the third power, it is clearly better to use many periodic cells with short bending magnets, rather than fewer cells with relatively long dipoles, if the aim is to keep the emittance small. The relation $\varepsilon_x \propto \Theta^3$ holds for all possible periodic magnet structures.

Since the magnet structure and the beam energy are fixed, the emittance in (6.55) is now only a function of the initial values β_0 and α_0 of the optical functions. The theoretical minimum allowed by these optical functions is then obtained in the usual way using the conditions

$$\frac{\partial \varepsilon_x}{\partial \alpha_0} = \mathcal{A}\frac{\partial}{\partial \alpha_0}\left(\frac{1+\alpha_0^2}{\beta_0}\frac{l}{20} - \frac{\alpha_0}{4} + \frac{\beta_0}{3l}\right) = \mathcal{A}\left(\frac{\alpha_0}{\beta_0}\frac{l}{10} - \frac{1}{4}\right) = 0 \tag{6.56}$$

and

$$\frac{\partial \varepsilon_x}{\partial \beta_0} = \mathcal{A}\left(-\frac{1+\alpha_0^2}{\beta_0^2}\frac{l^2}{20} + \frac{1}{3}\right) = 0 \tag{6.57}$$

with $\mathcal{A} = C_\gamma \gamma^2 \Theta^3$. Solving these two equations yields the initial values corresponding to the minimum possible emittance for given bending magnets. The solutions have the very simple form

$$\boxed{\begin{aligned} \beta_0 &= 2\sqrt{\frac{3}{5}}\, l = 1.549\, l \\ \alpha_0 &= \sqrt{15} = 3.873 \end{aligned}} \; . \tag{6.58}$$

Using this principle, *R. Chasman* and *K.G. Green* have suggested a magnet structure specially developed for synchrotron radiation sources with extremely small emittances [64]. This is shown in Fig. 6.8, together with the associated optical functions. The example shown here is taken from the planned Japanese project HiSOR [65], which has beam energies of up to $E_{\text{max}} = 1.5$ GeV. In this design the minimum possible number of quadrupoles are used. The 12 bending magnets each have a length of $l = 2.18$ m and a bending angle of $\Theta = 0.5236$ rad. These values give a minimum theoretical emittance of $\varepsilon_{\text{theor.}} = 6.68 \times 10^{-8}$ mrad. The actual emittance of the planned beam optics is somewhat larger, namely $\varepsilon_{\text{real}} = 7.89 \times 10^{-8}$ m rad. The difference arises from the use of different initial values, namely $\alpha_0 = 1.54$ and $\beta_0 = 2.82$ m, instead of the optimal values given by (6.58).

This modification is used because a system of beam optics developed in strict accordance with the conditions (6.58) would have an extremely high chromaticity. Mainly as a result of the large value of α_0, the beta function is strongly divergent outside the bending magnets, necessitating very strong quadrupoles for focusing and also leading to large beta functions. Together these effects result in high chromaticity, which cannot be compensated for by the use of sextupoles without drastically reducing the dynamic aperture. A further difficulty arises from the need for zero-dispersion sections to house wigglers and undulators, which means that there are few suitable places available for sextupoles.

Because of these complications, there are no existing or planned storage rings which strictly satisfy the conditions given in (6.58). Compromises must be found in order to achieve long beam lifetimes and stable, reliable operation, and these almost always entail some reduction in α_0. Fortunately the dependence of the minimum emittance upon α_0 is relatively flat, and so reducing its value only causes a very small increase in the emittance, as can be seen in the example shown.

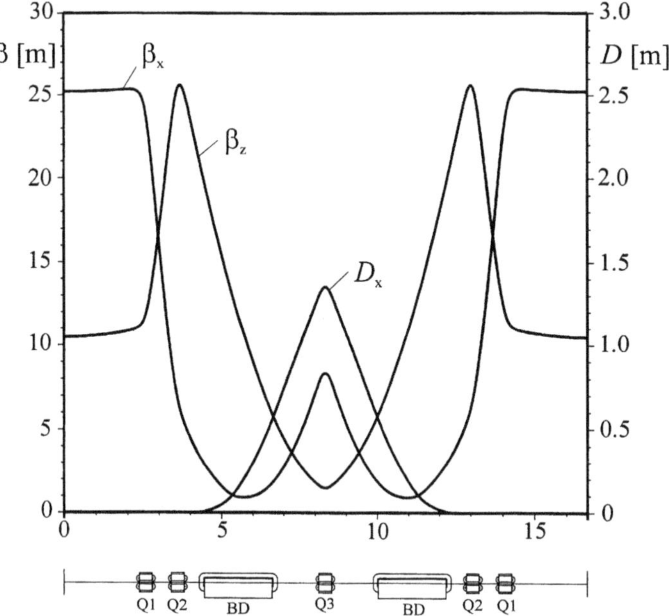

Fig. 6.8 One cell of a magnet structure based on the principle of R. Chasman and K.G. Green. The optics are symmetric about the centre of Q3 and there are dispersion-free sections at either end. The beta functions β_x and β_z are shown, along with the dispersion D_x.

Nowadays electron storage rings designed especially to produce synchrotron light are almost always based on the Chasman-Green principle, but with the following modification: in general more quadrupoles are used per cell than are shown in Fig. 6.8, to give greater flexibility in the beam optics and to allow a variety of choices of wigglers and undulators to be installed, depending on the particular requirements of the experiment. A modern example is the European Synchrotron Radiation Facility ESRF in Grenoble, which can reach beam energies of up to 6 GeV and contains a total of 32 straight sections, 30 of which can be used to house wigglers and undulators [45].

7 Luminosity

In Section 1.4.2 we saw that for a high-energy physics experiment at a colliding-beam storage ring, the rate of particle production is given by the simple expression

$$\dot{N}_{\mathrm{p}} = \sigma_{\mathrm{p}}\, \mathcal{L}. \tag{7.1}$$

Here σ_{p} is the interaction cross-section of the physical process under study. This is a property of nature and so is fixed. $\mathcal{L}$ is the **luminosity**, which describes how effectively the accelerator performs. The luminosity of course depends on the machine parameters, and so can be varied in different regions of the accelerator. The unit of luminosity is defined as

$$\mathcal{L}\left[\frac{1}{\mathrm{cm}^2\ \mathrm{s}}\right] = \mathcal{L}\left[\frac{10^{33}}{\mathrm{nb\ s}}\right] = \frac{\dot{N}_{\mathrm{p}}}{\sigma_{\mathrm{p}}}. \tag{7.2}$$

Nowadays processes with extremely small cross-sections are studied, namely $\sigma_{\mathrm{p}} \ll 1$ nb. This means that high luminosities are required in order to complete a measurement within a reasonable time. The total number of events collected during a data-taking run is

$$N_{\mathrm{p}} = \sigma_{\mathrm{p}} \int_{\mathrm{run}} \mathcal{L}\, dt = \sigma_{\mathrm{p}}\, \mathcal{I}. \tag{7.3}$$

The expression

$$\mathcal{I} = \int_{\mathrm{run}} \mathcal{L}\, dt \tag{7.4}$$

is called the **integrated luminosity**, often expressed in units of nb^{-1}. In order to get a feel for the quantities involved, let us consider as an example the luminosity produced by the storage ring DORIS II at energies of around $E = 5.3$ GeV. Average luminosities of $\langle\mathcal{L}\rangle = 2.27 \times 10^{31}\ \mathrm{cm}^{-2}\ \mathrm{s}^{-1}$ were achieved over long data-taking periods. The integrated luminosity per day is then

$$\begin{aligned} \int_{\mathrm{day}} \mathcal{L}\, dt &= \langle\mathcal{L}\rangle\, t_{\mathrm{day}} = \langle\mathcal{L}\rangle \times 24 \times 3600 \\ &= 1.961 \times 10^{36}\ \mathrm{cm}^{-2} = 1961\ \mathrm{nb}^{-1}. \end{aligned} \tag{7.5}$$

To calculate the luminosity we project the particles, which are all distributed in

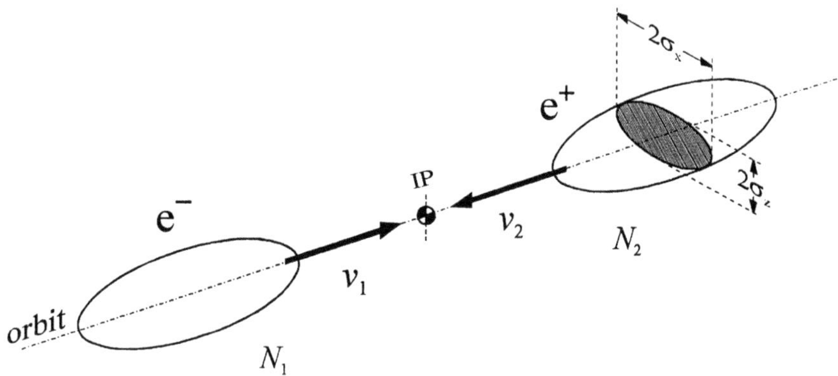

Fig. 7.1 Particle bunches colliding at the interaction point (IP).

the longitudinal direction within a bunch, onto a cross-sectional plane (Fig. 7.1). This reduces the problem to two dimensions. Since the charge distribution within a bunch is Gaussian in all directions, the surface density of positrons on this transverse plane is given by

$$n_2 = \frac{\partial^2 N_2}{\partial x\, \partial z} = \frac{N_2}{2\pi\sigma_x^*\sigma_z^*} \exp\left(-\frac{x^2}{2\sigma_x^{*2}} - \frac{z^2}{2\sigma_z^{*2}}\right). \tag{7.6}$$

Here N_2 is the total number of positrons in a bunch and $\sigma_{x,z}^*$ is the horizontal or vertical cross-section at the interaction point (IP). Because the collisions occur in a narrow region around the interaction point, in what follows we will be interested in the values of the beam cross-section and optical functions only at that point. For clarity we will follow the general convention and label values at the IP with *. The probability that an electron in a surface element $dA = dx\, dz$ of the beam cross-section in one bunch will collide with a positron in the other bunch is

$$dW = \sigma_\mathrm{p} \frac{n_2\, dx\, dz}{dA} = \sigma_\mathrm{p}\, n_2. \tag{7.7}$$

If we assume that the electron and positron beams have the same cross-section at the IP[1], then the number of electrons crossing the surface dA of the positron bunch per unit time is

$$d\dot{N}_1 = \frac{b f_\mathrm{rev} N_1}{2\pi\sigma_x^*\sigma_z^*} \exp\left(-\frac{x^2}{2\sigma_x^{*2}} - \frac{z^2}{2\sigma_z^{*2}}\right) dx\, dz, \tag{7.8}$$

where b equally-spaced bunches circulate with a frequency f_rev. N_1 is the number of electrons in a bunch. Hence we obtain the differential event rate

$$d\dot{N}_\mathrm{p} = \sigma_\mathrm{p}\, d\dot{N}_1\, n_2 = \sigma_\mathrm{p} \frac{b f_\mathrm{rev} N_1 N_2}{(2\pi)^2\sigma_x^{*2}\sigma_z^{*2}} \exp\left(-\frac{x^2}{\sigma_x^{*2}} - \frac{z^2}{\sigma_z^{*2}}\right) dx\, dz, \tag{7.9}$$

[1]This condition is automatically satisfied in storage rings in which the two particle beams both travel in the same beam pipe. There may be differences in machines with two separate rings, each with different beam optics.

which may be integrated using

$$\int_{-\infty}^{+\infty} \exp\left(-\frac{y^2}{\sigma^2}\right) dy = \sqrt{\pi}\sigma \tag{7.10}$$

to give the total rate

$$\dot{N}_{\mathrm{p}} = \sigma_{\mathrm{p}} \frac{b f_{\mathrm{rev}} N_1 N_2}{4\pi \sigma_x^* \sigma_z^*}. \tag{7.11}$$

Comparing this with (7.1) yields the expression for the luminosity

$$\mathcal{L} = \frac{b}{4\pi} \frac{N_1\, N_2}{\sigma_x^*\, \sigma_z^*} f_{\mathrm{rev}}, \tag{7.12}$$

which we have seen already in Section 1.4.2. Because they are easy to measure it is usually more convenient to use the average beam currents $I_i = N_i e f_{\mathrm{rev}} b$ rather than the number of particles. The luminosity then takes the form

$$\boxed{\mathcal{L} = \frac{1}{4\pi e^2 f_{\mathrm{rev}} b} \frac{I_1\, I_2}{\sigma_x^*\, \sigma_z^*}.} \tag{7.13}$$

We immediately see that the essential requirements for high luminosity are high currents and very small transverse beam dimensions. Note that the probability of collision in each bunch crossing is extremely small, and only a few particles in each bunch will interact. The beam cross-section is therefore usually chosen to be at least an order of magnitude smaller at the IP than at other points around the ring. At the IP the beta functions lie in the centimetre range, whereas elsewhere values up to tens of metres are common. In order to achieve such tiny beam dimensions, a special focusing system is needed on both sides of the IP. Nowadays this powerful focusing is relatively easy to achieve using the methods developed in Chapter 3. The key problem is the restriction of the beam currents, which is the subject of the following section.

7.1 Beam current restriction due to the space charge effect

Electrons lose a certain amount of energy ΔE_γ per revolution in the form of synchrotron radiation. This lost energy must be replaced by an average RF voltage $U = \Delta E_\gamma / e$. For an average beam current I_{beam} the power required is $P_{\mathrm{RF}} = I_{\mathrm{beam}}\, U$. The RF supply to an accelerator is always limited by cost, and this restricts the maximum attainable currents. Once this limit is reached, it can in principle be extended further by installing additional power generators and cavities.

Another limit can arise, due to beam instabilities in which the electromagnetic field induced by the beam in the vacuum chamber and accelerating sections then acts upon the beam itself. A type of feedback can occur, leading to rapidly-growing oscillations which result in partial or total beam loss. These fields grow

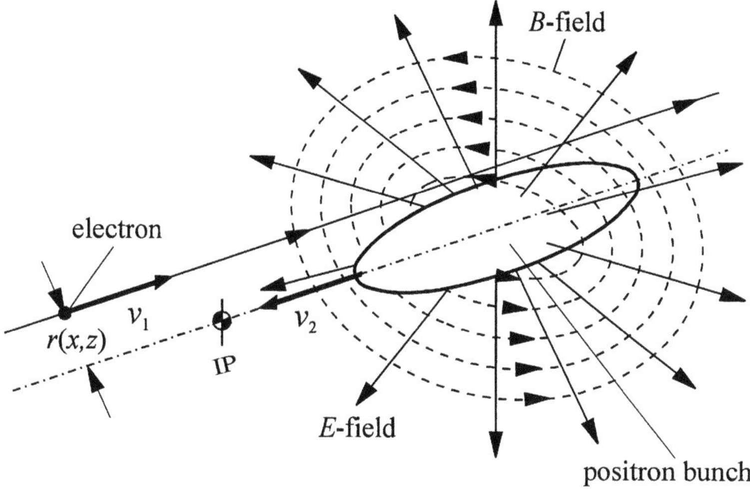

Fig. 7.2 Deflection of an electron due to the space charge of an oncoming bunch.

in proportion to the beam current and reach a critical strength when the current exceeds a certain threshold. Stable operation is possible only below this critical threshold, which in many cases can be raised by careful design of the vacuum chambers and by active and passive damping mechanisms.

The third and most significant current limitation is due to the **space charge effect**. During collision, the electromagnetic field around one bunch acts upon the particles in the other bunch, deflecting them out of orbit by an amount proportional to the beam current. Above a certain value the motion of the most strongly deflected particles becomes unstable and they are lost against the wall of the vacuum chamber. This current limitation in colliding beams due to space charge was first studied by F. Amman and D. Ritson [66] and is also known as the **Amman-Ritson effect**. To look at this effect quantitatively, let us consider a single electron in a bunch, travelling at a separation $\boldsymbol{r} = (x, z)$ from the orbit through an oncoming bunch. The electron is deflected by the field due to this bunch (Fig. 7.2).

In order to calculate the space charge forces we move to the centre of mass frame K' of the electron bunch. Here there is of course only an electric field $\boldsymbol{E}'$ produced by the electrons, which are at rest in this frame. If we move to the laboratory frame K, then the Lorentz transformation now gives both electric **and** magnetic fields. We thus separate the field components $B_\perp$, $E_\perp$ **perpendicular** to the direction of particle motion from the components $B_\parallel$, $E_\parallel$ which run **parallel** to it. The fields in the laboratory frame are then

$$\begin{aligned} \boldsymbol{E}_\perp &= \gamma \boldsymbol{E}'_\perp & \qquad \boldsymbol{E}_\parallel &= \boldsymbol{E}'_\parallel \\ \boldsymbol{B}_\perp &= \frac{\gamma}{c^2} \boldsymbol{v}_2 \times \boldsymbol{E}'_\perp & \qquad \boldsymbol{B}_\parallel &= 0. \end{aligned} \tag{7.14}$$

Only the transverse fields $B_\perp$ and $E_\perp$ are involved in the space charge effect, since they are the only components which change the focusing of the beam. As the electron travels through the positron bunch it feels a focusing force

$$\begin{aligned} \boldsymbol{F}_\perp &= -e\Big(\boldsymbol{E}_\perp + \boldsymbol{v}_1 \times \boldsymbol{B}_\perp\Big) \\ &= -e\left[\gamma \boldsymbol{E}'_\perp + \boldsymbol{v}_1 \times \left(\frac{\gamma}{c^2}\,\boldsymbol{v}_2 \times \boldsymbol{E}'_\perp\right)\right] \\ &= -e\Big(1+\beta_1\beta_2\Big)\,\boldsymbol{E}_\perp . \end{aligned} \tag{7.15}$$

Since the particles used in experiments are extremely relativistic, $\beta_1 = \beta_2 \approx 1$ and hence

$$\boldsymbol{F}_\perp = -2e\boldsymbol{E}_\perp . \tag{7.16}$$

The electrical and magnetic fields therefore contribute equally to the focusing. To continue the calculation let us again assume a Gaussian charge distribution inside the bunch and write the charge density in the centre of mass frame K' in the three-dimensional form

$$\rho'\,(x,z,s') = \frac{eN_2}{(2\pi)^{3/2}\sigma_x\sigma_z\sigma'_s}\exp\left(-\frac{x^2}{2\sigma_x^2} - \frac{z^2}{2\sigma_z^2} - \frac{(s'-s'_0)^2}{2\sigma_s'^2}\right). \tag{7.17}$$

Here it should be noted that $\sigma'_{x,z} = \sigma_{x,z}$ and $\sigma'_s = \gamma\sigma_s$. s'_0 is an arbitrary reference point on the beam axis. In the centre of mass frame the bunch length $\sigma'_s = \gamma\sigma_s$ is much greater than its transverse size, and so we can effectively regard the bunch as an infinitely long charge distribution, with the charge density varying only very slowly along the beam axis s'. It is then useful to rewrite the expression for the charge density (7.17) in the form

$$\rho'\,(x,z,s') = A(s')\exp\left(-\frac{x^2}{2\sigma_x^2} - \frac{z^2}{2\sigma_z^2}\right) \tag{7.18}$$

with

$$A(s') = \frac{eN_2}{(2\pi)^{3/2}\sigma_x\sigma_z\sigma'_s}\exp\left(-\frac{(s'-s'_0)^2}{2\sigma_s'^2}\right). \tag{7.19}$$

Using this charge density we can then calculate the field strength at the position of the electron, and hence determine the strength of the focusing force. If, however, we start from the general case of a beam with an oval cross-section at the IP, i.e. $\sigma_x \neq \sigma_z$, then the Gaussian distribution function in (7.18) makes the computation of the integral very time-consuming. We will therefore make a considerable simplification and assume here that the beam is circular. This is certainly not true in general, but it allows us to explain more simply the important physical ideas behind the method. If we insert the value $\sigma = \sigma_x = \sigma_z$ for the transverse cross-section of the beam into (7.18), then using $r^2 = x^2 + z^2$ we obtain

$$\rho'(r,s') = A(s')\exp\left(-\frac{r^2}{2\sigma^2}\right). \tag{7.20}$$

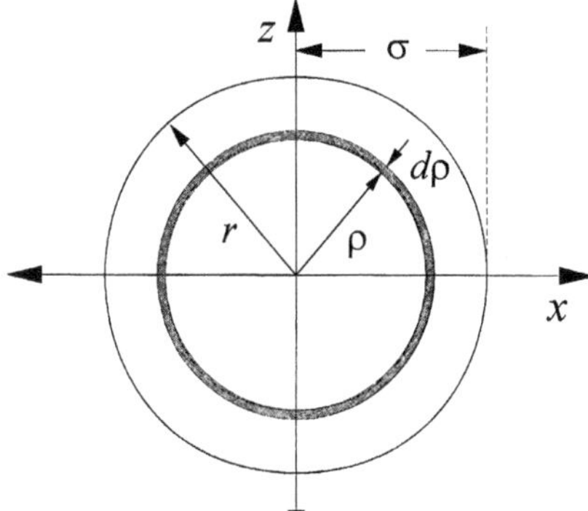

Fig. 7.3 Calculating the charge within a cylinder of radius r.

Since we have assumed the beam is circular it is convenient to work in polar coordinates. Using Fig. 7.3 we may then directly calculate the charge

$$dq = \rho'(r, s')\, 2\pi\, \varrho\, d\varrho\, ds' \tag{7.21}$$

contained within a cylindrical shell of radius ϱ and thickness $d\varrho$. From this we may immediately calculate the charge within a cylinder of length $\Delta s'$ and radius r to be

$$\Delta q(s') = 2\pi\, A(s')\, \Delta s' \int_0^r \exp\left(-\frac{\varrho^2}{2\,\sigma^2}\right) \varrho\, d\varrho. \tag{7.22}$$

Integrating by substitution we obtain

$$\Delta q(s') = 2\pi\, A(s')\, \sigma^2 \left[1 - \exp\left(-\frac{r^2}{2\,\sigma^2}\right)\right] \Delta s'. \tag{7.23}$$

Applying **Gauss's theorem** over the surface of the cylinder gives

$$E'_\perp(r)\, 2\pi\, r\, \Delta s' = \frac{\Delta q(s')}{\varepsilon_0}, \tag{7.24}$$

which yields the electrical field strength at the surface

$$E'_\perp(r) = \frac{\Delta q(s')}{2\pi\varepsilon_0\, r\, \Delta s'}. \tag{7.25}$$

Inserting (7.19) and (7.23) into this expression gives

$$E'_\perp(r) = \frac{eN_2}{(2\pi)^{3/2}\varepsilon_0\, r\, \sigma'_s} \exp\left(-\frac{(s' - s'_0)^2}{2\sigma'^2_s}\right) \left[1 - \exp\left(-\frac{r^2}{2\sigma^2}\right)\right]. \tag{7.26}$$

Using the relations $s' = \gamma\, s$ and $\sigma_s' = \gamma \sigma_s$ we now transform this field strength into the laboratory frame and obtain

$$E_\perp(r,s) = \gamma E_\perp'(r,s) = \frac{eN_2}{(2\pi)^{3/2}\varepsilon_0\,\sigma_s}\exp\left(-\frac{(s-s_0)^2}{2{\sigma_s}^2}\right)\frac{1}{r}\left[1-\exp\left(-\frac{r^2}{2\sigma^2}\right)\right]. \tag{7.27}$$

In what follows we will calculate the influence of the space charge effect to first order only. We therefore only consider electrons travelling at a very small distance from the beam axis as they pass through the positron bunch. If $r \ll \sigma$ we can expand the exponential as

$$\begin{aligned}\exp\left(-\frac{r^2}{2\sigma^2}\right) &= 1-\frac{r^2}{2\sigma^2}+\frac{1}{2!}\left(\frac{r^2}{2\sigma^2}\right)^2 - \ldots \\ &\approx 1-\frac{r^2}{2\sigma^2}.\end{aligned} \tag{7.28}$$

Under this approximation we finally obtain the following expression for the transverse electrical field strength

$$E_\perp(r,s) = \frac{eN_2}{(2\pi)^{3/2}\varepsilon_0\,\sigma_s}\exp\left(-\frac{(s-s_0)^2}{2{\sigma_s}^2}\right)\frac{r}{2\sigma^2}. \tag{7.29}$$

The electron is accelerated by this field and its transverse momentum changes by the amount

$$dp_\perp = F_\perp\,dt = F_\perp\,\frac{ds}{2c}. \tag{7.30}$$

The factor 2 in the ratio $ds/2c$ comes from the fact that the two relativistic bunches are travelling with equal velocity towards one another. If we now insert the force (7.16) into (7.30) and use the field strength (7.29), we obtain the momentum shift

$$\begin{aligned}dp_\perp &= -\frac{e\,E_\perp}{c}\,ds \\ &= -\frac{e^2\,N_2\,r}{4\pi\,\varepsilon_0\,c\,\sigma^2}\,\frac{1}{\sqrt{2\pi}\,\sigma_s}\exp\left(-\frac{(s-s_0)^2}{2{\sigma_s}^2}\right)ds.\end{aligned} \tag{7.31}$$

To obtain the total change in momentum we must integrate over the whole orbit. Since the Gaussian distribution dies away very rapidly for large $|s-s_0|$, we allow the integral to run to $\pm\infty$. The integral gives the simple result

$$\int_{-\infty}^{+\infty}\exp\left(-\frac{(s-s_0)^2}{2{\sigma_s}^2}\right)ds = \sqrt{2\pi}\,\sigma_s, \tag{7.32}$$

which finally yields the total change in transverse momentum

$$\Delta p_\perp(r) = -\frac{e^2\,N_2}{2\pi\,\varepsilon_0\,c}\,\frac{r}{2\sigma^2}. \tag{7.33}$$

The horizontal and vertical components are obtained simply by replacing r by x or z. The resulting changes in angle of the electron trajectory are

$$\begin{aligned}\Delta x' &= \frac{\Delta p_x}{p} = -\frac{e^2\,N_2}{2\pi\,pc\,\varepsilon_0}\,\frac{x}{2\sigma^2}\\ \Delta z' &= \frac{\Delta p_z}{p} = -\frac{e^2\,N_2}{2\pi\,pc\,\varepsilon_0}\,\frac{z}{2\sigma^2},\end{aligned} \tag{7.34}$$

where p is the particle momentum. Comparing this result with the change in angle $\Delta x' = k\,l\,x$ undergone by a particle at a separation x from the beam axis passing through a quadrupole of length l, we see that the space charge of the positron bunch has the same focusing effect on the electron as a quadrupole of integrated strength

$$k_r\,l = -\frac{e^2\,N_2}{2\pi\,pc\,\varepsilon_0}\,\frac{1}{2\sigma^2}. \tag{7.35}$$

Naturally the electron bunch also acts on the positrons in exactly the same way. This added quadrupole effect changes the focusing and hence also the tune $Q_{x,z}$ of the ring. The beam-beam interaction, due to the electromagnetic interaction between the bunches as they cross, has the same effect. If we remove the minus sign in (7.35) the electron trajectory will be bent towards the orbit, that is to say the beam will be focused, which according to (3.274) leads to the positive tune shift

$$\Delta Q_{x,z} = \frac{\beta^*_{x,z}}{4\pi}\,k_r\,l = \frac{e^2\,N_2}{8\pi^2\,pc\,\varepsilon_0}\,\frac{\beta^*_{x,z}}{2\sigma^2}. \tag{7.36}$$

Here the beta functions have the value $\beta^*_{x,z}$ in the interaction region, which we assume to be nearly constant over the length of the bunch. As the number of particles N_i in the bunch ($i = 1, 2$) increases, so the tune shift increases until the particles encounter a stronger optical resonance and are lost. This is the ultimate limiting effect of the space charge.

If we drop the simplifying assumption that the beam is circular and allow the horizontal and vertical beam dimensions to differ, i.e. $\sigma^*_x \neq \sigma^*_z$, then a rather lengthier calculation leads to effectively the same result. In (7.36) we must simply replace $2\sigma^2$ by the appropriate expression, namely

$$2\,\sigma^2 \quad \longrightarrow \quad \begin{cases} \sigma^*_x(\sigma^*_x + \sigma^*_z) & \text{horizontal} \\ \sigma^*_z(\sigma^*_x + \sigma^*_z) & \text{vertical.} \end{cases} \tag{7.37}$$

For colliding beams of arbitrary cross-section the tune shift due to the space charge effect is then

$$\begin{aligned}\Delta Q_x &= \frac{e^2\,N_2}{8\pi^2\,pc\,\varepsilon_0}\,\frac{\beta^*_x}{\sigma^*_x(\sigma^*_x + \sigma^*_z)}\\ \Delta Q_z &= \frac{e^2\,N_2}{8\pi^2\,pc\,\varepsilon_0}\,\frac{\beta^*_z}{\sigma^*_z(\sigma^*_x + \sigma^*_z)}.\end{aligned} \tag{7.38}$$

In what follows we will again replace the number of particles by the beam current using $N_2 = I_2/(b f_{\text{rev}} e)$ and express the beam cross-section as $\sigma^*_{x,z} = \sqrt{\varepsilon_{x,z}\beta^*_{x,z}}$. Recalling that $E = pc$ and $\varepsilon_0 = 1/\mu_0 c^2$, it then follows from (7.38) that

$$\Delta Q_x = \frac{\mu_0 e c^2 I_2}{8\pi^2 b f_{\text{rev}} E} \frac{\sqrt{\beta_x^*}}{\sqrt{\varepsilon_x}\left(\sqrt{\varepsilon_x \beta_x^*} + \sqrt{\varepsilon_z \beta_z^*}\right)}$$

$$\Delta Q_z = \frac{\mu_0 e c^2 I_2}{8\pi^2 b f_{\text{rev}} E} \frac{\sqrt{\beta_z^*}}{\sqrt{\varepsilon_z}\left(\sqrt{\varepsilon_x \beta_x^*} + \sqrt{\varepsilon_z \beta_z^*}\right)}. \tag{7.39}$$

For very large values of ΔQ the particles encounter optical resonances (Section 3.14) and are lost. This reduces the beam lifetime and causes the backgrounds in the particle detectors to increase drastically. Experience in electron–positron storage rings has shown that the tune shift should not be much greater than

$$\boxed{\Delta Q_{\text{max}} \approx 0.025.} \tag{7.40}$$

Values up to $\Delta Q = 0.04$ have been achieved in very carefully configured storage rings, but these conditions tend to be rather unstable and difficult to reproduce. According to (7.39), the space charge effect between two interacting beams thus restricts the maximum current per bunch to the value

$$I_{\text{max},x} = \frac{8\pi^2 b f_{\text{rev}} E \sqrt{\varepsilon_x}\left(\sqrt{\varepsilon_x \beta_x^*} + \sqrt{\varepsilon_z \beta_z^*}\right)}{\mu_0 e c^2 \sqrt{\beta_x^*}} \Delta Q_{\text{max},x}$$

$$I_{\text{max},z} = \frac{8\pi^2 b f_{\text{rev}} E \sqrt{\varepsilon_z}\left(\sqrt{\varepsilon_x \beta_x^*} + \sqrt{\varepsilon_z \beta_z^*}\right)}{\mu_0 e c^2 \sqrt{\beta_z^*}} \Delta Q_{\text{max},z} \tag{7.41}$$

At this point we must remember the vertical emittance ε_z. In an ideal planar machine this would in theory be zero, but in practice there is no such thing as a perfectly planar beam, and when calculating the luminosity we must assume a non-zero value. For one thing, the horizontal betatron oscillations are coupled to the vertical plane due to higher-order multipole fields or to slight rotations of the quadrupoles about the beam axis. Hence the beam always has a certain vertical extent. In addition, vertical disturbances of the orbit can produce vertical dispersion, which in the bending magnets gives rise to betatron oscillations due to quantum emission.

The term **emittance coupling** is usually used to describe all these effects which can increase the vertical beam size. It is expressed quantitatively through the coupling factor[2]

$$k = \frac{\varepsilon_z}{\varepsilon_x}. \tag{7.42}$$

If an ideal machine with zero vertical emittance has a theoretical horizontal value of ε_{x0}, then a finite coupling $0 < k < 1$ will result in emittances of

[2]In the USA the coupling is often defined as the ratio of the beam dimensions rather than of the emittances.

$$\varepsilon_x = \frac{\varepsilon_{x0}}{1+k} \qquad \text{and} \qquad \varepsilon_z = \frac{k\varepsilon_{x0}}{1+k}. \tag{7.43}$$

Inserting these expressions into (7.41), we obtain the upper limits to the current due to the space charge effect in both planes:

$$\begin{aligned} I_{\mathrm{max},x} &= \frac{8\pi^2 b f_{\mathrm{rev}} E \varepsilon_{x0}\left(\sqrt{\beta_x^*} + \sqrt{k\beta_z^*}\right)}{\mu_0 e c^2 (1+k)\sqrt{\beta_x^*}} \Delta Q_{\mathrm{max},x} \\ I_{\mathrm{max},z} &= \frac{8\pi^2 b f_{\mathrm{rev}} E \varepsilon_{x0}\sqrt{k}\left(\sqrt{\beta_x^*} + \sqrt{k\beta_z^*}\right)}{\mu_0 e c^2 (1+k)\sqrt{\beta_z^*}} \Delta Q_{\mathrm{max},z}. \end{aligned} \tag{7.44}$$

Naturally the current limits are not usually the same in the two planes. The vertical limit is generally lower and it is this which defines the maximum attainable beam current. Hence we will only use the second equation of (7.44) in what follows.

The maximum current depends strongly on the particle energy E, but this dependence is not yet manifest in (7.44) because the emittance is also energy-dependent. We therefore introduce a **normalized emittance** $\tilde{\varepsilon}$ which corresponds to the value of the emittance at $E_0 = m_0c^2$. For any value of the energy and hence $\gamma = E/E_0$, the emittance is then given by (6.48)

$$\varepsilon_{x,z} = \gamma^2 \tilde{\varepsilon}_{x,z}. \tag{7.45}$$

Replacing E by $\gamma m_0 c^2$ in (7.44) we obtain

$$\boxed{I_{\mathrm{max}} = \frac{8\pi^2 b m_0 f_{\mathrm{rev}} \gamma^3 \tilde{\varepsilon}_{x0}\sqrt{k}\left(\sqrt{\beta_x^*} + \sqrt{k\beta_z^*}\right)}{\mu_0 e (1+k)\sqrt{\beta_z^*}} \Delta Q_{\mathrm{max}}.} \tag{7.46}$$

Maximum luminosity is of course achieved when both colliding beams reach the maximum current allowed by the space charge limit, i.e. $I_1 = I_2 = I_{\mathrm{max}}$. Inserting the expression for the maximum current (7.46) into the definition of luminosity (7.13), we thus obtain the maximum possible luminosity for a particular system of beam optics at a given beam energy

$$\boxed{\mathcal{L}_{\mathrm{max}} = \frac{16\pi^3 b f_{\mathrm{rev}} m_0^2 \gamma^4 \tilde{\varepsilon}_{x0}\sqrt{k}\left(\sqrt{\beta_x^*} + \sqrt{k\beta_z^*}\right)^2}{\mu_0^2 e^4 (1+k)\sqrt{\beta_x^*}\ \beta_z^{*3/2}} \Delta Q_{\mathrm{max}}^2.} \tag{7.47}$$

For practical purposes it is more convenient to express the energy in units of GeV and the beam current in mA. It is also usual to express the luminosity in $\mathrm{cm^{-2}\,s^{-1}}$. With this in mind let us also define the emittance $\hat{\varepsilon}_{x0}$ normalized to the energy $E = 1$ GeV, so that the emittance at an arbitrary energy is given by

$$\varepsilon_{x0} = E^2\ \hat{\varepsilon}_{x0}, \qquad \text{with } E \text{ in GeV}. \tag{7.48}$$

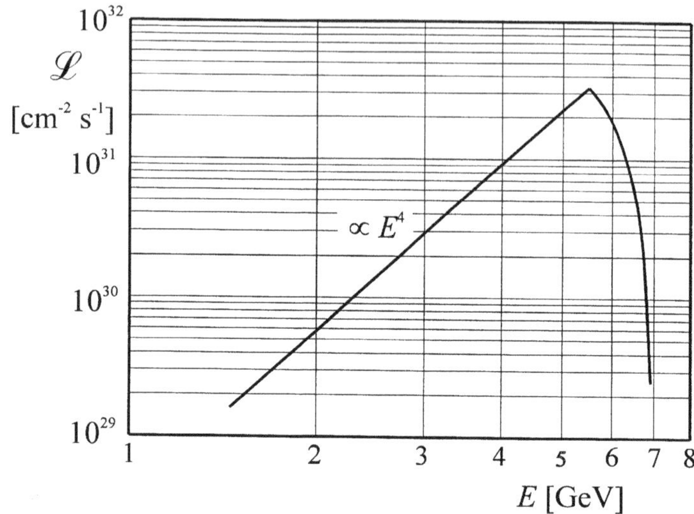

Fig. 7.4 Double logarithmic plot of the energy dependence of the maximum luminosity allowed by the space charge effect. Above a particular limit (in this example above 5.5 GeV) the RF supply is no longer powerful enough to maintain the required high currents and the luminosity falls off sharply.

If we now collect all the physical constants into one number, we obtain the following more convenient forms of (7.46) and (7.47):

$$I_{\max}\ [\mathrm{mA}] = 699.06 \frac{b f_{\mathrm{rev}} E^3 \hat{\varepsilon}_{x0} \sqrt{k}\left(\sqrt{\beta_x^*} + \sqrt{k\beta_z^*}\right)}{(1+k)\sqrt{\beta_z^*}} \Delta Q_{\max}, \qquad \text{with } E \text{ in GeV} \tag{7.49}$$

and

$$\mathcal{L}_{\max}\ [\mathrm{cm}^{-2}\,\mathrm{s}^{-1}] = 1.515\times 10^{32} \frac{b f_{\mathrm{rev}} E^4 \hat{\varepsilon}_{x0} \sqrt{k}\left(\sqrt{\beta_x^*} + \sqrt{k\beta_z^*}\right)^2}{(1+k)\sqrt{\beta_x^*}\ \beta_z^{*3/2}} \Delta Q_{\max}^2. \tag{7.50}$$

In the formulae (7.47) or (7.50) we see the very strong energy dependence of the luminosity, $\mathcal{L} \propto E^4$, which can also be seen from the curve in Fig. 7.4. Here it should be kept in mind that the current must also increase with the third power of the energy, as formula (7.46) shows. Once the maximum energy of the storage ring is reached it is no longer possible to increase the beam currents further, since the RF power supply is already operating at maximum capacity. Indeed since the rate of energy loss through synchrotron radiation increases with E^4, any further increase in beam energy necessitates a sharp reduction in beam currents. The luminosity falls off sharply as a result, as can be seen in Fig. 7.4.

In the luminosity formula (7.47) the emittance coupling k is in principle a free parameter. This raises the question of whether there is an optimal value of the coupling for a given system of beam optics, that is to say for given beta functions

β_x^* and β_z^*. We can answer this question if we assume that the maximum tune shift is the same in both planes, i.e. $\Delta Q_{\max,x} = \Delta Q_{\max,z}$. The optimal luminosity is then achieved when the current limitations due to the space charge effect are reached simultaneously in the two planes. We simply set the limiting currents $I_{\max,x}$ and $I_{\max,z}$ to be equal in (7.44) to obtain the optimal emittance coupling

$$k_{\text{opt}} = \frac{\beta_z^*}{\beta_x^*}. \tag{7.51}$$

Inserting this into (7.44) and using $\varepsilon_{x0} = \gamma^2 \tilde{\varepsilon}_{x0}$ we obtain the maximum current

$$I_{\max} = \frac{8\pi^2 b f_{\text{rev}} m_0 \gamma^3 \tilde{\varepsilon}_{x0}}{\mu_0 e} \Delta Q_{\max}. \tag{7.52}$$

In this situation the beam cross-section at the interaction point is

$$\sigma_x^* \sigma_x^* = \frac{\beta_x^* \beta_z^*}{\beta_x^* + \beta_z^*} \varepsilon_{x0}, \tag{7.53}$$

which along with the maximum current (7.52) gives a maximum luminosity from (7.13) of

$$\mathcal{L}_{\max} = \frac{16\pi^3 b f_{\text{rev}} m_0^2 \gamma^4 \tilde{\varepsilon}_{x0} \left(\beta_x^* + \beta_z^*\right)}{\mu_0^2 e^4 \beta_x^* \beta_z^*} \Delta Q^2. \tag{7.54}$$

The beam energy is not a parameter which can be tuned to optimize the luminosity, since it is prescribed by the physics of the process under study. In order to achieve the highest possible luminosities, the emittance ε_{x0} should first of all be large, limited only by the available aperture in the vacuum chamber. In addition, the maximum value of the tune shift $\Delta Q_{\max}$ can be optimized within certain limits by careful adjustment of the beams. Finally, the luminosity depends very strongly on the vertical beta function β_z^*, which can be made extremely small by suitable beam focusing, and in fact this method has proven to be by far the most effective.

7.2 The 'mini-beta' principle

As already mentioned, it is relatively easy to achieve very small vertical beta functions at the interaction point (IP) using the established principles of beam optics (Chapter 3). Values as low as $\beta_z^* < 3$ are used to operate experiments in electron storage rings. When developing a system of optics and a magnet structure, it is simplest to fix the beta functions at the IP and then arrange the magnets and the beam optics so that the optical functions remain within workable limits. In doing so it is important to remember that in order to fully measure a physical process we need to surround the IP with a particle detector which covers as much of the solid angle as possible. Such detectors, which contain a variety of counters and sub-detectors, are nowadays several metres in size in each dimension. To minimize any interference between the detector and the

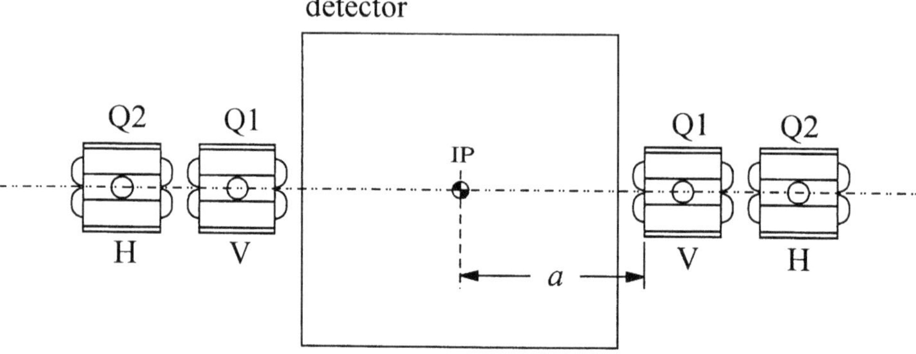

Fig. 7.5 Standard arrangement of focusing magnets around the interaction point (IP). A space is kept free for the particle detector, in order to avoid interference with the accelerator. In this arrangement the separation is $a > 2.5$ m.

accelerator, a space around the IP is kept completely free for the experiment, as shown in Fig. 7.5.

In this standard configuration with a separation of $a > 2.5$ m between the IP and the first vertically focusing magnets, beta functions of $\beta_z^* \approx 10$ cm are achievable, giving sufficient luminosities for current needs. However, any further reduction of the beta function below 10 cm creates major problems.

From the point of view of beam optics, the empty interaction region between the two quadrupoles Q1 is a pure drift section with a symmetry point at the IP. If the vertical beta function has the value β_z^* at the IP then, according to (3.151), by the beginning of the first quadrupole ($s = a$) it reaches the value

$$\beta_z(a) = \beta_z^* + \frac{a^2}{\beta_z^*}. \tag{7.55}$$

Extremely small amplitude functions at the IP will result in correspondingly large values at the first quadrupoles. Besides the aperture restriction, the chromaticity is again the major problem. The first quadrupoles Q1 must have a focal point at the IP and so must be relatively strong. If at the same time the beta function in the quadrupole is large then it makes a correspondingly large contribution to the chromaticity of

$$\Delta\xi = \frac{-1}{4\pi} \int_{\mathrm{Q1}} \beta_z(s)\, k\, ds. \tag{7.56}$$

Reducing β_z^* therefore strongly increases the chromaticity induced by the quadrupoles nearest to the interaction region. The lower limit of β_z^* is reached when the increase in chromaticity is no longer offset by the associated reduction in the dynamic aperture. A numerical example is given in Table 7.1. In this example, safe operation of the storage ring is no longer possible at values below $\beta_z^* = 10$ cm. In order to reach even smaller values and hence from (7.47) to achieve significantly higher luminosities, the sharp division between experiment and accelerator must be abandoned and the first vertically focusing quadrupole

Table 7.1 β_z^* dependence of the vertical beta function in the quadrupole Q1 and of the resulting contribution to the chromaticity. In this example the separation $a = 2.6$ m and Q1 has a strength $k = 0.6\ \mathrm{m}^{-2}$ and a length $l = 1$ m.

β_z^* [m]	$\beta_z(a)$ [m]	$\Delta\xi_z$
1.00	6.8	-0.32
0.50	13.5	-0.65
0.30	22.5	-1.08
0.10	67.6	-3.23
0.05	135.2	-6.46
0.03	225.3	-10.77 (!)

moved considerably closer to the interaction point. This means that the beams are focused earlier and the beta functions are prevented from growing to critical values, keeping the chromaticity within manageable bounds. Placing the first quadrupoles about 1 metre from the IP allows minimum beta functions as low as $\beta_z^* \approx 3$ cm. Because of these small values, this technique of arranging the magnets near the interaction point is termed the **'mini-beta' principle.**

This concept was first tried successfully in the DORIS storage ring, and the luminosity increased by more than an order of magnitude compared to the standard layout. The arrangement used within the ARGUS particle detector is shown in Fig. 7.6. The two 'mini-beta quadrupoles' are integrated into the iron yoke of the detector, reducing their separation from the IP to $a = 1.28$ m. To allow the measurement of particle momenta the detector is equipped with a large solenoid which generates a magnetic field parallel to the beam axis with a strength of $B = 0.8$ T. Of course the quadrupoles must not be exposed to this field, which would completely saturate them and considerably reduce their focusing power. They are therefore surrounded by compensating coils which produce a field exactly equal and opposite to that in the detector, thus keeping them in a region of effectively zero field.

A possibility currently being considered is to use superconducting quadrupoles placed even closer to the IP — separations of $a = 0.5$ to 0.7 m have been discussed — and hence reduce the beta function at the IP to values of $\beta_z^* = 1$ cm and below. It must be borne in mind, however, that according to (7.55) this would result in a clear variation in the beta function over the bunch length of a few centimetres. If the bunch length is only $\sigma_s = 1$ cm, for example, then for $\beta_z^* = 1$ cm the value at the head and tail of the bunch is already $\beta_z(\sigma_s) = 2$ cm. The effective beta function which determines the space charge limitation is therefore considerably larger than β_z^*. It clearly does not make sense to go to even smaller values. A rule of thumb for the smallest reasonable beta function at the IP is

$$\beta_z^* > 1.5 \times \sigma_s. \tag{7.57}$$

Shorter bunches can be achieved by using higher voltages in the accelerating sections, although this of course means higher costs.

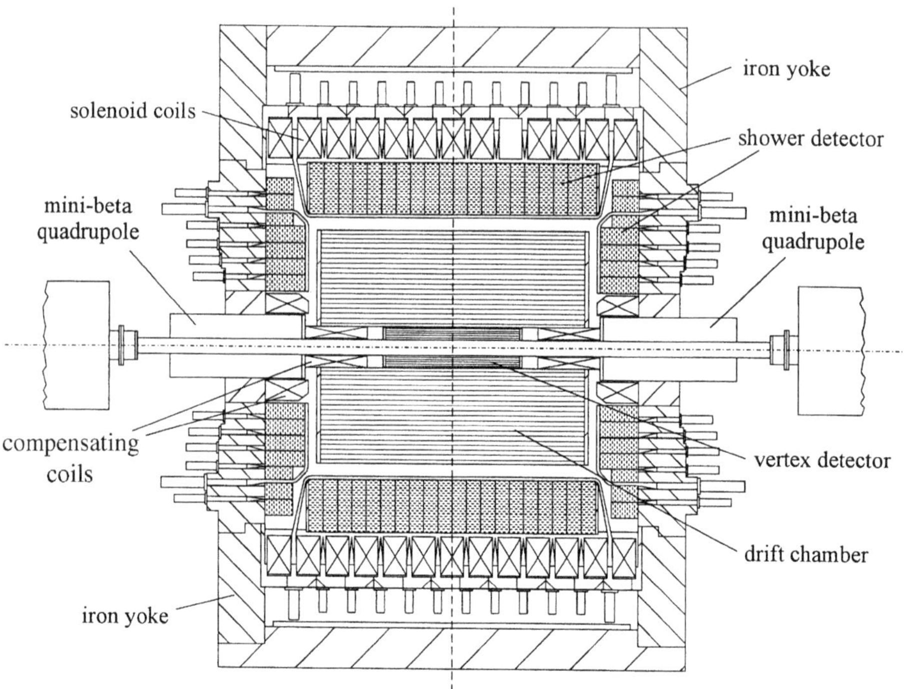

Fig. 7.6 'Mini-beta' arrangement of the DORIS storage ring in the ARGUS particle detector. The two vertically focusing 'mini-beta quadrupoles' are integrated into the yoke of the detector magnet. Special compensating coils maintain a field-free region around the quadrupoles.

8 Wigglers and undulators

It was mentioned in Chapter 2 that particularly intense and tightly collimated synchrotron radiation can be produced by special magnets consisting of a periodic series of short bending magnets (Fig. 2.7). Such magnets are termed 'wigglers' and 'undulators'. In Section 8.2 we will discuss when a magnet of this kind is a wiggler and when it is an undulator, but to begin with we will not make any distinction between the two types and will talk generally about W/U magnets.

8.1 The wiggler or undulator field

The field of a W/U magnet is periodic along the beam axis, with a period length λ_u. The potential may thus be written in the form

$$\varphi(s,z) = f(z)\cos\left(2\pi\frac{s}{\lambda_u}\right) = f(z)\cos(k_u s). \tag{8.1}$$

For simplicity it is assumed here that the magnet extends to infinity in the x direction, i.e. $\varphi(x) = \text{const}$. The unknown function $f(z)$ describes the vertical distribution of the field. Once again, this potential must satisfy the Laplace equation

$$\nabla^2\varphi(s,z) = 0. \tag{8.2}$$

From this it immediately follows that

$$\frac{d^2 f(z)}{dz^2} - f(z)k_u^2 = 0 \tag{8.3}$$

with the solution

$$f(z) = A\sinh(k_u z). \tag{8.4}$$

Inserting this into (8.1), we obtain the potential

$$\varphi(s,z) = A\sinh(k_u z)\cos(k_u s) \tag{8.5}$$

and the vertical field component

$$B_z(s,z) = \frac{\partial\varphi}{\partial z} = k_u A\cosh(k_u z)\cos(k_u s). \tag{8.6}$$

To determine the constant of integration A, let us start with the flux density B_0 in the middle of the poletip (Fig. 8.1). This point has the coordinates $(s,z) = (\lambda_u/4\,,\, g/2)$. It then follows from (8.6) that

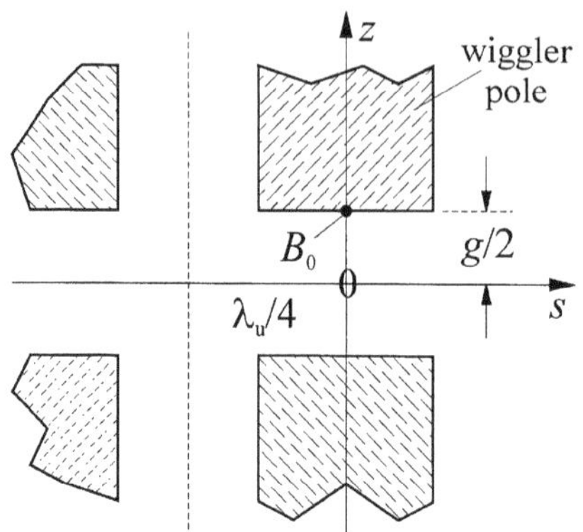

Fig. 8.1 Determination of the constants of integration from the field value B_0 in the middle of the poletip.

$$B_0 = B_z\left(0\,,\,\frac{g}{2}\right) = k_\mathrm{u} A \cosh\left(k_\mathrm{u}\frac{g}{2}\right) = k_\mathrm{u} A \cosh\left(\pi\frac{g}{\lambda_\mathrm{u}}\right) \tag{8.7}$$

and

$$A = \frac{B_0}{k_\mathrm{u}\cosh\left(\pi\dfrac{g}{\lambda_\mathrm{u}}\right)}. \tag{8.8}$$

Inserting this into (8.6) yields

$$B_z(s,z) = \frac{B_0}{\cosh\left(\pi\dfrac{g}{\lambda_\mathrm{u}}\right)}\cosh(k_\mathrm{u} z)\cos(k_\mathrm{u} s) \tag{8.9}$$

and similarly

$$B_s(s,z) = \frac{B_0}{\cosh\left(\pi\dfrac{g}{\lambda_\mathrm{u}}\right)}\sinh(k_\mathrm{u} z)\sin(k_\mathrm{u} s). \tag{8.10}$$

Along the beam axis the periodically varying field thus has the peak value

$$\tilde{B} = \frac{B_0}{\cosh\left(\pi\dfrac{g}{\lambda_\mathrm{u}}\right)}, \tag{8.11}$$

which critically depends on the ratio g/λ_u. This dependence is shown by the curve in Fig. 8.2. If, for a given period length λ_u, the gap height g between the poles of the magnet increases, then the field at the beam falls rapidly. For very short period lengths it is therefore necessary to make the gap between the magnet poles small as well. The minimum allowed size of this gap is of course determined by the transverse size of the beam. In what follows we will only really be interested in the field along the axis ($z \equiv 0$). From (8.9) and (8.11) we can write this field in the simple form

$$B_z(s) = \tilde{B}\cos(k_\mathrm{u} s). \tag{8.12}$$

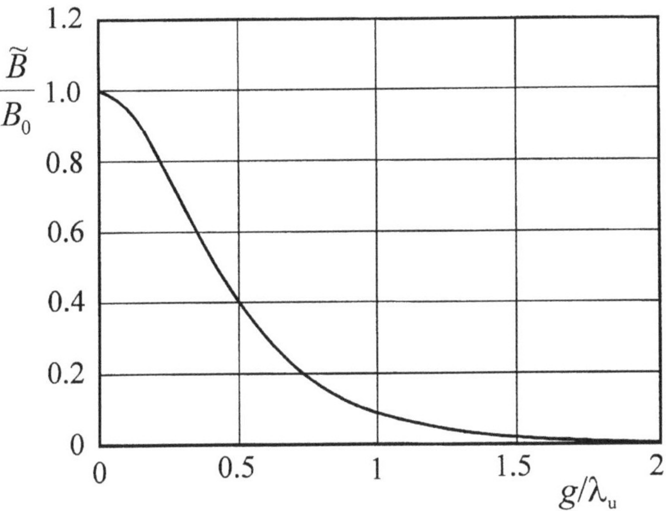

Fig. 8.2 Dependence of the peak field along the beam axis on the ratio g/λ_u.

Most W/U magnets are designed to produce approximately this field shape.[1] For more information about the theory, use, and design of W/U magnets the reader is referred to the review articles by H. Winick *et al.* [67, 68, 69] and G. Brown *et al.* [70].

There are essentially three different designs of W/U magnets: **electromagnets**, **permanent magnets**, and **hybrid magnets**. These are sketched in Fig. 8.3. The simplest type of W/U magnet uses conventional soft iron poles excited by current-carrying coils (Fig. 8.3 a). The magnet strength can be easily varied by changing the applied current, giving maximum fields of up to $\tilde{B} \approx 2$ T. However, this design only allows relatively long periods of at least $\lambda_u \approx 25$ cm. For smaller periods the space available for the coils between the poles becomes ever narrower and the current density required becomes ever higher. The ohmic heating increases rapidly and it becomes difficult to cool the coils.

In order to achieve field values along the beam of significantly more than 2 T, superconducting magnets must be used. Normally high-field magnets of this type only have a few periods and achieve field strengths of up to $\tilde{B} \approx 6$ T along the central plane. This results in an electron trajectory with a very small bending radius R. Equation (2.27) gives the critical wavelength

$$\lambda_c = \frac{4\pi}{3}\frac{R}{\gamma^3} = \frac{4\pi c(m_e c^2)^3}{3eE^2}\frac{1}{\tilde{B}}. \tag{8.13}$$

For a given beam energy it is thus possible to use a high field $\tilde{B}$ to push the critical wavelength of the synchrotron radiation spectrum down to smaller values. Of course, this is especially effective if superconductors are used. For this

[1] Asymmetric W/U magnets with a more complicated field shape are also sometimes used.

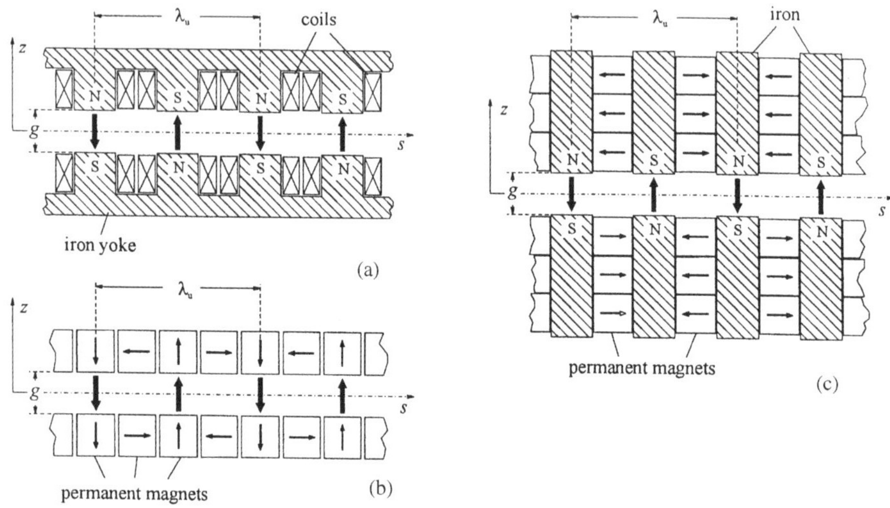

Fig. 8.3 The various designs used to construct W/U magnets. (a) shows a cross-section through a classical electromagnet with conventional iron poles; (b) illustrates the production of a periodic field by an arrangement of alternating permanent magnets and (c) shows a 'hybrid' magnet, a combination of iron poles excited by permanent magnets.

reason, such superconducting W/U magnets are often called **wavelength shifters**. Magnets of this kind are routinely used in electron storage rings [71, 72].

Nowadays W/U magnets with period lengths of only a few cm are used. This is not possible using electromagnets, for the reasons given above. In recent years permanent magnets have instead been employed with great success, with samarium–cobalt ($SmCo_5$) widely used. This has a remnant field of $B_r \approx 0.8 - 1.0$ T. In a magnetic field samarium–cobalt behaves almost as a non-magnetic material with $\mu_r = 1$, and so it may be used inside other magnets without any problems.

W/U magnets are constructed from square-ended blocks of this material in which the field runs parallel to the face of the block. Four such magnet blocks are needed to form each W/U pole (Fig. 8.3 b). The sinusoidal field is produced by arranging them such that the field direction of each individual block, shown by the arrows, is rotated by 90° relative to the previous one. Very long magnets with many periods may be constructed in this way, making use of the fact that the field in the gap remains constant if the dimensions of the magnet are scaled linearly.

Since the pole strength is constant for a permanent magnet of fixed geometry, the field strength of a W/U magnet built in this way can only be altered by varying the size of the gap g, as (8.11) shows. To allow this, the two rows of permanent magnet poles are mounted on supports which may be moved in or out by stepper motors. This mechanism must be very sturdy because the forces acting between the magnet poles are relatively strong, often as great as several

10^4 N. The first samarium–cobalt W/U magnet was built and successfully used at the Lawrence Berkeley Laboratory [73]. Nowadays this type of magnet is commercially available and can be built to the customer's specification.

A very interesting and flexible variant is the **hybrid magnet**, illustrated in Fig. 8.3 (c). Here the poles mostly consist of highly permeable iron, and the permanent magnet blocks are arranged with alternating polarity between them. If several such magnet blocks are arranged on top of one another, the flux from each one accumulates in the iron poles, resulting in considerably higher fields than can be achieved using the permanent magnets alone. If we assume a magnet with poles made of highly-permeable iron, using samarium–cobalt with a remnant field of 0.9 T as a magnetic material, the peak field in a hybrid magnet is given to a good approximation by the expression

$$B_0\ [T] \ = \ 3.33 \exp\left[-\frac{g}{\lambda_\mathrm{u}}\left(5.47 - 1.8\,\frac{g}{\lambda_\mathrm{u}}\right)\right]. \tag{8.14}$$

Field strengths of up to $B_0 > 2$ T can be achieved using this technology.

The W/U magnets are installed in specially reserved straight sections of the storage ring (Chapter 2), often termed **insertions**. If there is a non-zero dispersion in such a section, then according to (6.47) the function $\mathcal{H}$ is also non-zero, as explained in Section 6.4. Switching on the W/U magnet, which is of course a bending magnet, thus has the undesirable effect of increasing the beam emittance. Consequently, the insertions must have zero dispersion.

Furthermore, the W/U magnet must not cause any bending of the beam orbit, even though its field strength can vary widely during machine operation. Otherwise, any variation in the magnetic field would also change the beam position, which is unacceptable for experiments with high position resolution. For this reason, every W/U magnet must satisfy the condition

$$\int\limits_{\mathrm{W/U}} B_z(s)\,ds = \tilde{B}\int\limits_{s_1}^{s_2}\cos(k_\mathrm{u}s)\,ds = 0 \tag{8.15}$$

as strictly as possible. This is the case when

$$s_1 = 0 \qquad \text{and} \qquad s_2 = n\lambda_\mathrm{u} + \frac{\lambda_\mathrm{u}}{2} \tag{8.16}$$

with $n = 1, 2, \ldots$. The particle trajectory is shown in Fig. 8.4. The condition (8.16) is fulfilled by inserting a half-length pole at the beginning and end of the W/U magnet.[2] Since the poles are never all identical, they must be corrected individually with thin iron sheets after the magnet has been completed. This must be done to ensure that the condition (8.15) is satisfied over every individual period as well as for the whole magnet.

[2]It is also possible to use half-strength poles but it is better to adjust the length, especially for permanent magnets.

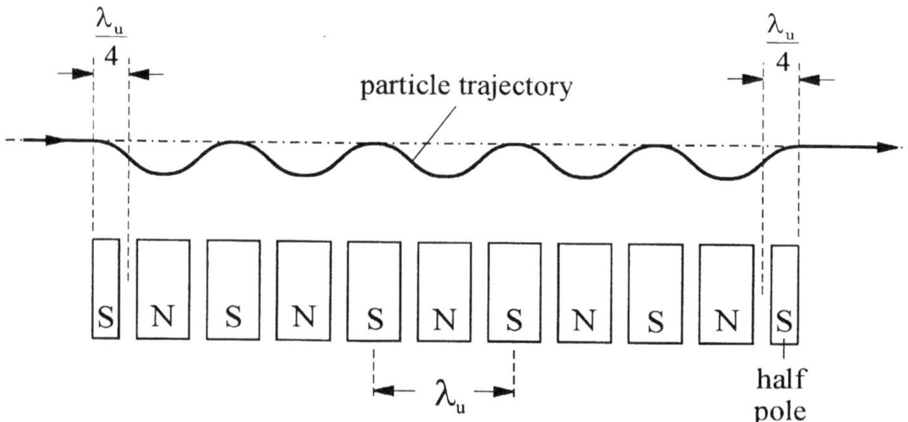

Fig. 8.4 Arrangement of a W/U magnet in a straight section (insertion) of a storage ring. The condition $\int B_z(s)ds = 0$ is achieved by means of the half-poles at the beginning and end of the magnet.

8.2 Equation of motion in a wiggler or undulator

As it moves through the W/U magnet, the electron is acted upon by the Lorentz force

$$\boldsymbol{F} = \dot{\boldsymbol{p}} = m_{\mathrm{e}}\gamma\dot{\boldsymbol{v}} = e\,\boldsymbol{v} \times \boldsymbol{B}. \tag{8.17}$$

Since the magnet is assumed to have infinite horizontal extent, the magnetic field has no B_x component. In addition, we neglect the vertical component of the electron's velocity. Both these approximations are adequately satisfied in most cases, allowing us to write

$$\boldsymbol{B} = \begin{pmatrix} 0 \\ B_z \\ B_s \end{pmatrix} \qquad \text{and} \qquad \boldsymbol{v} = \begin{pmatrix} v_x \\ 0 \\ v_s \end{pmatrix} \tag{8.18}$$

and it follows from (8.17) that

$$\dot{\boldsymbol{v}} = \frac{e}{m_{\mathrm{e}}\gamma} \begin{pmatrix} -v_s B_z \\ -v_x B_s \\ v_x B_z \end{pmatrix}. \tag{8.19}$$

If we again neglect the vertical component of the velocity ($v_z \equiv 0$), then using $\dot{x} = v_x$ and $\dot{s} = v_s$ we obtain the following set of equations describing the coupled motion in the s-x plane:

$$\boxed{\begin{aligned} \ddot{x} &= -\dot{s}\,\frac{e}{m_{\mathrm{e}}\gamma}\,B_z(s) \\ \ddot{s} &= \dot{x}\,\frac{e}{m_{\mathrm{e}}\gamma}\,B_z(s) \end{aligned}}\,. \tag{8.20}$$

The electrons travel along the s-axis through the W/U magnet and encounter the periodic field $B_z(s)$, which induces a horizontal oscillatory motion. The resulting horizontal particle velocity v_x, in combination with the same magnetic field, causes a periodic change in the longitudinal velocity. This is superimposed upon the original particle motion along the s-axis and so has a further effect on the horizontal bending. The problem of solving this coupled set of equations can, however, be considerably simplified, since for relativistic particles the velocity v_s along the s-axis totally dominates. We may start by directly calculating the horizontal motion from this component, assuming v_s to be constant and ignoring the modulation of the longitudinal motion. This modulation may then be determined in a second step by calculating it directly from v_x as a perturbation.

To form a picture of the basic features of particle motion in W/U magnets, let us begin by calculating only the horizontal particle motion. Since to a good approximation $\dot{x} = v_x \ll c$ and $\dot{s} = v_s = \beta c = \text{const}$, we only need to consider the first equation of (8.20) in this case. Using (8.12) it follows that

$$\ddot{x} = -\frac{\beta c e \tilde{B}}{m_e \gamma} \cos(k_u s). \tag{8.21}$$

As in the beam optics calculation, we again replace the time derivative by the spatial derivative with respect to s via $\dot{x} = x' \beta c$ and $\ddot{x} = x'' \beta^2 c^2$, and obtain

$$x'' = -\frac{e\tilde{B}}{m_e \beta c \gamma} \cos(k_u s) = -\frac{e\tilde{B}}{m_e \beta c \gamma} \cos\left(2\pi \frac{s}{\lambda_u}\right). \tag{8.22}$$

This equation may be solved by simple integration, in which we arbitrarily set the constants of integration to zero, since the particular initial conditions do not concern us. It then follows from (8.22) with $\beta = 1$ that

$$\begin{aligned} x'(s) &= \frac{\lambda_u e \tilde{B}}{2\pi m_e \gamma c} \sin(k_u s) \\ x(s) &= \frac{\lambda_u^2 e \tilde{B}}{4\pi^2 m_e \gamma\, c} \cos(k_u s). \end{aligned} \tag{8.23}$$

The maximum angle, which the particle reaches as it crosses the ideal orbit,

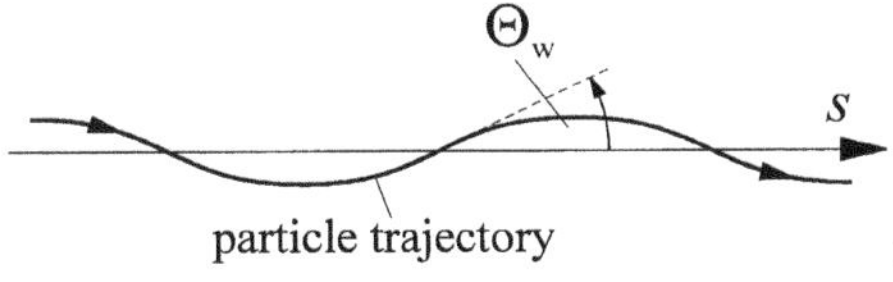

Fig. 8.5 Particle trajectory in a W/U magnet.

is Θ_w (Fig. 8.5). For $s = n\lambda_u$ with $n = 0, 1, 2, \ldots$ it then follows directly from (8.23) that

$$\Theta_w = x'_{\max} = \frac{1}{\gamma} \frac{\lambda_u e \tilde{B}}{2\pi m_e c}. \tag{8.24}$$

The dimensionless quantity

$$\boxed{K = \frac{\lambda_u e\tilde{B}}{2\pi m_e c}} \tag{8.25}$$

is called the **wiggler** or **undulator parameter**. The maximum trajectory angle thus has the form

$$\Theta_w = \frac{K}{\gamma}. \tag{8.26}$$

In a W/U magnet with $K = 1$, the maximum angle the electron trajectory makes with respect to the orbit is $\Theta_w = 1/\gamma$. This exactly corresponds to the natural opening angle of the synchrotron radiation (Chapter 2). Until now we have not drawn any distinction between wigglers and undulators, since they are essentially built in the same way. The difference lies in the bending strength, which we can now quantify using the convenient parameter K. Following the general convention, we use the definition:

$$\text{W/U magnet} = \begin{cases} \textbf{undulator} & \text{if } K \leq 1 \quad \text{i.e.} \quad \Theta_w \leq 1/\gamma \\ \textbf{wiggler} & \text{if } K > 1 \quad \text{i.e.} \quad \Theta_w > 1/\gamma. \end{cases} \tag{8.27}$$

In undulators the bending is very weak, so all the radiation is emitted virtually in parallel, with a very small opening angle. Wigglers, which are considerably stronger, emit a fan of radiation. This fan is of course particularly broad in the extremely strong superconducting wigglers inserted as 'wavelength shifters'. Here the wiggler parameter may be as high as $K \approx 100$ or above.

After this rather general introduction we now return to the set of equations (8.20), from which we wish to calculate the particle motion in a co-moving reference frame. Since the longitudinal velocity has a periodic modulation due to the changes in motion caused by the v_x component, we will choose a fixed average velocity $\bar{v}_s = \langle \dot{s} \rangle$ for the co-moving system. To calculate this we start with (8.23) and (8.25) and write the horizontal particle motion in the form

$$x'(s) = \frac{K}{\gamma}\sin(k_u s) = \Theta_w \sin(k_u s). \tag{8.28}$$

Using $\dot{x} = \beta c x'$ and $s = \beta c t$ gives

$$\dot{x}(t) = \beta c\Theta_w \sin(\omega_u t) = \beta c\frac{K}{\gamma}\sin(\omega_u t) \qquad \text{with} \qquad \omega_u = k_u \beta c. \tag{8.29}$$

For a finite horizontal velocity $\dot{x}$, the longitudinal velocity component $\dot{s}$ is the projection onto the orbit of the arbitrary but constant particle velocity βc, as shown in Fig. 8.6. This may be calculated according to $\dot{s}^2 = (\beta c)^2 - \dot{x}^2$, and

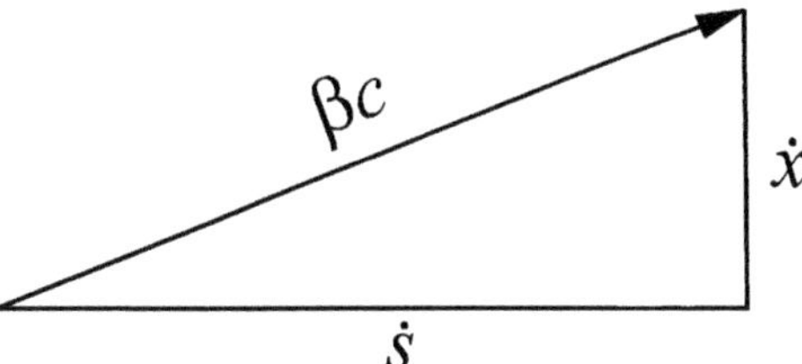

Fig. 8.6 Projection of the particle velocity onto the orbit.

using $\beta^2 = 1 - 1/\gamma^2$ we finally obtain

$$\dot{s}(t) = c\sqrt{1 - \left(\frac{1}{\gamma^2} + \frac{\dot{x}^2}{c^2}\right)}. \tag{8.30}$$

The expression in brackets under the root is very small compared to 1, so to a good approximation we may expand the root in the form

$$\begin{aligned}\dot{s}(t) &= c\left[1 - \frac{1}{2}\left(\frac{1}{\gamma^2} + \frac{\dot{x}^2}{c^2}\right)\right] \\ &= c\left[1 - \frac{1}{2\gamma^2}\left(1 + \frac{\gamma^2}{c^2}\dot{x}^2\right)\right].\end{aligned} \tag{8.31}$$

We now insert the explicit expression for the horizontal velocity from (8.29) and transform using $\sin^2(x) = \frac{1}{2}(1 - \cos 2x)$. The velocity projection along the beam axis thus becomes

$$\dot{s}(t) = c\left\{1 - \frac{1}{2\gamma^2}\left[1 + \frac{\beta^2 K^2}{2}\left(1 - \cos(2\omega_{\mathrm{u}} t)\right)\right]\right\}, \tag{8.32}$$

which consists of the **average velocity**

$$\bar{\dot{s}} = c\left\{1 - \frac{1}{2\gamma^2}\left[1 + \frac{\beta^2 K^2}{2}\right]\right\} \tag{8.33}$$

and a small perturbing oscillation

$$\Delta\dot{s}(t) = \frac{c\beta^2 K^2}{4\gamma^2}\cos(2\omega_{\mathrm{u}} t). \tag{8.34}$$

The average relative velocity

$$\beta^* = \frac{\bar{\dot{s}}}{c} = 1 - \frac{1}{2\gamma^2}\left[1 + \frac{K^2}{2}\right] \tag{8.35}$$

follows from (8.33), where to a good approximation we may set $\beta = 1$. From (8.29) and (8.33) to (8.35) we obtain the components of the particle velocity in the s-x plane

$$\begin{aligned}\dot{x}(t) &= \beta c\frac{K}{\gamma}\sin(\omega_{\mathrm{u}} t) \\ \dot{s}(t) &= \beta^* c + \frac{c\beta^2 K^2}{4\gamma^2}\cos(2\omega_{\mathrm{u}} t).\end{aligned} \tag{8.36}$$

From these expressions the coordinates of the particle in the laboratory frame may be calculated by simple integration. The initial conditions are of no relevance, and may be set to $x(0) = s(0) = 0$. With $\omega_\mathrm{u} = k_\mathrm{u}\beta c$ and $\beta = 1$ we obtain

$$\begin{aligned} x(t) &= -\frac{K}{k_\mathrm{u}\gamma}\cos(\omega_\mathrm{u}t) \\ s(t) &= \beta^* ct + \frac{K^2}{8k_\mathrm{u}\gamma^2}\sin(2\omega_\mathrm{u}t). \end{aligned} \tag{8.37}$$

This motion is most striking when viewed from the electron centre of mass frame K^*, which moves with the average velocity β^* relative to the rest frame. We thus apply the transformation

$$\begin{aligned} x^* &= x \\ s^* &= \gamma\,(s - \beta ct) \end{aligned} \tag{8.38}$$

to (8.37) and obtain the equations for the particle motion viewed in a co-moving reference frame

$$\begin{aligned} x^*(t) &= -\frac{K}{k_\mathrm{u}\gamma}\cos(\omega_\mathrm{u}t) \\ s^*(t) &= \frac{K^2}{8k_\mathrm{u}\gamma^2}\sin(2\omega_\mathrm{u}t). \end{aligned} \tag{8.39}$$

These describe a closed path in the shape of a figure '8', as shown in Fig. 8.7. For very small values of K the extent along the s^*-axis is very small, and essentially the only effect is a simple horizontal oscillation. As K increases the horizontal amplitude increases linearly, but the modulation of the longitudinal motion increases quadratically. Thus the '8' becomes ever broader.

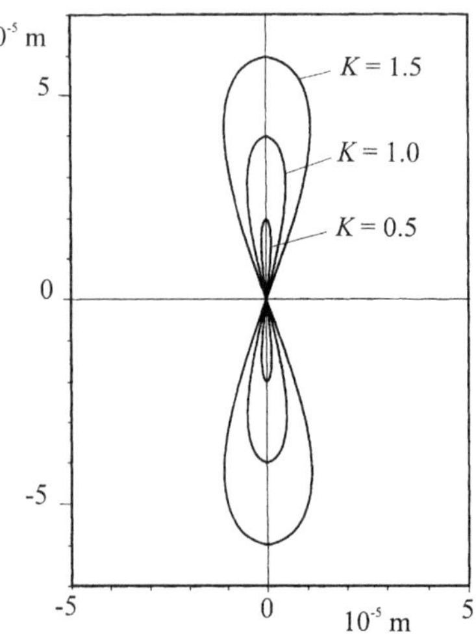

Fig. 8.7 Particle motion in an undulator viewed in a co-moving reference frame.

8.3 Undulator radiation

As they pass through the undulator the electrons perform transverse oscillations as a result of the periodic magnetic field. The oscillations have a well-defined frequency, determined by the period length of the undulator. In the laboratory frame this frequency is given by

$$\Omega_\mathrm{w} = \frac{2\pi}{T} = \frac{2\pi\beta c}{\lambda_\mathrm{u}} = k_\mathrm{u}\beta c. \tag{8.40}$$

For $K \ll 1$ the modulation of the longitudinal particle velocity may be neglected, and we effectively have a pure oscillation in the x direction. In a co-moving frame with the average velocity β^* this frequency transforms according to

$$\omega^* = \gamma^*\,\Omega_\mathrm{w}. \tag{8.41}$$

In this reference frame the electron thus emits **monochromatic radiation** with a frequency ω^*. This monochromatic radiation, which overlays the spontaneous synchrotron radiation which is always present, is typical of undulators. For a sufficiently large number of periods it can be several orders of magnitude more intense than the normal synchrotron radiation. The production of this relatively narrow-band monochromatic radiation at high intensity is one of the special features of undulators. Wigglers with very large K values have almost none of this radiation. This is because the strong bending causes the radiation to be emitted in a broad fan, preventing the coherent superposition of the radiation produced in each individual period.

The undulator radiation can of course only be observed experimentally in the laboratory frame. Hence it is necessary to transform it from the co-moving frame into the laboratory frame, using the relativistic Doppler effect. We consider a photon of momentum p, emitted in the laboratory frame at an angle Θ_0 to the s-axis, as shown in Fig. 8.8. In the laboratory frame the energy and momentum of the photon may be written in the form

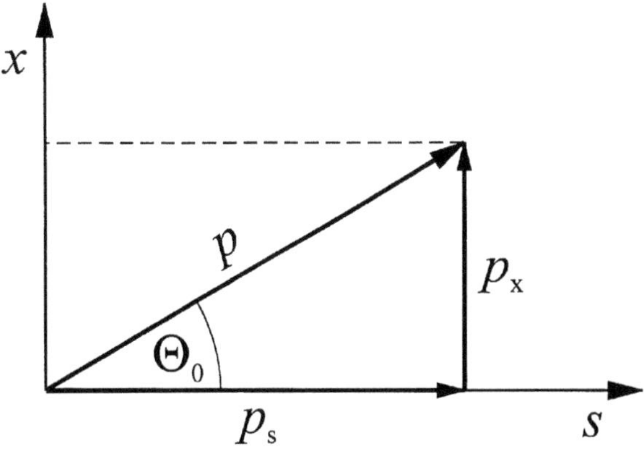

Fig. 8.8 Emission of a photon of momentum p in the laboratory frame.

$$\begin{aligned} E &= \hbar\omega \\ p &= \frac{\hbar\omega}{c}, \end{aligned} \tag{8.42}$$

where ω is the radiation frequency. Using these expressions we can immediately write down the photon four-momentum:

$$P_\mu = \begin{pmatrix} E/c \\ p_x \\ p_z \\ p_s \end{pmatrix} = \begin{pmatrix} E/c \\ p\sin\Theta_0 \\ 0 \\ p\cos\Theta_0 \end{pmatrix}. \tag{8.43}$$

This four-momentum transforms via a Lorentz transformation into a co-moving frame with velocity β^* and energy γ^* according to the relation

$$\begin{pmatrix} E^*/c \\ p_x^* \\ p_z^* \\ p_s^* \end{pmatrix} = \begin{pmatrix} \gamma^* & 0 & 0 & -\beta^*\gamma^* \\ 0 & 1 & 0 & 0 \\ 0 & 0 & 1 & 0 \\ -\beta^*\gamma^* & 0 & 0 & \gamma^* \end{pmatrix} \cdot \begin{pmatrix} E/c \\ p\sin\Theta_0 \\ 0 \\ p\cos\Theta_0 \end{pmatrix}. \tag{8.44}$$

From this we immediately obtain the expression for the transformed photon energy

$$\frac{E^*}{c} = \gamma^*\frac{E}{c} - \beta^*\gamma^* p\cos\Theta_0 = \gamma^*\frac{\hbar\omega_{\mathrm{w}}}{c}\Big(1 - \beta^*\cos\Theta_0\Big). \tag{8.45}$$

With $E^* = \hbar\omega^*$ this becomes

$$\frac{\hbar\omega^*}{c} = \gamma^*\frac{\hbar\omega_{\mathrm{w}}}{c}\Big(1 - \beta^*\cos\Theta_0\Big). \tag{8.46}$$

Solving this equation with respect to ω yields the required expression for the frequency transformation through the relativistic Doppler effect

$$\omega_{\mathrm{w}} = \frac{\omega^*}{\gamma^*(1 - \beta^*\cos\Theta_0)}. \tag{8.47}$$

We now insert the frequency ω^* from (8.41) into this transformation equation and obtain

$$\omega_{\mathrm{w}} = \frac{\Omega_{\mathrm{w}}}{1 - \beta^*\cos\Theta_0} \qquad \Longrightarrow \qquad \frac{\omega_{\mathrm{w}}}{\Omega_{\mathrm{w}}} = \frac{\lambda_{\mathrm{u}}}{\lambda_{\mathrm{w}}} = \frac{1}{1 - \beta^*\cos\Theta_0}, \tag{8.48}$$

where

$$\lambda_{\mathrm{w}} = \lambda_{\mathrm{u}}(1 - \beta^*\cos\Theta_0) \tag{8.49}$$

is the wavelength of the undulator radiation in the laboratory frame. In this expression we now replace β^* by the expression (8.35) and use the approximation $\cos\Theta_0 \approx 1 - \Theta_0^2/2$, which is valid here since $\Theta_0 \approx 1/\gamma \ll 1$. From (8.49) it then follows that

$$\begin{aligned}\lambda_{\mathrm{u}}(1-\beta^*\cos\Theta_0) &= \lambda_{\mathrm{u}}\left[1-\left(1-\frac{1+K^2/2}{2\gamma^2}\right)\left(1-\frac{\Theta_0^2}{2}\right)\right] \\ &= \lambda_{\mathrm{u}}\left[1-\left(1-\frac{\Theta_0^2}{2}-\frac{1+K^2/2}{2\gamma^2}+\cdots\right)\right] \\ &\approx \lambda_{\mathrm{u}}\left(\frac{\Theta_0^2}{2}+\frac{1+K^2/2}{2\gamma^2}\right).\end{aligned} \quad (8.50)$$

This approximation is almost always well satisfied. In conjunction with (8.49) it leads to the very important **coherence condition** for undulator radiation.

$$\boxed{\lambda_{\mathrm{w}} = \frac{\lambda_{\mathrm{u}}}{2\gamma^2}\left(1+\frac{K^2}{2}+\gamma^2\Theta_0^2\right).} \quad (8.51)$$

The wavelength of the radiation is principally determined by the undulator period λ_{u}, the beam energy γ, and the undulator parameter K. The wavelength becomes longer with increasing angle Θ_0, i.e. there is a red-shift behind the undulator as the distance from the beam axis increases.

For many experiments which use undulator radiation, the line width of the undulator spectrum is very important. This is principally determined by the number of undulator periods N_{u}, with the length of a period denoted by λ_{u}. The total length of the undulator is thus $L_{\mathrm{u}} = N_{\mathrm{u}}\,\lambda_{\mathrm{u}}$. The electrons enter the undulator at the point $s_0 - L_{\mathrm{u}}/2$ and continuously emit radiation with a frequency ω_{w} until they leave the undulator again at the point $s_0 + L_{\mathrm{u}}/2$ (Fig. 8.9). The point s_0 marks the centre of the undulator. A wave train of radiation is emitted, which may be described by the time-dependent function

$$u(\omega_{\mathrm{w}}, t) = \begin{cases} a\exp i\omega_{\mathrm{w}} t & \text{if } -\frac{T}{2} \le t \le \frac{T}{2} \\ 0 & \text{otherwise} \end{cases}, \quad (8.52)$$

viewed from a fixed position. The duration of the wave train is given by $T = N_{\mathrm{u}}\,\lambda_{\mathrm{w}}/c$. A finite-length wave such as this does not have a sharply defined

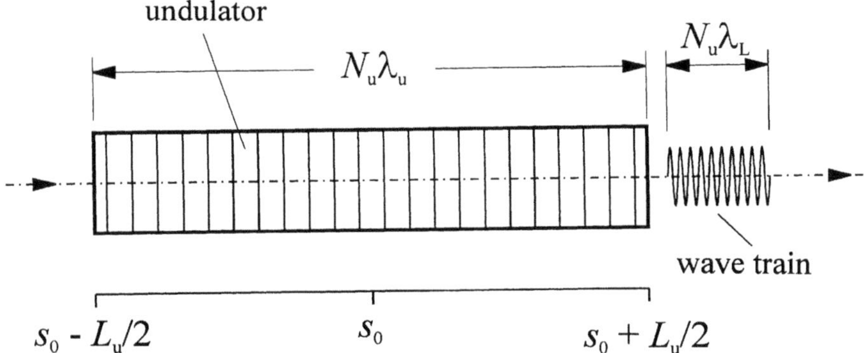

Fig. 8.9 Finite wave-train of coherent radiation from an undulator.

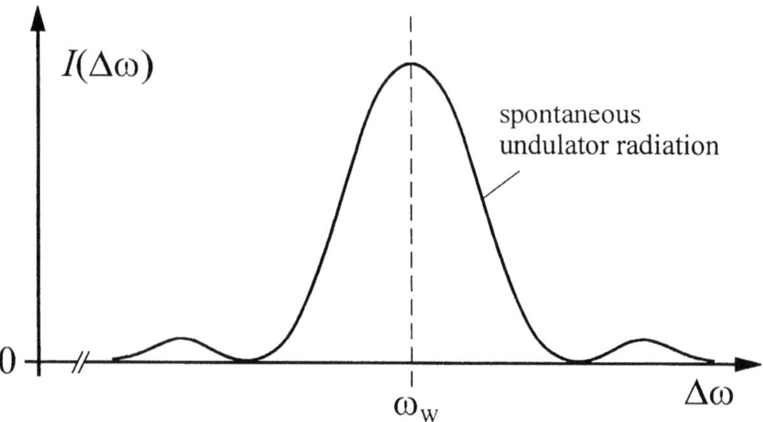

Fig. 8.10 Intensity distribution of the undulator radiation.

frequency, but instead consists of a continuous spectrum of partial waves whose amplitudes are given by the Fourier integral

$$A(\omega) = \frac{1}{\sqrt{2\pi}T} \int_{-\infty}^{+\infty} u(\omega_\mathrm{W}, t) \exp(-i\omega t)\, dt. \tag{8.53}$$

Inserting (8.52) into (8.53) and noting that the wave $u(\omega_\mathrm{W}, t)$ is only non-zero in the interval $[-T/2, +T/2]$, it follows that

$$A(\omega) = \frac{a}{\sqrt{2\pi}T} \int_{-T/2}^{+T/2} \exp\left[-i\,(\omega - \omega_\mathrm{W})\,t\right] dt = \frac{2a}{\sqrt{2\pi}T} \frac{\sin(\omega - \omega_\mathrm{W})\dfrac{T}{2}}{\omega - \omega_\mathrm{W}}. \tag{8.54}$$

Writing $\Delta\omega = \omega - \omega_\mathrm{W}$ with $\omega_\mathrm{W}\, T = 2\pi\, N_\mathrm{u}$ yields the partial wave amplitude in the form

$$A(\omega) = \frac{a}{\sqrt{2\pi}} \frac{\sin\left(\pi N_\mathrm{u} \dfrac{\Delta\omega}{\omega_\mathrm{W}}\right)}{\pi N_\mathrm{u} \dfrac{\Delta\omega}{\omega_\mathrm{W}}}. \tag{8.55}$$

The intensity of the coherent radiation from the undulator is proportional to the square of the partial wave amplitudes. The intensity distribution thus has the form

$$I(\Delta\omega) \;\propto\; \left[\frac{\sin\left(\pi N_\mathrm{u} \dfrac{\Delta\omega}{\omega_\mathrm{W}}\right)}{\pi N_\mathrm{u} \dfrac{\Delta\omega}{\omega_\mathrm{W}}}\right]^2, \tag{8.56}$$

which is illustrated in Fig. 8.10. The full width half maximum of the line can

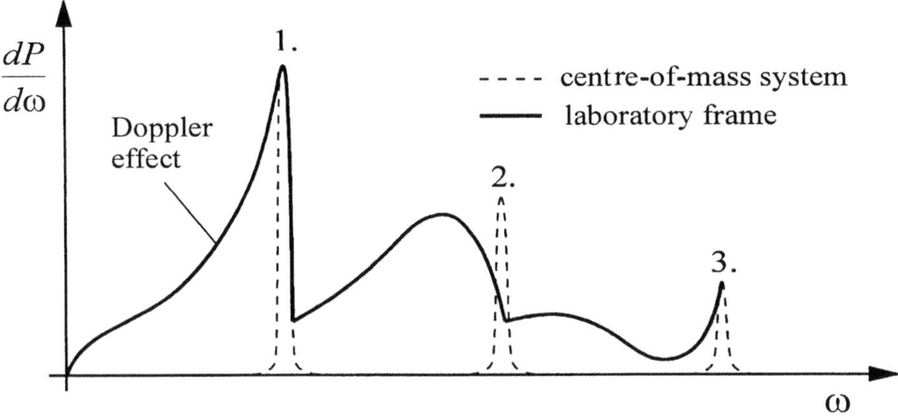

Fig. 8.11 A typical spectrum of coherent radiation emitted from an undulator. In the centre of mass frame only sharp lines are present, while in the laboratory frame the frequencies are broadened by the relativistic Doppler effect.

immediately be calculated from this relation using the condition

$$\left(\frac{\sin x}{x}\right)^2 = \frac{1}{2} \qquad \text{with} \qquad x = \pi N_\mathrm{u} \frac{\Delta\omega}{\omega_\mathrm{w}}. \tag{8.57}$$

With $x = 1.392$ we end up with the simple expression

$$\frac{2\Delta\omega}{\omega_\mathrm{w}} = \frac{2x}{\pi N_\mathrm{u}} = \frac{0.886}{N_\mathrm{u}} \approx \frac{1}{N_\mathrm{u}}. \tag{8.58}$$

The sharpness of the spectral lines radiated from an undulator thus increases linearly with the number of undulator periods N_u.

In reality we are not dealing with exactly sinusoidal particle motion and higher order oscillations are also present. As a result the spectrum also contains higher harmonics whose intensity decreases with increasing order. A typical undulator spectrum is shown in Fig. 8.11. In the centre of mass frame this spectrum consists of sharp lines, which are, however, broadened in the laboratory frame due to the relativistic Doppler effect. From (8.51) we see that this broadening is caused by the photons emitted at angles less than Θ_0. If we also want sharp lines in the laboratory frame, the transverse size of the photon beam must be restricted by the use of collimators. In the spectrum in Fig. 8.11, the broadband spontaneous synchrotron radiation is not shown, as it is generally several orders of magnitude weaker than the undulator radiation. As a result, undulators are nowadays one of the most important sources of coherent radiation in electron storage rings.

9 The free electron laser (FEL)

In a laser, a wave travels through an energy reservoir (which may be a solid, liquid, or gas) and induces stimulated emission, so that when it leaves the reservoir it has the same frequency but has been amplified. External energy is supplied to the medium, either by methods such as optical pumping or by chemical processes (Fig. 9.1). In the **free electron laser** the medium is replaced by a high energy electron beam. In this case the exchange of energy does not occur via quantum transitions of bound electrons but instead by electromagnetic interactions with the beam electrons, which are moving freely in a magnetic field. This is where the **free electron laser (FEL)** gets its name. Because there is no dependence on atomic energy levels, the FEL can in principle operate over a very broad range of wavelengths. There are two types of FEL: the **Compton** and the **Raman** FEL. In the Raman FEL the interaction between the electrons dominates, which is the case at low beam energies. In relativistic particle beams this interaction becomes negligible, and we instead have a Compton FEL. Since accelerators generally operate at relatively high particle energies, we will only discuss the Compton FEL.

There are three different ways to consider the FEL principle. The first involves the construction of the Hamiltonian in the moving centre of mass frame, and was first developed by A. Renieri *et al.* [74, 75, 76, 77]. The second approach is based on an analysis of the Lorentz force in the laboratory frame and leads to the pendulum equation. The foundation of this method was the work of W.B. Colson in particular [78, 79, 80]. This approach is very clear and pedagogical, and we will follow it in this chapter. C. Pellegrini and S. Krinsky have pursued a third approach, that of a coupled system of Vlasov and Maxwell equations, concentrating particularly on the high-gain region. [81, 82].

In discussing the FEL principle it is important to determine whether the gain G of the electromagnetic field is only a few percent per crossing (low-gain regime) or much larger ($G \gg 1$, i.e. the high-gain regime). In the first case the laser field may be regarded as essentially constant, which considerably simplifies the calculation of the gain. In the second case the rapid and generally non-linear growth of the field within the FEL must be taken into account, which can usually only be done using numerical methods.

The possibility of using an FEL as a particle accelerator was first suggested by R.B. Palmer [83]. In 1976 L.R. Elias *et al.* in Stanford performed the first suc-

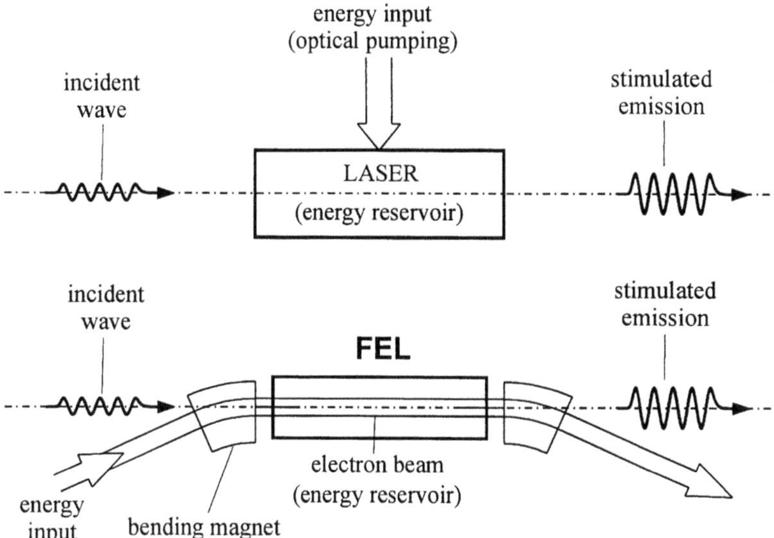

Fig. 9.1 Comparison of the classical laser (upper picture) with the FEL. The medium used as an energy reservoir in the laser is replaced by the beam of free electrons in the FEL.

cessful experiments with stimulated emission[84]. A year later the FEL principle was first extended experimentally to oscillator operation by D.A.G. Deacon *et al.* [85] Further studies were mainly performed at linear accelerators with relatively low beam energies. In storage rings only the special form of the optical klystron has so far been used. Further detailed reviews of the theory and development of the free electron laser may be found in [76, 92, 93, 94, 95].

9.1 Conditions for energy transfer in the FEL

In a free electron laser the electron and laser beam must travel along the same axis in order to maximize the length of the interaction region and so allow sufficient exchange of energy per beam crossing. The electric field $\boldsymbol{E}_\mathrm{L}$ and the velocity $\boldsymbol{v}$ of the electrons are then perpendicular to one another. The energy gained in this case is

$$\Delta W = -e\int \boldsymbol{E}_\mathrm{L}\cdot d\boldsymbol{s} = -e\int \boldsymbol{v}\cdot\boldsymbol{E}_\mathrm{L}\,dt = 0 \qquad (\boldsymbol{v}\perp\boldsymbol{E}_\mathrm{L}). \tag{9.1}$$

Direct exchange of energy between the electron and laser beams is thus not possible. An undulator is therefore inserted, which gives the beam a horizontal velocity component according to (8.29)

$$v_x = \dot{x} = \beta c\frac{K}{\gamma}\sin(\omega_\mathrm{u}t) \qquad \text{with} \qquad \omega_\mathrm{u} = k_\mathrm{u}\beta c. \tag{9.2}$$

The electric field of the FEL beam can couple to this horizontal component. A positive energy yield is, however, only possible if the phase between the undulator

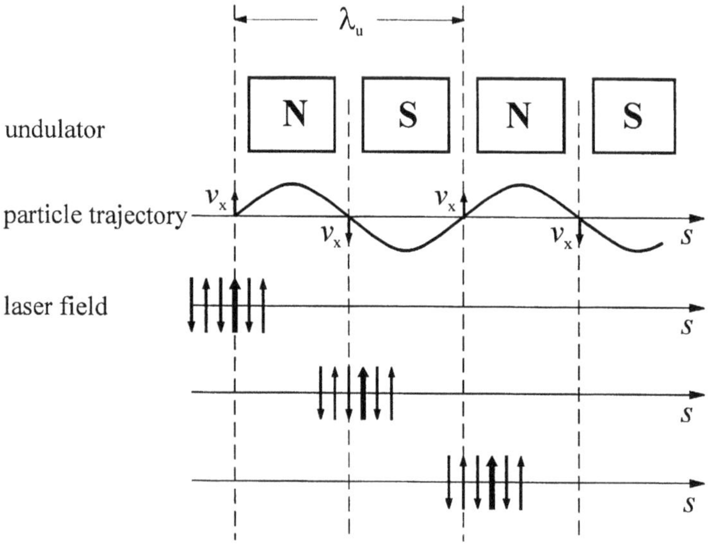

Fig. 9.2 Phase condition between the laser field and the electron beam oscillating in an undulator. Since the laser field has a higher velocity than the electrons there is a phase shift, shown for three positions of the laser field. The broad arrow always indicates the same phase.

period and the laser field satisfies certain conditions. We will introduce these with the help of Fig. 9.2, considering just a single electron. We assume that at a particular point in time the electron crosses the orbit and has a transverse velocity v_x. If this time and position coincides with a maximum of the electric field E_L of the laser wave travelling alongside the beam, and this field acts in the same direction as v_x, then the negatively charged electron will be decelerated by this field and hence energy will be transferred to the laser field.

After half an undulator period the electron again crosses the orbit and this time has the velocity component $-v_x$. Since the electron always moves more slowly than the electromagnetic field of the laser, and also since the bending in the undulator forces the electron to travel further, the laser field moves ahead of it. If the system is set up correctly this phase shift will be exactly $\Delta\Psi = \pi$, and the electric field acting on the electron will now have the value $-E_L$. In this case the electron again loses energy, which is transferred to the laser field. This process is repeated every time the electron crosses the orbit, until it reaches the end of the undulator. The result is thus an overall gain in energy by the laser field.

Having looked qualitatively at the energy transfer in the FEL, let us now consider the phase condition in quantitative detail. If $\boldsymbol{E}_L$ is the electric field of the laser wave, then the change in energy of the electrons travelling along with it is given by the general relation

$$\Delta W = -e \int \boldsymbol{E}_{\mathrm{L}} \cdot d\boldsymbol{s} = -e \int \boldsymbol{v} \cdot \boldsymbol{E}_{\mathrm{L}} \, dt. \tag{9.3}$$

To a good approximation, the x-component of the laser field may be regarded here as a plane wave of the form

$$E_{\mathrm{L},x} = E_{\mathrm{L},0} \cos(k_{\mathrm{L}} s - \omega_{\mathrm{L}} t + \varphi_0), \tag{9.4}$$

where $k_{\mathrm{L}} = 2\pi/\lambda_{\mathrm{L}}$ is the wavenumber, ω_{L} the frequency of the laser wave and φ_0 is an arbitrary initial phase. The horizontal component of the electron velocity

$$v_x = c\frac{K}{\gamma} \sin(k_{\mathrm{u}} s) \tag{9.5}$$

may be obtained for $\beta = 1$ directly from (8.29) for the undulator parameter K and the undulator wavenumber k_{u}. Inserting (9.4) and (9.5) into (9.3), the change in the electron energy is given by

$$\begin{aligned} \Delta W &= -\frac{ceE_{\mathrm{L},0}K}{\gamma} \int \cos(k_{\mathrm{L}} s - \omega_{\mathrm{L}} t + \varphi_0) \sin(k_{\mathrm{u}} s) \, dt \\ &= -\frac{ceE_{\mathrm{L},0}K}{2\gamma} \int \Big\{ \sin\Big[(k_{\mathrm{L}} + k_{\mathrm{u}}) s - \omega_{\mathrm{L}} t + \varphi_0\Big] \\ &\qquad - \sin\Big[(k_{\mathrm{L}} - k_{\mathrm{u}}) s - \omega_{\mathrm{L}} t + \varphi_0\Big] \Big\} \, dt. \end{aligned} \tag{9.6}$$

On average, it is only possible to transfer energy between the electron and the laser field if the phase

$$\Psi_{\pm} = (k_{\mathrm{L}} \pm k_{\mathrm{u}})\bar{s} - \omega_{\mathrm{L}} t + \varphi_0 \tag{9.7}$$

between the electron and the electromagnetic wave of the FEL only varies very slowly with time and remains virtually constant within the undulator. Thus the time derivative

$$\frac{d\Psi_{\pm}}{dt} = (k_{\mathrm{L}} \pm k_{\mathrm{u}})\dot{\bar{s}} - \omega_{\mathrm{L}} \approx 0 \tag{9.8}$$

must vanish. Here $\dot{\bar{s}}$ is the average velocity of the electron along the s-axis. With $\omega_{\mathrm{L}} = k_{\mathrm{L}} c$ and $\dot{\bar{s}} = \beta^* c$ it follows that

$$\frac{1}{c}\frac{d\Psi_{\pm}}{dt} = (k_{\mathrm{L}} \pm k_{\mathrm{u}})\, \beta^* - k_{\mathrm{l}} = 0. \tag{9.9}$$

Inserting the average relative velocity from (8.35) and realizing that $k_\mathrm{u} \ll k_\mathrm{L}$, (9.9) yields

$$\begin{aligned} 0 &= (k_\mathrm{L} \pm k_\mathrm{u}) \left[1 - \frac{1}{2\gamma^2}\left(1 + \frac{K^2}{2}\right)\right] - k_\mathrm{L} \\ &\approx -\frac{k_\mathrm{L}}{2\gamma^2}\left(1 + \frac{K^2}{2}\right) \pm k_\mathrm{u}. \end{aligned} \tag{9.10}$$

It is clear that only the solution with the positive sign can exist, namely

$$k_\mathrm{u} = \frac{k_\mathrm{L}}{2\gamma^2}\left(1 + \frac{K^2}{2}\right). \tag{9.11}$$

If we now replace k_L and k_u by the corresponding wavelengths, we finally obtain the coherence condition familiar from (8.51)

$$\lambda_\mathrm{L} = \frac{\lambda_\mathrm{u}}{2\gamma^2}\left(1 + \frac{K^2}{2}\right) \tag{9.12}$$

for a radiation angle $\Theta_0 = 0$. In the energy equation (9.6) only the phase

$$\Psi_+ = (k_\mathrm{L} + k_\mathrm{u})\bar{s} - \omega_\mathrm{L} t + \varphi_0 \approx \mathrm{const} \tag{9.13}$$

contributes to the energy exchange, while the term with the phase

$$\Psi_- = (k_\mathrm{L} - k_\mathrm{u})s - \omega_\mathrm{L} t + \varphi_0 \tag{9.14}$$

oscillates rapidly and so on average has no effect on the total transferred energy. Θ_0 does not appear at all in formula (9.12) because for simplicity we have only considered the problem in one dimension. Transverse effects, such as the finite sizes of the electron and laser beams and overlapping of the two beams, are neglected here. However, it turns out that the most important features of the FEL are very well described by this simplified treatment.

9.2 Equation of motion for electrons in the FEL (pendulum equation)

Electrons of energy $\gamma = E/m_\mathrm{e}c^2$ interact with the electromagnetic field of the FEL and undergo a change in energy $\Delta\gamma$ and a phase shift $\Delta\Psi$ relative to the laser field. The functions $\Delta\gamma(s)$ and $\Delta\Psi(s)$ within the undulator give a complete description of the particle motion through the FEL. The relative change in energy per path element $ds = c\,dt$ may be written as

$$\frac{d\gamma}{ds} = \frac{dW}{c\,dt}\frac{1}{m_\mathrm{e}c^2}. \tag{9.15}$$

Inserting the differential energy dW from (9.6) gives

$$\frac{d\gamma}{ds} = -\frac{eE_{\mathrm{L},0}K}{2\gamma m_{\mathrm{e}}c^2}\left\{\sin\left[(k_{\mathrm{L}}+k_{\mathrm{u}})\,s-\omega_{\mathrm{L}}t+\varphi_0\right] - \sin\left[(k_{\mathrm{L}}-k_{\mathrm{u}})\,s-\omega_{\mathrm{L}}t+\varphi_0\right]\right\}. \tag{9.16}$$

Rearranging this expression using the relation

$$\sin x = -\Re\Big(i\,\exp(i\,x)\Big)$$

we obtain

$$\frac{d\gamma}{ds} = \frac{eE_{\mathrm{L},0}K}{2\gamma m_{\mathrm{e}}c^2}\Re\left(i\,\exp\left\{i\Big[(k_{\mathrm{L}}+k_{\mathrm{u}})\,s(t)-\omega_{\mathrm{L}}+\varphi_0\Big]\right\} - i\,\exp\left\{i\Big[(k_{\mathrm{L}}-k_{\mathrm{u}})\,s(t)-\omega_{\mathrm{L}}+\varphi_0\Big]\right\}\right). \tag{9.17}$$

Using (8.32) to (8.34), the position variable $s(t)$ may be calculated from

$$\dot{s}(t) = \beta^* c + \frac{cK^2}{4\gamma^2}\cos(2\omega_{\mathrm{u}}t). \tag{9.18}$$

Simple integration yields

$$s(t) = \beta^* ct + \frac{cK^2}{8\gamma^2\omega_{\mathrm{u}}}\sin(2\omega_{\mathrm{u}}t), \tag{9.19}$$

where we again set $\beta = 1$ because the velocities involved are highly relativistic. In what follows we also insert the average velocity $\bar{s} = \beta^* ct$. Combining (9.19) and (9.17), and noting that $k_{\mathrm{L}} \gg k_{\mathrm{u}}$, we may write

$$\frac{d\gamma}{ds} = \frac{eE_{\mathrm{L},0}K}{2\gamma m_{\mathrm{e}}c^2}\Re\left(i\,\exp\left[i\,\frac{ck_{\mathrm{L}}K^2}{8\gamma^2\omega_{\mathrm{u}}}\sin(2k_{\mathrm{u}}\bar{s})\right] \times\left\{\exp\left\{i\big[(k_{\mathrm{L}}+k_{\mathrm{u}})\bar{s}-\omega_{\mathrm{L}}+\varphi_0\big]\right\} - \exp\left\{i\left[(k_{\mathrm{L}}-k_{\mathrm{u}})\bar{s}-\omega_{\mathrm{L}}+\varphi_0\right]\right\}\right\}\right). \tag{9.20}$$

Here we have used $2k_{\mathrm{u}}\bar{s} = 2\omega_{\mathrm{u}}t$ in the argument of the sine function. To rearrange this expression we use the identity from Abramowitz-Stegun [96]

$$\exp\Big(i\,x\,\sin\Phi\Big) = \sum_{n=-\infty}^{+\infty} J_n(x)\,\exp\Big(i\,n\Phi\Big), \tag{9.21}$$

where $J_n(x)$ is the nth-order Bessel function. We obtain

$$\frac{d\gamma}{ds} = \frac{eE_{\mathrm{L},0}K}{2\gamma m_{\mathrm{e}}c^2}\Re\left(i\sum_{n=-\infty}^{+\infty} J_n\left(\frac{ck_{\mathrm{L}}K^2}{8\gamma^2\omega_{\mathrm{u}}}\right)\exp(i\,2nk_{\mathrm{u}}\bar{s})\right.$$
$$\left.\times\left\{\exp\left\{i\left[(k_{\mathrm{L}}+k_{\mathrm{u}})\bar{s}-\omega_{\mathrm{L}}t+\varphi_0\right]\right\}-\exp\left\{i\left[(k_{\mathrm{L}}-k_{\mathrm{u}})\bar{s}-\omega_{\mathrm{L}}t+\varphi_0\right]\right\}\right\}\right). \tag{9.22}$$

We now collect together the individual terms in this expression with the same wavenumber nk_{u}, replacing the index n by another integer index N. The energy change per wavelength is now

$$\begin{aligned}\frac{d\gamma}{ds} &= \frac{eE_{\mathrm{L},0}K}{2\gamma m_{\mathrm{e}}c^2}\sum_{N=-\infty}^{+\infty}\left[J_{\frac{N-1}{2}}(N\eta)-J_{\frac{N+1}{2}}(N\eta)\right] \\ &\qquad\times\Re\left\{i\exp\left\{i\left[(k_{\mathrm{L}}+Nk_{\mathrm{u}})\bar{s}-\omega_{\mathrm{L}}t+\varphi_0\right]\right\}\right\} \\ &= -\frac{eE_{\mathrm{L},0}K}{2\gamma m_{\mathrm{e}}c^2}\sum_{N=-\infty}^{+\infty}\left[J_{\frac{N-1}{2}}(N\eta)-J_{\frac{N+1}{2}}(N\eta)\right] \\ &\qquad\times\sin\left[(k_{\mathrm{L}}+Nk_{\mathrm{u}})\bar{s}-\omega_{\mathrm{L}}t+\varphi_0\right]\end{aligned} \tag{9.23}$$

with

$$\eta = \frac{ck_{\mathrm{L}}K^2}{8N\gamma^2\omega_{\mathrm{u}}} = \frac{k_{\mathrm{L}}K^2}{8N\gamma^2 k_{\mathrm{u}}}. \tag{9.24}$$

By analogy with the undulator parameter K we introduce the dimensionless parameter

$$K_{\mathrm{L}} = \frac{eE_{\mathrm{L},0}}{k_{\mathrm{u}}m_{\mathrm{e}}c^2}, \tag{9.25}$$

which describes the effect of the laser field and leads to the expression

$$\begin{aligned}\frac{d\gamma}{ds} &= -\frac{k_{\mathrm{u}}K_{\mathrm{L}}K}{2\gamma}\sum_{N=-\infty}^{+\infty}\left[J_{\frac{N-1}{2}}(N\eta)-J_{\frac{N+1}{2}}(N\eta)\right] \\ &\qquad\times\sin\left[(k_{\mathrm{L}}+Nk_{\mathrm{u}})\bar{s}-\omega_{\mathrm{L}}t+\varphi_0\right].\end{aligned} \tag{9.26}$$

In order to achieve a non-zero overall energy transfer over the path through the undulator, the phase

$$\Psi = (k_{\mathrm{L}}+Nk_{\mathrm{u}})\bar{s}-\omega_{\mathrm{L}}t+\varphi_0 \tag{9.27}$$

between the Nth harmonic of the electron oscillation and the laser wave must be effectively constant over time. This was shown above for the first harmonic ($N=1$), and we now extend this to all harmonics. We thus have the condition

$$\frac{d\Psi}{dt} = (k_{\mathrm{L}}+Nk_{\mathrm{u}})\dot{\bar{s}}-\omega_{\mathrm{L}} = (k_{\mathrm{L}}+Nk_{\mathrm{u}})\beta^* c-\omega_{\mathrm{L}} = 0, \tag{9.28}$$

into which we insert (8.35). Since $k_{\mathrm{L}} \gg Nk_{\mathrm{u}}$, we can reduce this expression to the form

$$\frac{d\Psi}{dt} = -\frac{ck_\mathrm{L}}{2\gamma^2}\left(1+\frac{K^2}{2}\right) + cNk_\mathrm{u} = 0 \tag{9.29}$$

in the same way as in the derivation of (9.12). It immediately follows that

$$k_\mathrm{u} = \frac{k_\mathrm{L}}{2N\gamma^2}\left(1+\frac{K^2}{2}\right), \tag{9.30}$$

or if we again replace the wavenumbers by the corresponding wavelengths

$$\lambda_\mathrm{L} = \frac{\lambda_\mathrm{u}}{2N\gamma^2}\left(1+\frac{K^2}{2}\right). \tag{9.31}$$

The is the generalized coherence condition for any harmonic N. Wavelengths are always positive definite quantities, and we see immediately from equation (9.31) that the case $N \leq 0$ is not physically meaningful. Moreover, in most cases only one particular harmonic is considered in an FEL, rather than the sum of all possible harmonics. Thus from (9.26) we finally obtain the change in energy per path element for the Nth harmonic

$$\boxed{\begin{aligned}\left(\frac{d\gamma}{ds}\right)_N = & -\frac{k_\mathrm{u}K_\mathrm{L}K}{2\gamma}\left[J_{\frac{N-1}{2}}(N\eta) - J_{\frac{N+1}{2}}(N\eta)\right] \\ & \times \sin\left[(k_\mathrm{L}+Nk_\mathrm{u})\bar{s} - \omega_\mathrm{L}t + \varphi_0\right] \\ & \text{with} \quad N = 1,\ 3,\ 5,\ \ldots\end{aligned}} \tag{9.32}$$

The change in energy $d\gamma/ds$ is proportional to the field parameter K_L and so from (9.25) is also proportional to the laser field strength $E_{\mathrm{L},0}$. Clearly the exchange of energy can only occur when a laser field is present. This is precisely the same as the key process of **stimulated emission** in the laser.

In order to describe the complete motion of the electrons in the FEL, we also need to know the change in the phase per path element. According to (9.26), the phase between the Nth harmonic of the electron oscillation and the laser wave is

$$\Psi = (k_\mathrm{L} + Nk_\mathrm{u})\,\bar{s} - \omega_\mathrm{L}t + \varphi_0. \tag{9.33}$$

The time derivative of this phase was already calculated in (9.29), from which we immediately obtain

$$\frac{d\Psi}{ds} = \frac{1}{c}\frac{d\Psi}{dt} = Nk_\mathrm{u} - \frac{k_\mathrm{L}}{2\gamma^2}\left(1+\frac{K^2}{2}\right). \tag{9.34}$$

The **resonance energy** is defined as the energy at which there is no phase shift relative to the laser field, i.e. $d\Psi/ds = 0$. Using this condition we may immediately calculate the resonance energy from (9.34), giving

$$\gamma_\mathrm{r}^2 = \frac{k_\mathrm{L}}{2Nk_\mathrm{u}}\left(1+\frac{K^2}{2}\right). \tag{9.35}$$

Inserting this expression for the resonance energy back into (9.34) yields the change in phase per path element

$$\frac{d\Psi}{ds} = N k_{\mathrm{u}} \left(1 - \frac{\gamma_{\mathrm{r}}^2}{\gamma^2}\right). \tag{9.36}$$

The FEL is always operated at electron energies very close to the resonance energy. It is thus useful to use the difference $\Delta\gamma$ from the resonance energy as an energy variable with

$$\gamma = \gamma_{\mathrm{r}} + \Delta\gamma \qquad \text{and} \qquad \Delta\gamma \ll \gamma_{\mathrm{r}}. \tag{9.37}$$

Under this condition we have

$$1 - \frac{\gamma_{\mathrm{r}}^2}{\gamma^2} = \frac{(\gamma_{\mathrm{r}} + \Delta\gamma)^2 - \gamma_{\mathrm{r}}^2}{\gamma_{\mathrm{r}}^2} \approx 2\frac{\Delta\gamma}{\gamma_{\mathrm{r}}} \tag{9.38}$$

and from (9.36) we finally obtain

$$\boxed{\frac{d\Psi(s)}{ds} = 2\frac{N k_{\mathrm{u}}}{\gamma_{\mathrm{r}}} \Delta\gamma(s).} \tag{9.39}$$

It is also useful to relate the total energy transfer to the resonance energy and to consider only the energy difference $\Delta\gamma$ from (9.37). It then follows from the energy relation (9.32) and the phase function (9.33) that

$$\frac{d\Delta\gamma}{ds} = \frac{d\gamma}{ds} - \frac{d\gamma_{\mathrm{r}}}{ds} = -\frac{k_{\mathrm{u}} K_{\mathrm{L}} K}{2\gamma_{\mathrm{r}}} \sqrt{F(N\eta)}\, \sin\Psi(s) \tag{9.40}$$

with

$$F(N\eta) = \left[J_{\frac{N-1}{2}}(N\eta) - J_{\frac{N+1}{2}}(N\eta)\right]^2. \tag{9.41}$$

In a uniform undulator the resonance energy γ_{r} is constant and so the derivative $d\gamma_{\mathrm{r}}/ds$ vanishes. The change in energy per path element is then

$$\boxed{\frac{d\Delta\gamma(s)}{ds} = -\frac{k_{\mathrm{u}} K_{\mathrm{L}} K}{2\gamma_{\mathrm{r}}} \sqrt{F(N\eta)}\, \sin\Psi(s).} \tag{9.42}$$

We now differentiate (9.39) once again with respect to s and insert it into (9.42), giving

$$\frac{d^2\Psi(s)}{ds^2} = 2\frac{N k_{\mathrm{u}}}{\gamma_{\mathrm{r}}} \frac{d\Delta\gamma(s)}{ds} = -\frac{N k_{\mathrm{u}}^2 K_{\mathrm{L}} K}{\gamma_{\mathrm{r}}^2} \sqrt{F(N\eta)}\, \sin\Psi(s). \tag{9.43}$$

This relation leads directly to the general **equation of a pendulum**

$$\boxed{\Psi''(s) + \Omega_{\mathrm{L}}^2 \sin\Psi(s) = 0} \tag{9.44}$$

with a frequency

$$\boxed{\Omega_{\mathrm{L}}^2 = \frac{N k_{\mathrm{u}}^2 K_{\mathrm{L}} K}{\gamma_{\mathrm{r}}^2} \sqrt{F(N\eta)}.} \tag{9.45}$$

The length of an oscillation along the s-axis is $L_{\mathrm{L}} = 2\pi/\Omega_{\mathrm{L}}$ and $f_{\mathrm{L}} = c/L_{\mathrm{L}}$ is the frequency with which the electron oscillates in the potential of the laser field.

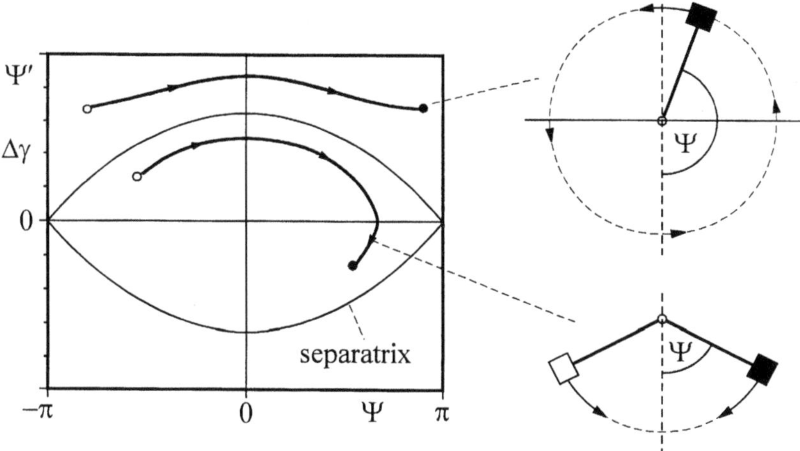

Fig. 9.3 Electron motion in the FEL field. The phase shift relative to the laser wave is plotted along the horizontal axis, and the energy difference from the resonance energy is plotted vertically. Two different types of electron motion are plotted and may be visualized in terms of the corresponding pendulum-like motion.

This corresponds to the synchrotron frequency in circular accelerators. Since $\Delta\gamma \propto \Psi'$ according to (9.39), the functions $\Delta\gamma(s)$ and $\Delta\Psi(s)$ are orthogonal. Hence we can depict the electron motion in a Ψ-$\Delta\gamma$ diagram, as shown in Fig. 9.3. At small amplitudes and with strong laser fields the electron performs stable oscillations, which to a good approximation would be represented by an ellipse in this diagram. As the amplitude increases the frequency decreases and reaches zero in the limiting case $\Psi_{\text{max}} \to \pi$ for $\Delta\gamma = 0$. This represents the limit between stable particle oscillation and the unstable region in which the particles can no longer be contained within the potential of the laser field. This limit corresponds to the separatrix for synchrotron oscillations.

9.3 Amplification of the FEL (low gain approximation)

Let the energy gain in the laser field be

$$\Delta W_{\text{L}} = -m_{\text{e}}c^2\Delta\gamma. \tag{9.46}$$

The minus sign comes from the fact that the energy loss $\Delta\gamma$ from the electrons leads to a corresponding gain in energy in the laser field. The energy stored in the laser field is

$$W_{\text{L}} = \frac{\varepsilon_0}{2}\, E_{\text{L},0}^2\, V, \tag{9.47}$$

where V is the volume occupied by the laser field. The amplification of the FEL due to a single electron is defined as

$$G_1 = \frac{\Delta W_{\text{L}}}{W_{\text{L}}} = -\frac{2m_{\text{e}}c^2}{\varepsilon_0 E_{\text{L},0}^2 V}\,\Delta\gamma. \tag{9.48}$$

We now replace $\Delta\gamma$ by the expression (9.39) and obtain

$$G_1 = -\frac{m_e c^2 \gamma_r}{\varepsilon_0 E_{L,0} V N k_u} \Delta\Psi'. \tag{9.49}$$

Here

$$\Delta\Psi' = \Psi'_{\text{end}} - \Psi'_{\text{start}} \tag{9.50}$$

is the difference in Ψ' that the electron undergoes from the beginning of the undulator (Ψ'_{start}) to the end (Ψ'_{end}). Using the definitions of the frequency Ω_L in (9.45) and the field parameter K_L in (9.25) we can rewrite (9.49) in the form

$$G_1 = -\frac{e^2 N k_u K^2}{\varepsilon_0 V m_e c^2 \gamma_r^3} F(N\eta) \frac{\Delta\Psi'}{\Omega_L^4}. \tag{9.51}$$

Virtually all the electrons in a bunch contribute to the laser amplification, and we will assume that they are homogeneously distributed within the bunch. We thus need to sum the amplification induced by each electron and average over all initial phases of the electrons relative to the laser wave as they enter the undulator. This gives the total amplification of the FEL

$$G = -\frac{e^2 N k_u K^2}{\varepsilon_0 m_e c^2} \frac{n_b}{\gamma_r^3} F(N\eta) \frac{\langle\Delta\Psi'\rangle}{\Omega_L^4} \tag{9.52}$$

for an electron density $n_b = n/V$ in a bunch and with

$$\langle\Delta\Psi'\rangle = \frac{1}{n}\sum_{i=1}^{n} \Delta\Psi'_i. \tag{9.53}$$

To average over the phase $\Psi(s)$ through the undulator and hence obtain the value of $\langle\Delta\Psi'\rangle$ we must solve the equation of motion (9.44)

$$\Psi''(s) + \Omega_L^2 \sin\Psi(s) = 0.$$

Multiplying by $2\Psi'(s)$ gives

$$2\Psi'\Psi'' + 2\Omega_L^2 \Psi' \sin\Psi = 0, \tag{9.54}$$

which may be integrated directly to give the result

$$\Psi'^2 - 2\Omega_L^2 \cos\Psi = C, \tag{9.55}$$

where C is the constant of integration. We again start with the case of a single electron, which enters the undulator with an initial phase Ψ_{start}. It then follows from (9.55) that

$$\Psi'^2 - {\Psi'_{\text{start}}}^2 = 2\,\Omega_L^2 (\cos\Psi - \cos\Psi_{\text{start}}). \tag{9.56}$$

Using (9.39) we can express the initial condition in the form

$$\Psi'_{\text{start}} = 2\frac{Nk_{\text{u}}}{\gamma_{\text{r}}}(\gamma_{\text{start}} - \gamma_{\text{r}}) \tag{9.57}$$

and obtain

$$\Psi'^2 = \frac{4N^2k_{\text{u}}^2}{\gamma_{\text{r}}}(\gamma_{\text{start}} - \gamma_{\text{r}})^2 + 2\,\Omega_{\text{L}}^2(\cos\Psi - \cos\Psi_{\text{start}}). \tag{9.58}$$

To simplify the calculation let us define the variable

$$w = \frac{2\pi N N_{\text{u}}}{\gamma_{\text{r}}}(\gamma_{\text{start}} - \gamma_{\text{r}}), \tag{9.59}$$

where N is the harmonic number and N_{u} is the number of undulator periods of length λ_{u}. Integrating the equation of motion (9.44) once gives

$$\boxed{\Psi'(s) = \frac{2w}{N_{\text{u}}\lambda_{\text{u}}}\sqrt{1 + \frac{N_{\text{u}}^2\lambda_{\text{u}}^2\Omega_{\text{L}}^2}{2w^2}\Big[\cos\Psi(s) - \cos\Psi_{\text{start}}\Big]}.} \tag{9.60}$$

Once again this equation cannot in general be solved analytically, and one must make approximations or use numerical methods.

For relatively weak laser fields $E_{\text{L},0}$ the field parameter K_{L} is very small, and so too is the frequency Ω_{L} according (9.45). From (9.42) the amplification due to each electron passing through the undulator is also small. Thus the intensity of the laser field only changes slightly and may be assumed to be effectively constant during each pass. This is the so-called **low-gain regime** of the FEL. Here the laser field is so weak that virtually all the electrons lie outside the separatrix (Fig. 9.3) and do not perform stable oscillations in the laser field.

In the low-gain regime the second term under the square root in (9.60) is small compared to 1 and so the square root may be expanded as

$$\sqrt{1+x} = 1 + \frac{1}{2}x - \frac{1}{8}x^2 + \ \ldots$$

giving

$$\begin{aligned}\Psi'(s) \;&=\; \frac{2w}{N_{\text{u}}\lambda_{\text{u}}}\Bigg\{1 + \frac{1}{2}\,\frac{N_{\text{u}}^2\lambda_{\text{u}}^2\Omega_{\text{L}}^2}{2w^2}\Big[\cos\Psi(s) - \cos\Psi_{\text{start}}\Big] \\ &\quad - \frac{1}{8}\left(\frac{N_{\text{u}}^2\lambda_{\text{u}}^2\Omega_{\text{L}}^2}{2w^2}\right)^2\Big[\cos\Psi(s) - \cos\Psi_{\text{start}}\Big]^2 + \ \ldots\Bigg\}\end{aligned} \tag{9.61}$$

Let us calculate the average value of Ψ' using this equation to first order. To do so we need to determine the phase $\Psi(s)$ in the argument of the cosine, which we calculate from (9.61) to zeroth order. We then have

$$\Psi_0'(s) = \frac{2w}{N_u \lambda u} \qquad \Longrightarrow \qquad \Psi(s) = \frac{2w}{N_u \lambda u}\, s. \tag{9.62}$$

The phase difference between the beginning and end of the undulator is then

$$\Delta\Psi = \frac{2w}{N_u \lambda u} L_u = 2w, \tag{9.63}$$

where $L_u = N_u \lambda u$ is the total length of the undulator. From (9.61) we thus obtain the difference to first order

$$\Delta\Psi_1' = \Psi'(s_0 + L_u) - \Psi'(s_0) = \frac{N_u \lambda_u \Omega_L^2}{2w} \Big[\cos(2w + \Psi_{\text{start}}) - \cos \Psi_{\text{start}} \Big], \tag{9.64}$$

where s_0 denotes the beginning of the undulator. As the electrons are uniformly distributed within the bunch and the bunch is very long compared to the wavelength of the laser, the initial phases of the electrons are also uniformly distributed over the range $0 \leq \Psi_{\text{start}} \leq 2\pi$. We must therefore average over all initial phases and obtain

$$\langle \Delta\Psi_1' \rangle = \frac{N_u \lambda_u \Omega_L^2}{2w} \frac{1}{2\pi} \int_0^{2\pi} \Big[\cos(2w + \Psi_{\text{start}}) - \cos \Psi_{\text{start}} \Big] d\Psi_{\text{start}} = 0. \tag{9.65}$$

To first order the energy contributions from all the electrons average out and there is no net energy exchange between the electron beam and the laser field. Amplification of the FEL is thus only a higher-order effect.

To confirm this let us repeat the calculation, this time to second order. This requires us to expand the argument $\Psi(s)$ of the cosine to first order. We go back to (9.61) and, using (9.62), write

$$\Delta\Psi_1'(s) = \Psi'(s) - \Psi'(s_0) = \frac{N_u \lambda_u \Omega_L^2}{2w} \left[\cos\left(\frac{2w}{N_u \lambda_u} s + \Psi_{\text{start}} \right) - \cos \Psi_{\text{start}} \right]. \tag{9.66}$$

From this we again obtain the phase difference

$$\Delta\Psi_1 = \frac{N_u \lambda_u \Omega_L^2}{2w} \left\{ \int_0^{L_u} \cos\left(\frac{2w}{N_u \lambda_u} s + \Psi_{\text{start}} \right) ds - N_u \lambda_u \cos \Psi_{\text{start}} \right\} \tag{9.67}$$

by integrating over the length of the undulator. The phase difference is then given to first order by

$$\Delta\Psi_1 = \frac{N_u^2 \lambda_u^2 \Omega_L^2}{4w^2} \Big[\sin(2w + \Psi_{\text{start}}) - \sin \Psi_{\text{start}} - 2w \cos \Psi_{\text{start}} \Big]. \tag{9.68}$$

From (9.61), the difference to second order in the first derivative of the phase $\Psi(s)$ has the form

$$\Delta\Psi_2' = \frac{N_\mathrm{u}^3\lambda_\mathrm{u}^3\Omega_\mathrm{L}^4}{16w^3}\left\{\frac{8w^2}{N_\mathrm{u}^2\lambda_\mathrm{u}^2\Omega_\mathrm{L}^2}\Big[\cos(\Delta\Psi_1+2w+\Psi_\mathrm{start})-\cos\Psi_\mathrm{start}\Big]\right.$$
$$\left.-\Big[\cos(2w+\Psi_\mathrm{start})-\cos\Psi_\mathrm{start}\Big]^2\right\} \quad (9.69)$$

We can further simplify this expression by noting that in the low-gain regime the changes in phase are very small, i.e. $\Delta\Psi_1 \ll 1$. We may thus write

$$\cos(\Delta\Psi_1+2w+\Psi_\mathrm{start})-\cos\Psi_\mathrm{start}$$
$$\approx \cos(2w+\Psi_\mathrm{start})-\cos\Psi_\mathrm{start}-\Delta\Psi_1\sin(2w+\Psi_\mathrm{start}). \quad (9.70)$$

Inserting this expression together with the phase (9.68) into (9.69) we obtain

$$\Delta\Psi_2' = \frac{N_\mathrm{u}^3\lambda_\mathrm{u}^3\Omega_\mathrm{L}^4}{16w^3}\left\{\frac{8w^2}{N_\mathrm{u}^2\lambda_\mathrm{u}^2\Omega_\mathrm{L}^2}\Big[\cos(2w+\Psi_\mathrm{start})-\cos\Psi_\mathrm{start}\Big]\right.$$
$$-2\sin(2w+\Psi_\mathrm{start})\Big[\sin(2w+\Psi_\mathrm{start})-\sin\Psi_\mathrm{start}-2w\cos\Psi_\mathrm{start}\Big]$$
$$\left.-\Big[\cos(2w+\Psi_\mathrm{start})-\cos\Psi_\mathrm{start}\Big]^2\right\}. \quad (9.71)$$

In order to determine the effect of all the electrons on the laser field we must again average over all initial phases Ψ_start. The following initial values are used:

$$\begin{aligned}
\langle\cos(2w+\Psi_\mathrm{start})-\cos\Psi_\mathrm{start}\rangle &= 0\\
\langle\sin^2(2w+\Psi_\mathrm{start})\rangle &= \langle\cos^2(2w+\Psi_\mathrm{start})\rangle = \langle\cos^2\Psi_\mathrm{start}\rangle = \frac{1}{2}\\
\langle\sin(2w+\Psi_\mathrm{start})\sin\Psi_\mathrm{start}\rangle &= \langle\cos(2w+\Psi_\mathrm{start})\cos\Psi_\mathrm{start}\rangle = \frac{1}{2}\cos(2w)\\
\langle\sin(2w+\Psi_\mathrm{start})\cos\Psi_\mathrm{start}\rangle &= \frac{1}{2}\sin(2w).
\end{aligned} \quad (9.72)$$

Equation (9.71) then simplifies to

$$\langle\Delta\Psi_2'\rangle = -\frac{N_\mathrm{u}^3\lambda_\mathrm{u}^3\Omega_\mathrm{L}^4}{8}\,\frac{1}{w^3}\Big[1-\cos(2w)-w\sin(2w)\Big]. \quad (9.73)$$

It is easy to show that

$$\frac{d}{dw}\left(\frac{\sin w}{w}\right)^2 = -\frac{1}{w^3}\Big[1-\cos(2w)-w\sin(2w)\Big]. \quad (9.74)$$

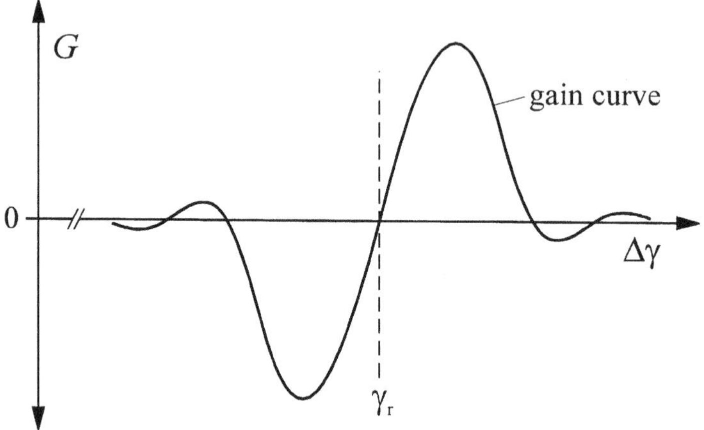

Fig. 9.4 Amplification or gain curve of the FEL.

Equation (9.73) thus becomes

$$\langle \Delta \Psi_2' \rangle = -\frac{N_u^3 \lambda_u^3 \Omega_L^4}{8} \frac{d}{dw} \left(\frac{\sin w}{w} \right)^2 . \tag{9.75}$$

We insert this average value into the amplification relation (9.52) and obtain the required FEL amplification factor in the low-gain regime for the Nth harmonic:

$$\begin{aligned} G_N &= -\frac{\pi e^2 N K^2 N_u^3 \lambda_u^2}{4\varepsilon_0 m_e c^2} \frac{n_b}{\gamma_r^3} F(N\eta) \frac{d}{dw} \left(\frac{\sin w}{w} \right)^2 \\ \text{with} & \\ F(N\eta) &= \left[J_{\frac{N-1}{2}}(N\eta) - J_{\frac{N+1}{2}}(N\eta) \right]^2 \\ \eta &= \frac{k_L K^2}{8N\gamma^2 k_u} \\ w &= \frac{2\pi N N_u}{\gamma_r} (\gamma_{\text{start}} - \gamma_r) \\ N &= 1, 3, 5, \ldots . \end{aligned} \tag{9.76}$$

This amplification function strongly depends on the injection energy of the electrons. Plotting G as a function of the injection energy $\Delta\gamma$ results in a curve, shown qualitatively in Fig. 9.4, which is typical of an FEL. At the resonance energy, $\gamma = \gamma_r$, the curve passes through zero, i.e. there is no amplification. Hence it is important to tune the electron energy to be slightly higher than that required by the coherence condition (9.31). If on the other hand $\Delta\gamma < 0$, the electron beam gains energy at the expense of the laser field, i.e. the particles are accelerated by the laser field. In this regime it is thus possible to accelerate particles by means of the 'inverse free electron laser' principle.

9.4 The Madey theorem

In Chapter 8 we showed that the intensity distribution of spontaneous undulator radiation is described by the function

$$I(\Delta\omega) \propto \left[\frac{\sin\left(\pi N_{\mathrm{u}}\dfrac{\Delta\omega}{\omega_{\mathrm{w}}}\right)}{\pi N_{\mathrm{u}}\dfrac{\Delta\omega}{\omega_{\mathrm{w}}}}\right]^2 . \tag{9.77}$$

In this representation the electron energy is constant and the intensity of the radiation is considered as a function of its frequency. It is of course also possible to keep the frequency or wavelength constant and vary the electron beam energy, which in principle will give a similar intensity distribution. To calculate this we return to the coherence condition (9.31), from which the frequency

$$\omega_{\mathrm{N}} = \frac{2\pi c}{\lambda_{\mathrm{w}}} = \frac{4\pi c N \gamma^2}{\lambda_{\mathrm{u}}(1+K^2/2)} \tag{9.78}$$

of the Nth harmonic of the undulator radiation immediately follows. Differentiating with respect to energy gives

$$\frac{d\omega}{d\gamma} = \frac{4\pi c N}{\lambda_{\mathrm{u}}(1+K^2/2)} 2\gamma = N\omega_{\mathrm{w}}\,\frac{2}{\gamma}, \tag{9.79}$$

where ω_{w} refers to the first harmonic of the undulator wave. The relative change in frequency can thus also be expressed in terms of the relative change in energy

$$\frac{\Delta\omega}{\omega_{\mathrm{w}}} = \frac{2N\Delta\gamma}{\gamma}. \tag{9.80}$$

Using $\Delta\gamma = \gamma - \gamma_{\mathrm{r}}$ we obtain the relation

$$\pi N_{\mathrm{u}}\frac{\Delta\omega}{\omega_{\mathrm{w}}} = \frac{2\pi N N_{\mathrm{u}}}{\gamma}(\gamma - \gamma_{\mathrm{r}}). \tag{9.81}$$

This is exactly the same as the definition of the variable w in (9.59), if we set the electron energy γ in the undulator equal to the initial energy. Any difference may certainly be neglected, since in the low-gain regime the change in electron energy per pass through the undulator is very small. The intensity distribution of the undulator radiation line may thus also be written in the form

$$I(\Delta\gamma) \propto \left(\frac{\sin w}{w}\right)^2 . \tag{9.82}$$

Comparing this expression with the amplification function (9.76) of an FEL we see that this amplification function is proportional to the differential of the intensity distribution of the **spontaneous** undulator radiation along the beam axis. This fundamental relation in the theory of free electron lasers was first demonstrated in a very general form by J.M.J. Madey [86], and is known as the **Madey theorem**. This relation is illustrated in Fig. 9.5.

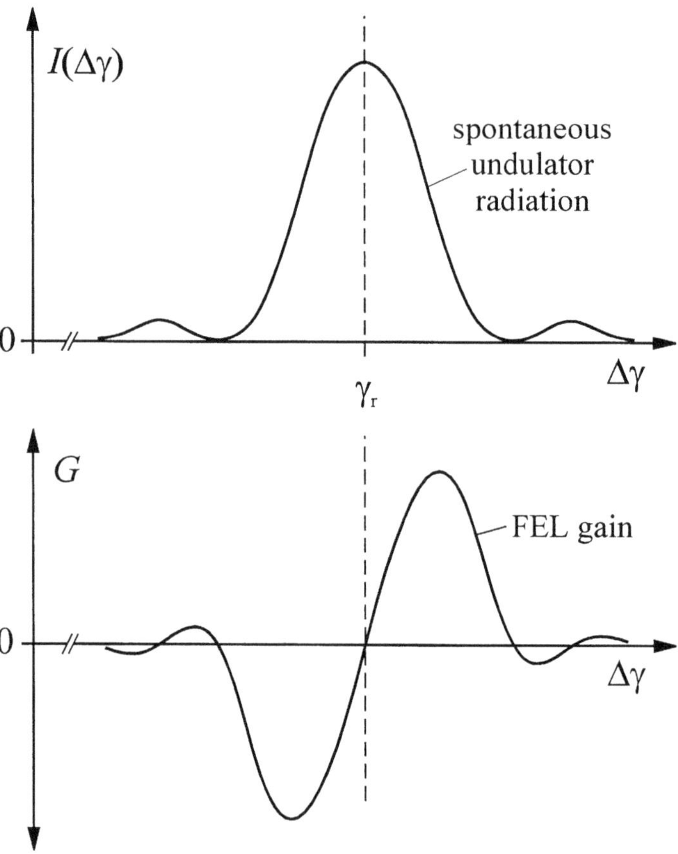

Fig. 9.5 Illustration of the Madey theorem. The lower amplification curve of the FEL is obtained by differentiating the intensity function of the undulator radiation line at a fixed wavelength λ.

9.5 FEL amplification in the high-gain regime

For very strong FEL fields the separatrix shown in Fig. 9.3 is very broad along the direction of the energy axis. A relatively large number of electrons therefore lie inside this broadened separatrix and perform stable pendulum-like oscillations. Very strong amplification results and one talks of a high-gain regime. The laser field can no longer be assumed to be constant inside the undulator, and in calculating the FEL amplification one must take into account the evolution of the electromagnetic wave. The frequency Ω_L of the electron motion in the FEL-field is very high in this regime, and the approximations we made in the low-gain regime in order to solve the root in

$$\Psi'(s) = \frac{2w}{N_\mathrm{u}\lambda_\mathrm{u}}\sqrt{1 + \frac{N_\mathrm{u}^2\lambda_\mathrm{u}^2\Omega_\mathrm{L}^2}{2w^2}\left[\cos\Psi(s) - \cos\Psi_\mathrm{start}\right]} \qquad (9.83)$$

are no longer valid. As a result it is no longer possible to integrate this relation analytically, and numerical methods such as the Runge-Kutta method must be employed to find the solution. It then becomes possible to calculate the course of individual electrons through the laser field for arbitrary initial values γ_{start} and Ψ_{start}. Here we must take into account the fact that the particles in the electron beam are approximately evenly distributed along the length of the bunch, i.e. as they enter the FEL undulator they completely fill the available phase range per period $-\pi \leq \Psi \leq +\pi$. We now assume that the incoming particles all have the same energy γ_{start}, so that they form a horizontal line in the Ψ-$\Delta\gamma$ plane (Fig. 9.6). The electron beam with n particles thus has the initial energy

$$E_{\mathrm{start}} = n\gamma_{\mathrm{start}} m_{\mathrm{e}} c^2 = n(\gamma_{\mathrm{r}} + \Delta\gamma_{\mathrm{start}}) m_{\mathrm{e}} c^2, \tag{9.84}$$

with $\gamma_{\mathrm{start}} = \gamma_{\mathrm{r}} + \Delta\gamma$. Since we have assumed that the electrons are uniformly distributed, we can take each small phase volume containing an electron and associate it with a particular phase, i.e. the ith electron has the initial phase $\Psi_{\mathrm{start},i}$. The relative initial energy, denoted by $\Delta\gamma_{\mathrm{start},i}$, is the same for all electrons. Using the equations of motion (9.39) and (9.42) we may now calculate the path of each electron and hence also its final energy $\Delta\gamma_{\mathrm{end},i}$.

As we have seen, this calculation may only be performed numerically and, since the number of electrons in the beam is very large, would require unrealistically long computation times. We thus take large numbers of electrons lying close together in phase space and combine them into a single **macro-particle**, and then calculate the FEL amplification using a relatively small number (100 to 1000) of such macro-particles. This calculation may then be performed within a reasonable amount of computing time.

The final energy of the total assembly of electrons (or macro-particles) as they leave the FEL undulator is

$$E_{\mathrm{end}} = m_{\mathrm{e}} c^2 \sum_{i=1}^{N} (\gamma_{\mathrm{r}} + \Delta\gamma_{\mathrm{end},i}) = E_{\mathrm{start}} + m_{\mathrm{e}} c^2 \sum_{i=1}^{N} \Delta\gamma_{\mathrm{end},i}. \tag{9.85}$$

Defining

$$\overline{\Delta\gamma} = \frac{1}{N} \sum_{i=1}^{N} \Delta\gamma_{\mathrm{end},i} \tag{9.86}$$

as the average final energy of all the electrons, the net change in energy of the entire electron beam as it leaves the FEL is then given by

$$\Delta E = E_{\mathrm{start}} - E_{\mathrm{end}} = N m_{\mathrm{e}} c^2 (\Delta\gamma - \overline{\Delta\gamma}). \tag{9.87}$$

If we start at the resonance energy (i.e. $\Delta\gamma = 0$) then it is clear from Fig. 9.6 (a) that for every electron with an initial phase $+\Psi$ there is another with the phase $-\Psi$, which by symmetry will exactly compensate for it in the energy sum. In this situation the total change of energy of the laser field is

$$\Delta E = 0 \qquad \text{if} \qquad \Delta\gamma_{\mathrm{start}} = 0 \tag{9.88}$$

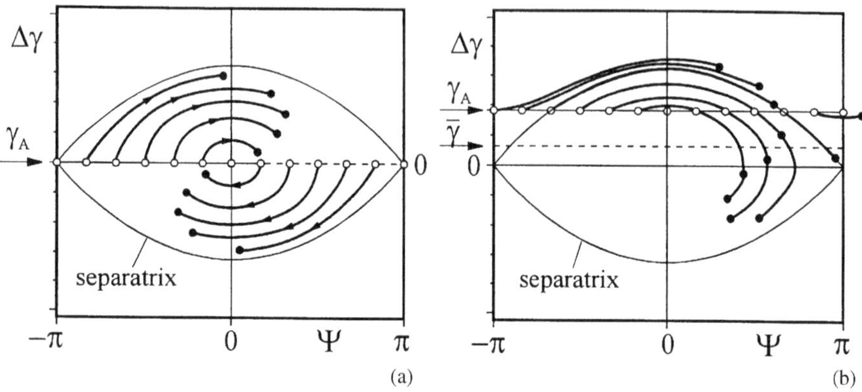

Fig. 9.6 Motion of electrons with a flat phase distribution as they pass through the FEL. (a) shows the motion when $\gamma_{\text{start}} = \gamma_{\text{r}}$ and (b) the motion when $\gamma_{\text{start}} > \gamma_{\text{r}}$.

and there is no FEL amplification. This is the same result as in the low-gain regime. The situation changes, however, if the particles are injected at an energy γ_{start} which is higher than the resonance energy. Then the average energy $\overline{\gamma} = \gamma_{\text{r}} + \overline{\Delta\gamma}$ of the outgoing electrons at the end of the FEL undulator is lower than the injection energy, as Fig. 9.6(b) shows. The laser field thus gains energy.

The calculation may be repeated for various initial energies to obtain, in each case, the average energy of the electrons as they leave the FEL. The relation

$$\frac{\gamma_{\text{start}} - \overline{\gamma}}{\gamma_{\text{start}}} = G \tag{9.89}$$

again leads to the amplification or gain curve of FEL, as plotted in Fig. 9.4.

In general the undulator is so short and the frequency Ω_{L} so low that only a small part of a full ellipse in phase space will be covered. As the undulator length increases, the energy transfer at first rises but then declines again and reaches a certain saturation level when the trajectory of the particle in phase space becomes closed. This saturated state is, however, only reached in high gain FELs, if at all.

9.6 The FEL amplifier and FEL oscillator

As we have seen above, coherent electromagnetic radiation can be amplified by means of the FEL principle, using a high-energy electron beam as an energy reservoir. The wave enters an undulator, with the wavelength λ and the undulator parameter K chosen to satisfy the coherence condition (9.12). In order to amplify laser field, the energy γ of the electron beam must be tuned to slightly above the resonance energy γ_{r}. The basic layout of this kind of **FEL amplifier** is shown in the upper diagram of Fig. 9.7. This amplifier is of little practical use, however, since in general the amplification achieved in one pass through the undulator is well below $G < 1$. The gain in energy

$$\Delta E = N m_{\text{e}} c^2 (\Delta\gamma_{\text{start}} - \overline{\Delta\gamma}) \tag{9.90}$$

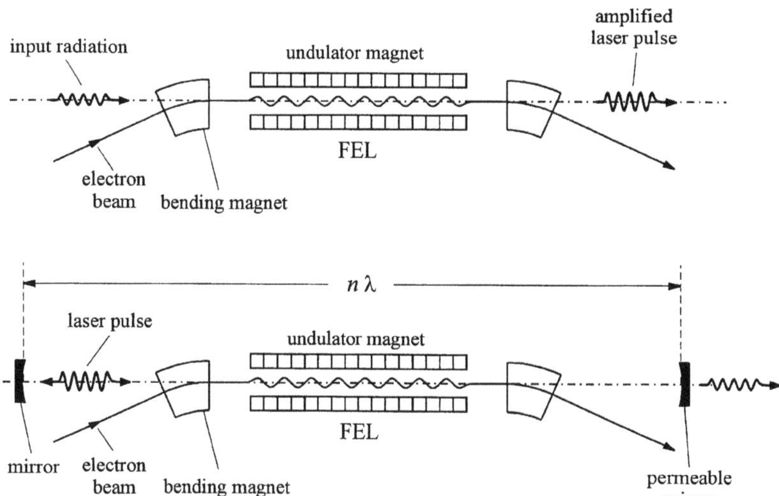

Fig. 9.7 The FEL amplifier (*upper diagram*) and FEL oscillator with optical resonator.

is relatively small in standard undulators of limited length. It is only possible to achieve sufficient amplification to reach FEL saturation and deliver high output power in the so-called high-gain regime by using very long undulators with many periods [87].

The FEL principle becomes considerably more efficient if an optical resonator is used, in which the laser wave is reflected many times between two spherical mirrors and energy is transferred each time the laser field passes through the electron beam. This arrangement is shown in the lower diagram in Fig. 9.7. The repeated recirculation of the FEL wave results in a type of feedback, and so this system is called an **FEL oscillator**. Here the energy transfer per pass is

$$\Delta E_{\mathrm{osc}} = N m_{\mathrm{e}} c^2 (\Delta\gamma_{\mathrm{start}} - \overline{\Delta\gamma}) - 2\Delta E_{\mathrm{loss}}. \tag{9.91}$$

The losses arise due to the fact that the mirrors are not perfect reflectors, with a factor 2 due to the fact that the wave is reflected off both mirrors on each pass through the oscillator. If the energy of the FEL wave stored in the optical resonator is E_{L} then the loss from the mirrors is

$$\Delta E_{\mathrm{loss}} = E_{\mathrm{L}}(1 - R), \tag{9.92}$$

where R is the average reflectance of the mirror at the wavelength used. In the visible and infrared regions there are mirrors with a reflectance of only just less than 1. In this case the amplification only needs to reach a few percent in order for the FEL amplifier to be usable. The first time the electron beam passes through the FEL-undulator it emits undulator radiation which, because the coherence condition has been carefully satisfied, has the same wavelength as the laser radiation. This undulator radiation is now reflected between the mirrors and steadily gains in intensity, because of the FEL amplification, until saturation

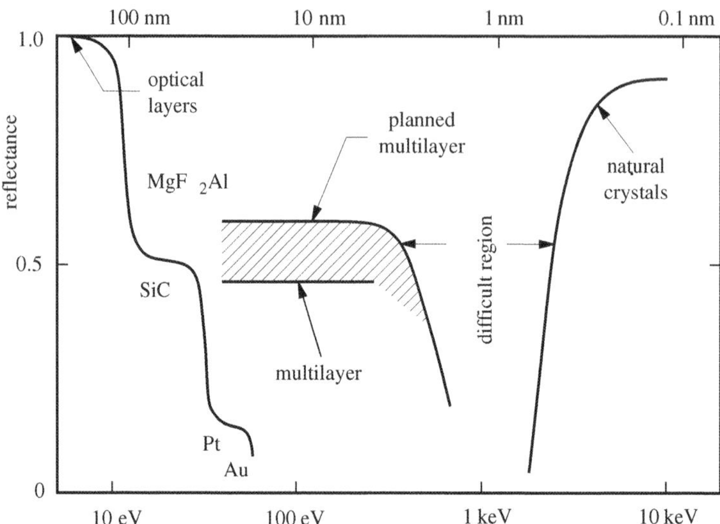

Fig. 9.8 Reflectance of various mirror materials for wavelengths below the visible range.

is reached. Part of this radiation is allowed to exit, either by making one or both mirrors slightly transparent or by using an additional deflector mirror, and is then transported to an experiment.

With an appropriate choice of undulator parameter and beam energy it is in principle possible to satisfy the coherence condition for relatively short wavelengths of well below $\lambda < 100$ nm. The FEL principle thus offers the possibility of producing intensive coherent radiation ranging from the UV region up to soft X-rays. There is, however, the problem that for wavelengths below $\lambda \approx 100$ nm, the reflectance of the currently available mirrors falls off sharply, as shown by the curve in Fig. 9.8. In this wavelength region low-gain FELs may no longer be safely used, since an amplification of $G \gg 1$ is required per pass through the undulator. This necessitates the use of very long undulators ($l > 10$ m) and places extremely stringent demands on the beam quality. The amplification increases with the number of electrons within the phase volume, or in other words with the particle density within the bunch. This means that very low beam emittance and very short bunches are needed.

9.7 The optical klystron

As a rule the total phase range $-\pi < \Psi < +\pi$ is filled with particles and the amplification factor is given by the average energy transfer of all the particles, a non-negligible fraction of which actually take energy out of the laser field (Section 9.3). If, however, we only fill the phase range $0 < \Psi + \pi$ with particles, all of which give energy to the FEL field, and leave the range $-\pi < \Psi < 0$ empty, then the amplification per wavelength is considerably higher than in a normal FEL. This principle was suggested by N.A. Vinokurov and A.N. Skrinsky and

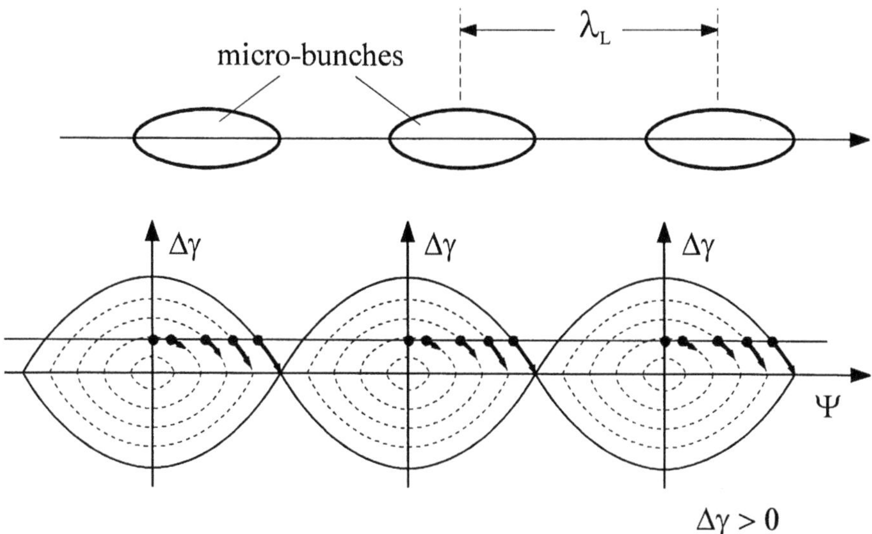

Fig. 9.9 Phase space distribution of electrons in the laser field of the optical klystron.

is known as the **optical klystron** [88]. Here it is necessary to divide the electron beam up into micro-bunches which travel with a separation equal to the laser wavelength λ_L and are separated by regions of very low particle density (Fig. 9.9).

The practical implementation of the optical klystron is illustrated in Fig. 9.10. The arrangement consists of two identical undulators, arranged one after another with a dispersive section between them. This device functions in a similar way to the microwave klystron (Fig. 5.11), hence the name used for this type of FEL.

The first undulator has the task of modulating the energy of the electron beam using the laser wave. We refer back to Fig. 9.6(a), which illustrates the motion of the electrons in the FEL. Initially all particles have the same energy γ_{start}, but because of their interaction with the laser field they leave the undulator with varying energies. The average energy is the same as before, but the final energy of each individual particle varies according to its initial phase in the FEL field. This is exactly what happens in the first undulator of the optical klystron.

In an RF klystron the bunching then takes place in a pure drift section, due to the varying velocities of the particles. In the optical klystron, however, one is dealing with relativistic particles which all have practically the same velocity. Short bending magnets are thus used to produce a dispersive section, in which particles of different energies follow trajectories of different lengths. This section can also be produced within the normal periodic structure of the undulator, simply by exciting the poles more strongly for a few periods. The electrons leave the dispersive section grouped into the required micro-bunches, as shown in Fig. 9.9. The actual amplification then takes place in the second undulator of the optical klystron. This procedure is so effective that even relatively short systems can reach higher values than normal FELs of the same length. The disadvantage

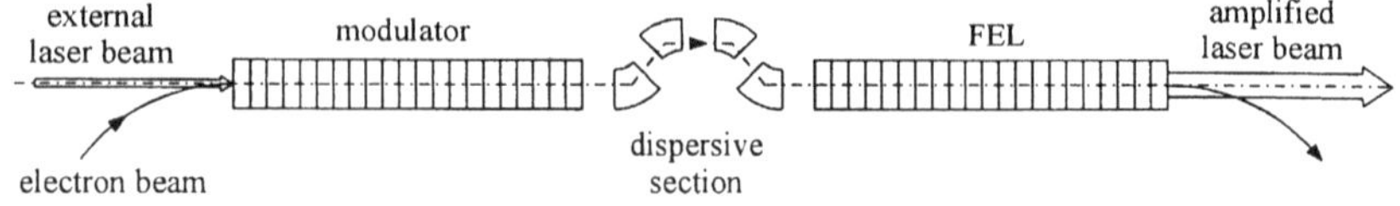

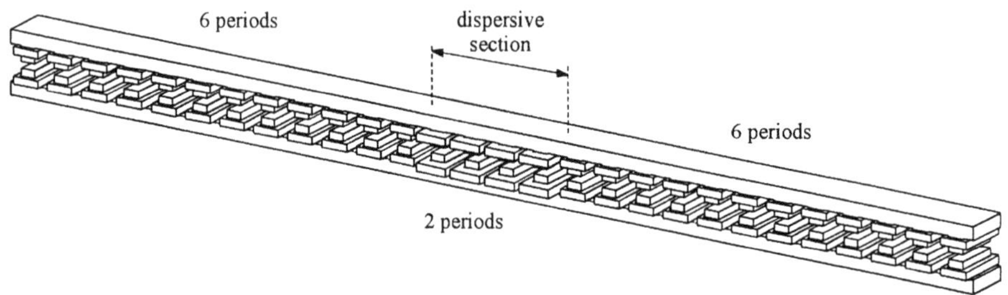

Fig. 9.10 Basic layout of an optical klystron. The first undulator modulates the particle energy, so that micro-bunches are produced in the dispersive section which follows. The actual FEL amplification then takes place in the second undulator. The dispersive section can also be produced within a few periods of a single undulator by exciting the poles in this region more strongly.

of optical klystrons, however, is their considerably lower power output.

Optical klystrons may again be used as amplifiers and oscillators. This principle has been successfully demonstrated at the ACO storage ring in Orsay (France) [89, 90] and VEPP III in Novosibirsk (Russia) [91], with laser frequencies essentially in the visible region.

9.8 Time structure of the FEL radiation

The quality and usefulness of the laser radiation depends not only on its intensity and the range of wavelengths available, but also on its time structure. The shortest time interval is the length of an individual micropulse produced from one electron bunch. This directly determines the line width of the radiation, an important quantity. The length of the laser pulse is determined by two effects. The distance d between the mirrors of the optical resonator is very large compared to the wavelength λ_L of the laser radiation (e.g. $d \approx 10$ m and $\lambda_\mathrm{L} \approx 10^{-7}$ m). Thus there are very many resonator modes of the wavelength $\lambda_{\mathrm{L},j}$, with integer j in the region $j \approx 10^8$, which satisfy the resonance condition

$$d = j\,\lambda_{\mathrm{L},j}. \tag{9.93}$$

In order for energy to be transferred from the electron beam to the laser wave, the wavelength must fulfil the coherence condition

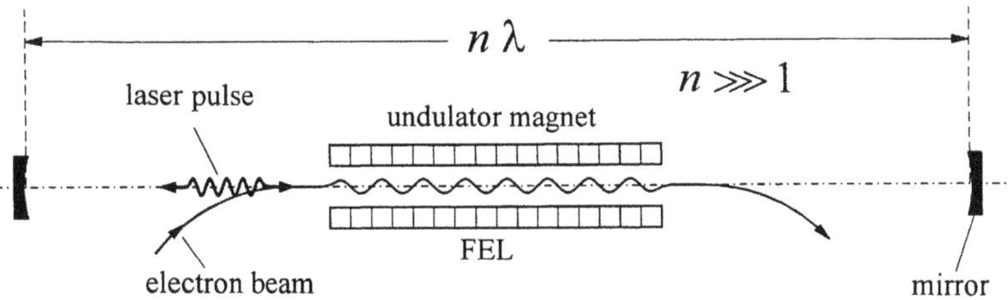

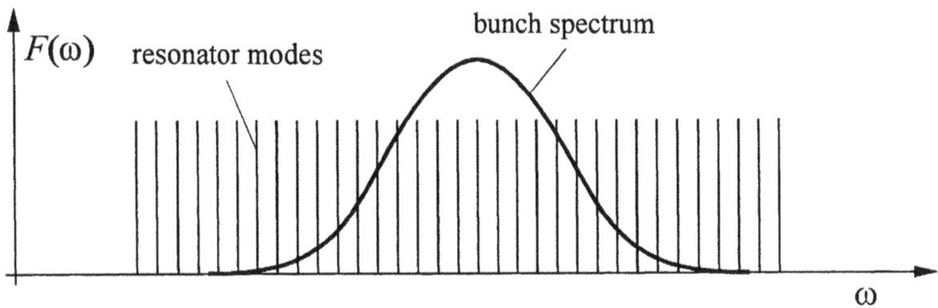

Fig. 9.11 Excitation of multiple modes in the FEL optical resonator due to the finite energy spread in the bunch.

$$\lambda_{\mathrm{L},j} = \frac{\lambda_{\mathrm{u}}}{2\gamma_j^2}\left(1 + \frac{K^2}{2}\right). \tag{9.94}$$

The electrons do not, of course, all have a single sharp energy but instead follow a Gaussian distribution across a broad energy range, due to quantum fluctuations in the emission of synchrotron radiation and the resulting synchrotron oscillations. This means that there are particles in the beam with energies γ_j corresponding to a large number of resonator modes (Fig. 9.11). As a result, a great many modes with wavelengths $\lambda_{\mathrm{L},j}$ and frequencies $\omega_{\mathrm{L},j} = 2\pi c/\lambda_{\mathrm{L},j}$ are amplified by the FEL in the optical resonator. These all lie relatively close together and overlap according to

$$I(t) = \sum_j a_j e^{i\,\omega_j t}. \tag{9.95}$$

This gives a short pulse with a typical duration of between $\tau_{\mathrm{L}} = 1$ ps and $\tau_{\mathrm{L}} = 10$ ps. The energy spread of the line then follows immediately from the uncertainty relation

$$\Delta E = \frac{h}{\tau_{\mathrm{L}}}. \tag{9.96}$$

If the FEL is used in a linac, then as well as the very short micropulse duration there is the considerably longer timescale of the pulse rate of the accelerator, which usually has a repetition frequency of a few hundred Hz. If, on the other hand, the FEL is used in a storage ring, then the next largest unit of time is the revolution time in the accelerator, which is of the order of MHz. However, the

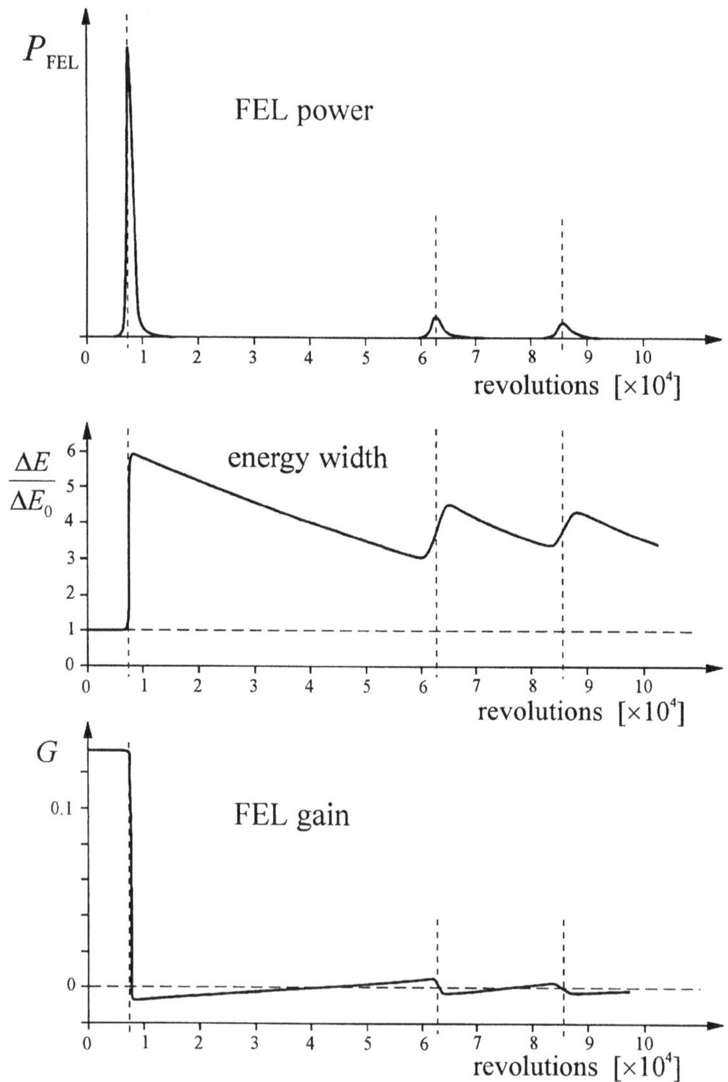

Fig. 9.12 Form of the FEL power, energy spread, and FEL amplification as a function of the number of revolutions in a storage ring.

FEL wave radiated per revolution certainly does not have a constant intensity, but instead has a relatively complicated behaviour over the timescale of a few revolutions, as in the example shown in Fig. 9.12.

When the FEL starts up, the amplitude of the radiation grows exponentially for the first few revolutions. As a result the field strength increases and the energy spread in the particle beam grows through interactions of the electrons with the laser field. A continually increasing fraction of the electrons end up with such a strong energy deviation that they can no longer contribute to the FEL amplification. The amplification falls off rapidly and finally drops below zero, at which point the laser effect ends and the FEL power decays away rapidly.

The electron beam, which now has a broad spread of energies, is then cooled by damping of the synchrotron oscillations and the energy spread is reduced. As the energy spread decreases, the FEL amplification grows again until the laser threshold is reached and the FEL radiation again grows exponentially. The energy spread then increases again until the amplification once again falls below zero. The peak power output of the FEL is much lower in the second pulse than in the first because the quality of the beam is much poorer by the start of this second pulse.

This process is repeated several times until a state of equilibrium is reached. The intensity at equilibrium is well below that of the first pulse. A. Renieri has found an approximation for the FEL power output in this equilibrium state [97], given by

$$P_{\mathrm{L}} \approx \frac{1}{2N_{\mathrm{u}}} P_{\mathrm{syn}}, \tag{9.97}$$

where P_{syn} is the total power radiated through synchrotron radiation and N_{u} is the number of undulator periods. This leads to the Renieri criterion, which states that in equilibrium the radiated FEL power is proportional to the total power W_0 emitted as synchrotron radiation.

10 Diagnostics

The beam circulating inside a closed vacuum chamber is not visible from outside. Furthermore, no-one is usually allowed near the accelerator while it is in operation. In order to establish first of all whether a beam is present in the machine, and then to measure the physical parameters of that beam, the accelerator must therefore be equipped with a wide range of measuring instruments, often termed **monitors**. In this chapter we will discuss the most common beam monitors used in accelerators and introduce various diagnostics used to measure the key beam parameters.

10.1 Observation of the beam and measurement of the beam current

10.1.1 The fluorescent screen

The simplest way to observe a charged particle beam is to use a fluorescent screen, either fixed at the end of a beam transport line or mounted on a movable support which can be inserted into the path of the beam when required (Fig. 10.1). Zinc sulphide (ZnS) has proven to be an effective fluorescent material. Mixed with sodium silicate or some other suitable binder it is applied in thin layers onto glass, ceramic, or metal. The resulting screens emit green light with a high light yield. However, they cannot be used in cases where an extremely high vacuum is required. In addition they only have a limited lifetime and burn out at the beam spot after extended exposure.

To avoid this problem, thicker screens made of Al_2O_3 doped with chrome and called 'chromox-screens' are more commonly used nowadays [98]. These screens give off a predominantly red light and are characterized by their high tolerance to exposure to the beam. They also have an extremely low degassing rate and so may also be used in ultra high vacuum (UHV).

The emitted light is viewed using a television camera in the control room. Nowadays very compact CCD cameras are available for this purpose. Since these cameras are relatively susceptible to radiation damage, care must be taken during installation to ensure that they are well protected by lead shielding and are positioned where the radiation level is as low as possible. It is useful to mark the fluorescent screen with calibrated cross-hairs, allowing the position and size of the beam to be measured with reasonable accuracy.

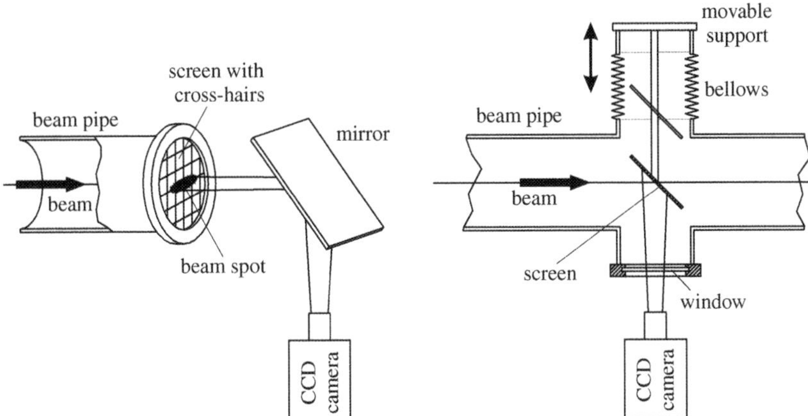

Fig. 10.1 The fluorescent monitor. The figure on the *left* shows a fixed version used, for example, at the end of a linac, while the diagram on the *right* shows a screen which may be moved in and out of the beam line.

As well as determining the beam position, a fluorescent screen can in principle be used to measure the beam profile and intensity. In practice, however, there are restrictions. First of all, there is not always a linear relation between the light yield and the beam intensity, which reduces the accuracy of the measurement. Furthermore, usually only a very small region of the screen is illuminated by the beam, with the rest remaining dark. The brightness adjustment of the camera is usually tuned for the regions of average light yield, so that the relatively few light pixels are overemphasized. As a result one must either adjust the tuning to pick out only the peak values or illuminate the fluorescent screen externally to lighten the background. This reduces the difference between the lightest and darkest regions enough for both to be within the range detectable by the camera. This latter method is simple to implement and also allows a much improved view of the cross-hairs on the screen.

Fluorescent screens have a relatively long afterglow ranging from several milliseconds to more than a second. They are thus not able to resolve the time structure of the beam, which typically lies in the region of nanoseconds and below, due to the generally very high frequency of the accelerator.

10.1.2 The Faraday cup

The simplest way to measure the current or intensity of a beam of charged particles is to completely absorb the beam in a block of conducting material and find the amount of captured charge by measuring the resulting current. This was the first method used to measure particle beams. Although the principle is very simple, it does have problems and significant limits. However, it does allow the intensity to be quantitatively determined to high accuracy, and so is often used for calibrations. One possible design, often called the **'Faraday cup'**, is shown in Fig. 10.2.

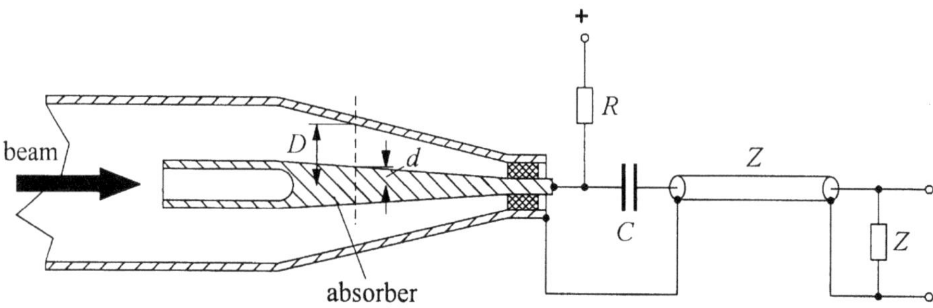

Fig. 10.2 Outline of a Faraday cup with a coaxial structure.

For an exact measurement it is necessary to completely absorb all the particles in the block of material. Since at high energies the penetration depth is large, the block of material must be very thick. Moreover, at high beam energies the amount of energy transferred to the block is naturally also very high, leading to strong heating of the absorber. Above all, however, there is the problem of multiple scattering in the absorber, resulting in very strong transverse broadening of the beam and hence in particle losses. A further uncertainty in the measurement comes from the production of secondary particles by pair production, the rate of which increases strongly at high energies. As a result, the use of the Faraday cup is restricted to measuring the current of low-energy beams.

To minimize the error due to the emission of secondary electrons, the absorber is raised to a positive potential, in order to prevent electron emission. Alternatively the absorber can be surrounded by a negatively charged cage.

As well as providing a fairly precise absolute measurement of the beam current, Faraday cups offer a further advantage: if carefully designed they can achieve a very high bandwidth of many GHz, allowing the measurement of the time structure of the particle current. This is of particular interest in linacs operating in the S-band region. A prerequisite for high bandwidth is a consistent coaxial structure in which the ratio of the diameter D of the outer conductor to the diameter d of the inner conductor, and with it the impedance

$$Z = \frac{1}{2\pi}\sqrt{\frac{\mu_0}{\varepsilon_0}}\ln\frac{D}{d} \tag{10.1}$$

remains constant from the beginning of the cup to the readout electronics connected to the end of the coaxial cable. This avoids the production of reflections, which can strongly distort the time-varying voltage signal. Distortions can also arise due to the fact that the bandwidth of this device is nevertheless only finite, resulting in a measured current pulse $I_{\mathrm{meas}}(t)$ which is markedly longer than the original current pulse in the beam, especially for very short particle bunches with pulse lengths in the sub-nanosecond region. In this case the area under the pulses $\int I_{\mathrm{meas}}(t)\,dt$ is a useful measure of the number of particles absorbed.

10.1.3 The wall current monitor

The two monitors considered so far either absorb the beam totally (Faraday cup) or at least partially, as well as causing a transverse broadening of the beam (fluorescent screen). These types of monitors are thus only usable in linacs or at the end of beam transport lines running to an experiment. They are not suitable for measuring the current of a beam circulating in a cyclic accelerator. Here monitors must be employed which measure the current without disturbing the beam. An example is the **wall current monitor** [99].

Outside the metal vacuum chamber through which the beam travels there is no measured magnetic field, i.e. if one travels around a closed path outside the vacuum chamber and measures the magnetic field along it one finds

$$\oint \boldsymbol{B}_{\text{external}} \cdot d\boldsymbol{r} = 0. \tag{10.2}$$

Within the vacuum chamber, however, one finds that around the beam

$$\frac{1}{\mu_0} \oint \boldsymbol{B}_{\text{beam}} \cdot d\boldsymbol{r} = I_{\text{beam}}, \tag{10.3}$$

i.e. there must be a wall current I_{wall} flowing in the vacuum chamber, where

$$I_{\text{beam}} = -I_{\text{wall}}. \tag{10.4}$$

The beam current may thus be determined by instead measuring the current in the wall of the vacuum chamber. To do this, the chamber wall is broken at a

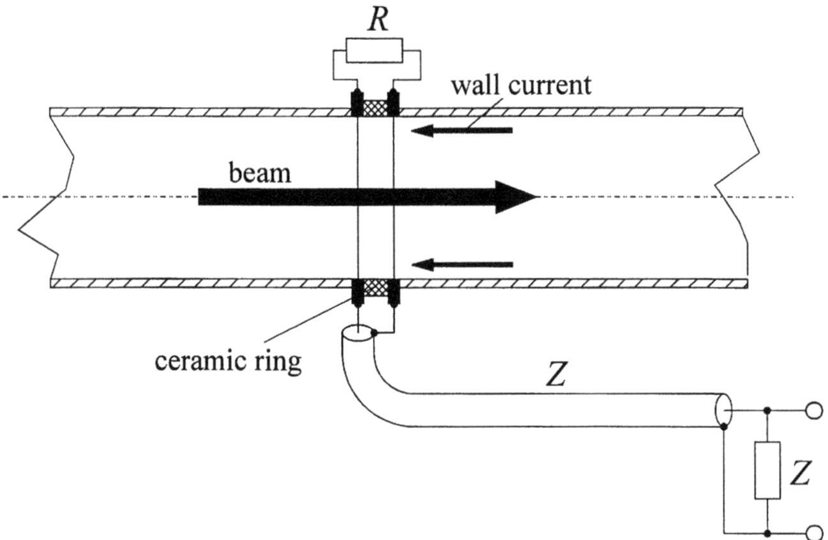

Fig. 10.3 Layout of the wall current monitor.

certain point and the gap filled with an insulating material, usually ceramic. The gap is bridged by an ohmic resistance R, across which the voltage

$$U_{\rm R} = R\, I_{\rm beam} \quad = \quad -R\, I_{\rm wall}. \tag{10.5}$$

is developed. This voltage can be measured directly. The layout of the wall current monitor is sketched in Fig. 10.3. If instead of only one resistor a large number are used, connected in parallel around the vacuum chamber, and if care is taken to keep the inductance low, then this current monitor can also achieve very high bandwidths of several GHz. In practice is has proven successful to choose a resistance of $R \approx 1\Omega$.

10.1.4 The beam transformer

A relativistic bunch of length σ_s containing N particles of elementary charge e, travelling past an observer at time t_0, produces a beam current pulse

$$I_{\rm beam}(t) \;=\; \frac{Ne}{\sqrt{2\pi}\,\tau}\exp\left(-\frac{(t-t_0)^2}{2\tau^2}\right) \qquad \text{with} \qquad \tau \;=\; \frac{\sigma_s}{c}. \tag{10.6}$$

A magnetic field is induced around the beam which varies very rapidly with time. If we now surround the beam with a ring-shaped iron core (assumed ideal) with a mean radius $R_{\rm core}$, as shown in Fig. 10.4, then inside the core the field

$$B(t) \;=\; \frac{\mu_0\mu_{\rm r}}{2\pi\, R_{\rm core}}\, I_{\rm beam}(t) \tag{10.7}$$

is produced. A coil is wound evenly around this iron core, which has a cross-sectional area A. A voltage

$$U_{\rm ind}(t) \;=\; n\, A\, \dot{B}(t) \;=\; \frac{\mu_0\mu_{\rm r}\, n\, A}{2\pi\, R_{\rm core}}\, \dot{I}_{\rm beam}(t) \tag{10.8}$$

is induced in the coil, which has a total of n windings. This arrangement acts like a transformer, in which the beam is the primary winding and the inductive coil is the secondary winding. Hence it is known as a **beam transformer** [100]. An older name, **Rogowski coil**, is also sometimes used.

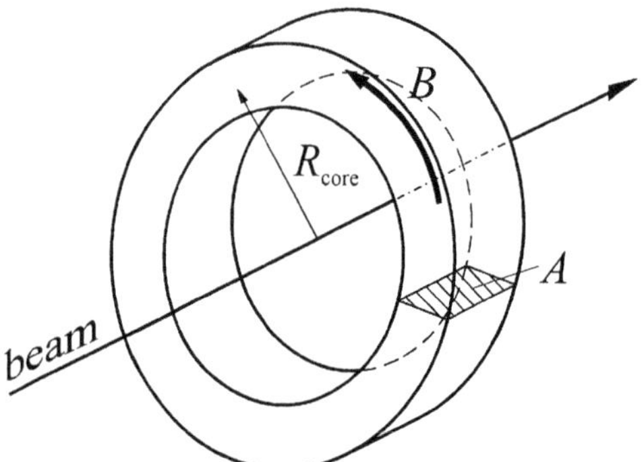

Fig. 10.4 Ideal iron core around a particle beam.

From (10.8) we expect a signal that is proportional to the first time derivative of the beam current. In fact the relationship is more complicated, because the iron core is not ideal. The losses which occur in the coil grow proportionally with the square of the frequency, which means that the high frequency components are strongly suppressed. The inductive coil also has a stray capacitance, which again weakens the high frequency signals in particular. A wound tape core consisting of thin layers of iron is often used, because it has a high permeability μ_r and hence gives a high degree of sensitivity. The eddy currents in such a core are low, but still non-negligible. Overall this current monitor behaves like a low-pass filter with a threshold frequency of a few tens of MHz, considerably lower than the spectral range of the bunch of several GHz. A real beam transformer can thus be simply described by the equivalent circuit shown in Fig. 10.5.

The output voltage of the beam transformer is then, to a good approximation,

$$U_{\text{out}}(t) = \frac{1}{C_T} \int I(t)\, dt, \tag{10.9}$$

where the current $I(t)$ through the inductive coil is given by the relation

$$\dot{I}(t) + \frac{1}{C_T R_T} I(t) = \frac{\dot{U}_{\text{ind}}}{R_T}. \tag{10.10}$$

As already mentioned, the bandwidth of a real beam transformer is very small compared to the frequency spectrum of the beam current. In other words, the time constant $C_T R_T$ is long compared to the duration τ of the bunch pulses. Hence the contribution from the term $I(t) / C_T R_T$ is almost zero and may be neglected. From (10.10) it then follows that

$$\dot{I}(t) \approx \frac{\dot{U}_{\text{ind}}}{R_T} \qquad \Longrightarrow \qquad I(t) \approx \frac{U_{\text{ind}}}{R_T} + C_1. \tag{10.11}$$

For simplicity we set the constant of integration $C_1 = 0$. According to (10.9) the secondary voltage takes the following form

$$U_{\text{out}}(t) = \frac{1}{C_T} \int I(t)\, dt \approx \frac{1}{C_T R_T} \int U_{\text{ind}}\, dt. \tag{10.12}$$

Referring again to (10.8) and once again setting the constant of integration to zero, it immediately follows that

$$U_{\text{out}}(t) = \frac{1}{C_T R_T} \frac{\mu_0 \mu_r\, n\, A}{2\pi\, R_{\text{core}}} I_{\text{beam}}(t). \tag{10.13}$$

This relation tells us that the time dependence of the output voltage $U_{\text{out}}(t)$ is proportional to the beam current $I_{\text{beam}}(t)$. In fact by using an oscilloscope, for example, one again sees an almost Gaussian pulse and not a differentiated signal such as one might expect from (10.8). It turns out that the signal is only roughly proportional to the beam current, and even that is only true for relatively long

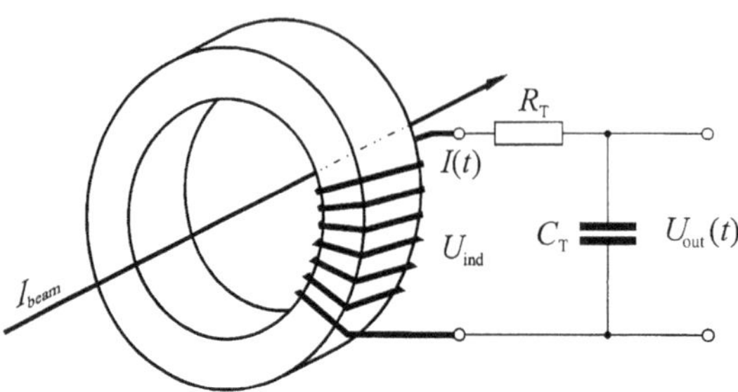

Fig. 10.5 Simplified equivalent circuit corresponding to a real beam transformer.

bunches with limited frequency components. For short bunches the signal $U_{\text{out}}(t)$ is considerably longer than the current pulse. In this case the simple equivalent circuit shown in Fig 10.5 is far from sufficient. Here, however, the area under the voltage pulse

$$\int U_{\text{out}}(t)\, dt \;\propto\; N e \tag{10.14}$$

can be used as a good measure of the number of particles. In any case, it is always necessary to calibrate the monitor, which may be done either using a test beam and a Faraday cup, for example, or by replacing the beam through the device by a wire and passing a known current through it.

Equation (10.14) does not actually contribute, since the transformer is not able to deliver a DC component. The zero line shifts until the integral vanishes. It is therefore necessary to relate the integral to the shift in the zero line. This effect can lead to errors, especially for beams with unevenly spaced bunches.

10.1.5 The current transformer

In order to measure the average current in a circulating beam or the DC current of a beam with no RF structure, as is sometimes found for proton or heavy-ion beams, the principle of the current transformer [101] is employed. The operation of this device is best described with the help of Fig. 10.6.

The current transformer again consists of an iron core around the beam, but with a primary coil wound around it. A periodically varying current $I(t) = I_0 \sin \omega t$ is passed through the coil. The coil is sufficiently strongly excited for the transformer to be driven around a significant part of the hysteresis loop (Fig. 10.7). Let us begin by assuming that no beam is present, i.e. $I_{\text{beam}}(t) = 0$. The hysteresis curve is then **symmetric** about zero. A periodic magnetic field $B(t)$ is generated in the iron core which, because of the non-linearity of the hysteresis curve, contains higher harmonics in addition to the frequency ω. As a consequence of the symmetry, only **odd-numbered** frequency components of the form $(2j-1)\omega$ are present, with $j = 1, 2, \ldots$.

Using a second winding, which acts as an induction coil, voltages of these frequencies can be selectively measured. By doubling the frequency of the signal generator the second harmonic 2ω is produced. This serves as a reference signal

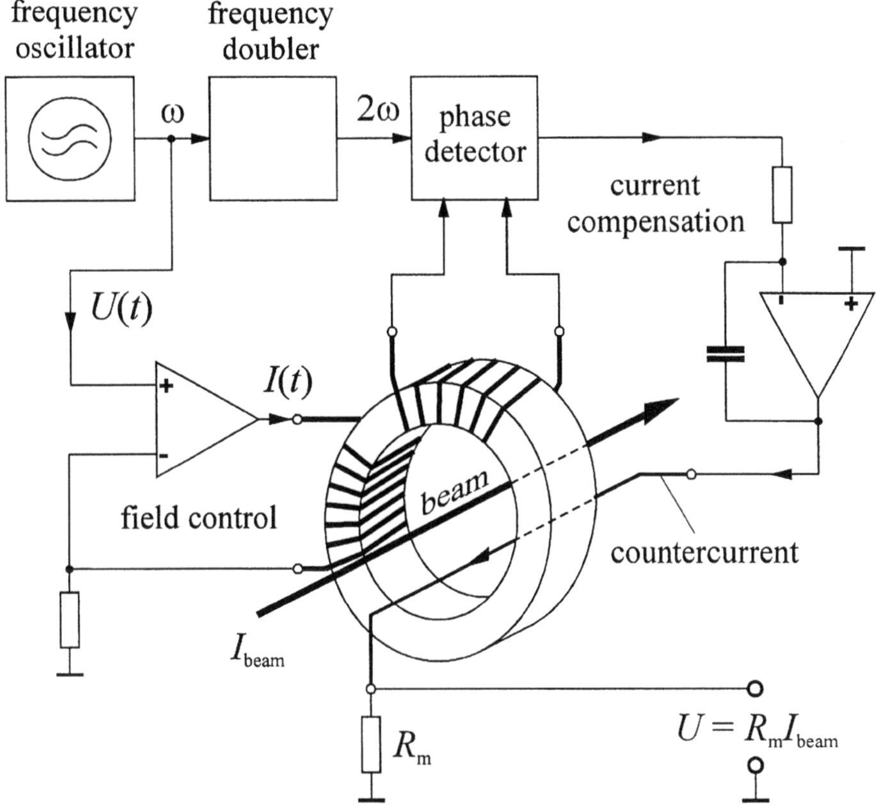

Fig. 10.6 The current transformer, used to measure the DC component of the beam current.

for the detector, which is very sensitive to phase shifts. The detector thus picks out these very small-bandwidth and largely noise-free components from the induced voltage. None are present in the symmetric case, and so the detector gives no output.

This situation changes if direct current from the beam also flows though the ring. The hysteresis curve is then shifted horizontally (Fig. 10.8) and so is no longer symmetric about the origin. As a result, even-numbered harmonics $2j\,\omega$ are now also present and the phase detector, which is sensitive to the second harmonic, gives a non-zero output.

This signal is used to produce a direct current in a wire running though the coil parallel to the beam, which exactly compensates for the effect of the beam current. The second harmonic thus disappears from the frequency spectrum. The compensating current is therefore practically identical to the beam current and can be measured very simply, for example through a resistor.

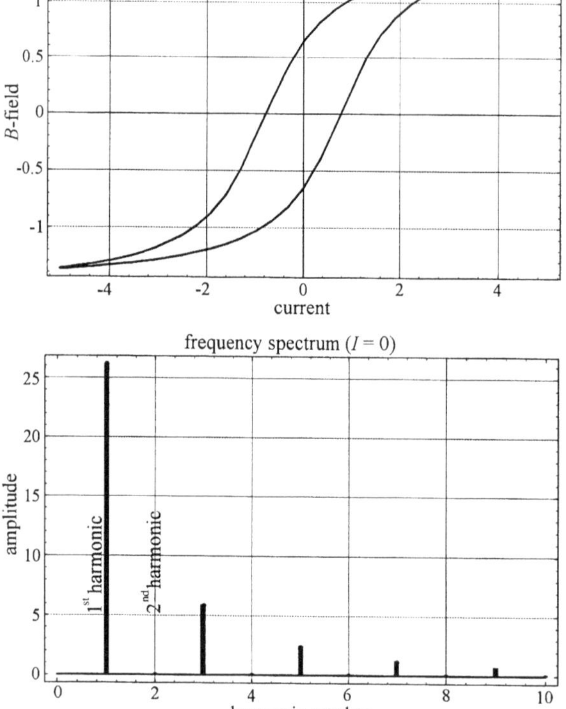

Fig. 10.7 Hysteresis curve and frequency spectrum of an iron core undergoing symmetric periodic excitation in the absence of any beam current ($I_{\text{beam}} = 0$).

10.1.6 The measurement cavity

A beam accelerated by a high-frequency field has a periodic time structure. The beam current may thus be written in the form

$$I(t) = \sum_{j=1}^{n} \left[a_j \cos(j\,\omega t) + b_j \sin(j\,\omega t)\right]. \tag{10.15}$$

The lowest frequency present, ω, is the revolution frequency in cyclic accelerators. The Fourier coefficients a_j and b_j are proportional to the beam current. If we filter out one harmonic (which in principle may be freely chosen) from this frequency spectrum and measure its intensity, we have a further measure of the beam intensity. This is most successful if a round cavity is used, operating in the TM_{010} mode at the chosen frequency (Fig. 10.9). This system closely resembles an accelerating cavity, as described in Chapter 5.

As the beam passes along the axis of the resonant cavity, the accompanying electric field generates a high-frequency wave. Once the transient compo-

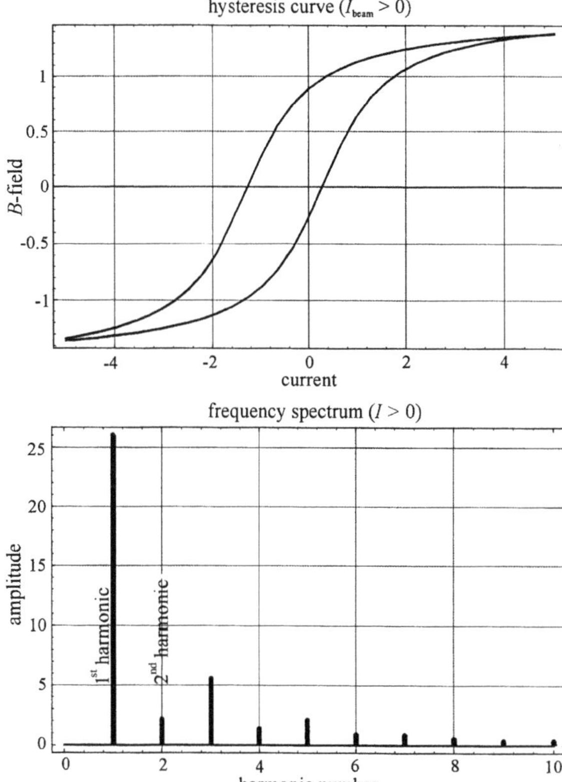

Fig. 10.8 Hysteresis curve and frequency spectrum of an iron core undergoing asymmetric periodic excitation in the presence of a beam current ($I_{\text{beam}} > 0$).

nents have died away, after many passes through the cavity, a stationary state is reached which depends on the beam current. This corresponds to the resonant excitation of an electric oscillator by a periodic current.

In order to make a measurement, part of the field energy must be extracted from the cavity and then measured electronically. This may be done either inductively using a coupling loop or capacitively across a coupling antenna. It does not matter in principle which method is chosen. As an example let us consider the capacitive coupling antenna in more detail. This carries the signal via a coaxial cable to a terminating resistance, which in practice is always given by the input impedance of the readout electronics which follow. This arrangement is illustrated in Fig. 10.10.

Due to the additional coupling capacitance C_{c} and the real load resistance R_{c}, the total capacitance and effective shunt impedance $R_{\text{s}}^{\text{eff}}$ of the oscillator change. Let us assume that the coupling is sufficiently weak, i.e. the value of the AC resistance $1/\omega\, C_{\text{c}}$ and the shunt impedance with no load R_{s} of the cavity are large compared to the load resistance R_{c}. We then have $\omega\, C_{\text{c}}\, R_{\text{c}} \ll 1$. Under

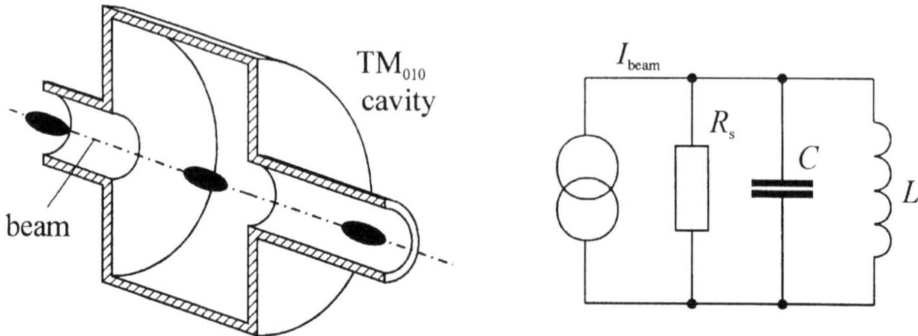

Fig. 10.9 Cavity for measuring beam current. Its electrical behaviour is described by the equivalent circuit.

these assumptions it follows that

$$R_s^{\text{eff}} = \frac{R_s}{1 + \omega^2 C_c^2 R_c R_s} \tag{10.16}$$

and the effective capacitance of the circuit

$$C^{\text{eff}} = C + C_c \approx C. \tag{10.17}$$

The detuning of the circuit is therefore small, but the quality factor (Q-value)

$$Q = \frac{\omega}{\Delta\omega} = \frac{R_s^{\text{eff}}}{\omega L} = R_s^{\text{eff}} \sqrt{\frac{C}{L}} \qquad (\omega = 1/\sqrt{L\,C}) \tag{10.18}$$

can be considerably reduced by the coupling, compared to the unloaded case. Rearranging yields

$$Q = \frac{R_s}{\omega L\,(1 + \omega^2 C_c^2 R_c R_s)} = \sqrt{\frac{C}{L}} \cdot \frac{R_s\, C\, L}{C\,L + C_c^2 R_c R_s}, \tag{10.19}$$

i.e. the Q-value decreases sharply as the coupling capacitance increases. This is, however, desirable. Cavity resonators typically have very high Q-values and correspondingly narrow, sharp resonance peaks. This means that even slight variations in temperature are sufficient to shift the resonance curve and distort the measurement, requiring the use of an expensive frequency control system. In this case, however, the coupling means that the resonator is considerably damped, and a suitably-sized cavity will have such a large bandwidth that the sensitivity to temperature variation becomes negligible. Frequency regulation is thus no longer necessary.

In the simplest case, the coupled signal can be rectified using a diode and then directly measured (Fig. 10.10). Here it must be remembered that diodes have a strongly non-linear behaviour, especially for low signals, and this measurement is

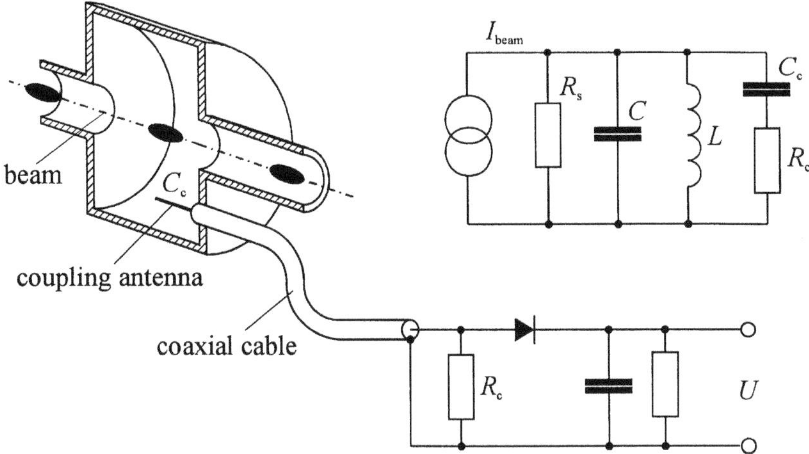

Fig. 10.10 Signal extraction from a measurement cavity using a capacitive coupling antenna. The equivalent circuit contains the coupling capacitance C_c and the load resistance R_c.

hence rather imprecise. For a high-precision measurement, a technique common in communications engineering must be used in which the signal is overlaid on a low-frequency reference signal, amplified, and then rectified. This allows a highly linear measurement over the entire range.

In cyclic machines, such as storage rings, a stationary and hence well-defined state is always achieved. In linacs with short beam pulses, however, the transient oscillations are more significant since of course they do not die away during a single pass of the beam. The measured output thus depends on the duration of the pulse rather than the number of particles and so is difficult to interpret. In this case the pickup cavity delivers a perfectly usable relative signal, but if absolute measurements are required then it must first be accurately calibrated using other methods, such as the Faraday cup.

10.2 Determination of the beam lifetime in a storage ring

A beam circulating in a storage ring decays in intensity due to collisions with residual gas atoms, occasional large energy losses through synchrotron radiation in the case of electrons, and non-linear resonances. In the simplest description the decline in intensity has the exponential form

$$I(t) = I_0 \exp\left(-\frac{t}{\tau_{\mathrm{beam}}}\right). \tag{10.20}$$

Here τ_{beam} is the so-called **lifetime** of the beam and the relation (10.20) may be regarded as a definition of beam lifetime. Differentiating this relation immediately yields

$$\frac{dI(t)}{dt} = -\frac{I_0}{\tau_{\mathrm{beam}}} \exp\left(-\frac{t}{\tau_{\mathrm{beam}}}\right) = \frac{I(t)}{\tau_{\mathrm{beam}}} \tag{10.21}$$

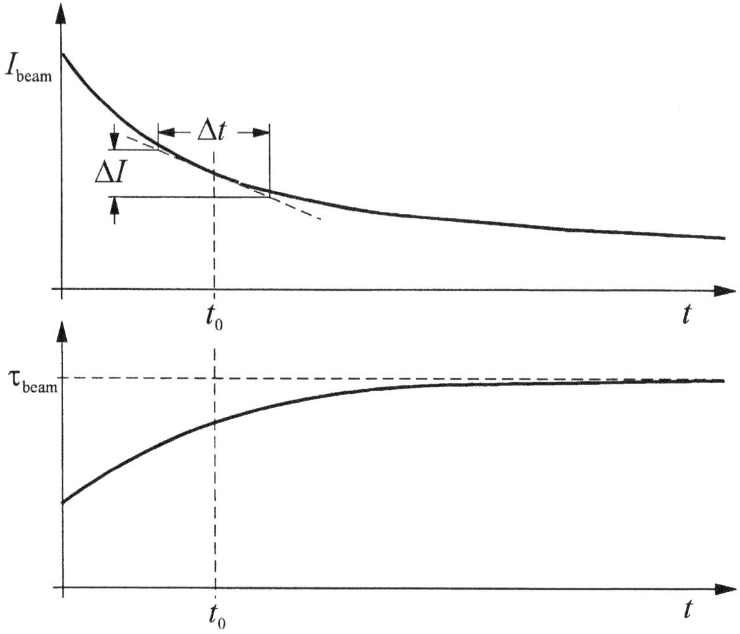

Fig. 10.11 Time dependence of the beam current and lifetime. The lower curve shows that the lifetime increases with decreasing current.

and hence

$$\tau_{\text{beam}} = \frac{I(t)}{dI/dt}. \tag{10.22}$$

In addition, the lifetime often does not remain constant during machine operation. For electron beams in particular the lifetime is often relatively short at the beginning, when the intensity is high, because the intense synchrotron radiation causes a high level of gas desorption on the surface of the vacuum chamber and so reduces the vacuum pressure. As the beam current decreases the vacuum improves and the lifetime increases. Hence the lifetime is itself a function of time (Fig. 10.11).

The instantaneous lifetime at a particular time t_0 is given by (10.22). Using one of the current monitors described in the preceding section, the current is continuously monitored, with measurements repeated at frequent intervals. A straight line is fitted to each measurement falling in the time interval Δt about t_0, and from this the gradient $\Delta I/\Delta t$ is obtained. The average instantaneous current $\langle I(t_0)\rangle$ is also calculated from the measured values. From (10.22) the instantaneous lifetime is then

$$\tau_{\text{beam}} = \frac{\langle I(t_0)\rangle}{\Delta I}\,\Delta t. \tag{10.23}$$

Since the beam lifetime can range from a few seconds to many hours, depending on the operating conditions, it is useful to be able to vary the time interval between measurements. For short lifetimes the beam current varies rapidly and only a few measurements are required for a reliable lifetime measurement. In this case a short time interval is chosen, which is often also necessary because the rapid decrease in the beam intensity does not allow a long measurement. For very long beam lifetimes of several hours, the individual current measurements must last sufficiently long for the statistical fluctuations in the measured values not to cause large errors in the lifetime measurement.

10.3 Measurement of the momentum and energy of a particle beam

10.3.1 The magnetic spectrometer

The simplest way to measure the momentum and hence also the energy of a particular charged particle is to measure the angle of deflection in a known magnetic field. As we saw in Chapter 3, the angle of deflection in the x-s plane is given by (Fig. 10.12)

$$d\alpha(\boldsymbol{r}) = \frac{e}{p} B_z(\boldsymbol{r})\, ds, \tag{10.24}$$

where the vertical field component at the position $\boldsymbol{r}$ is given by $B_z(\boldsymbol{r})$. Homogeneous magnetic fields are generally used, as these greatly simplify the measurement. The total bending angle α_{tot} as the particle passes through a particular magnet is obtained by integrating equation (10.24) along the particle path, i.e. beginning at a point of zero field before the magnet and then ending at another zero-field point behind the magnet. The particle momentum is then

$$p \;=\; \frac{e}{\alpha_{\text{tot}}} \int\limits_{\text{path}} B_z\, ds. \tag{10.25}$$

For a particle of known type the energy is then immediately given by

$$E \;=\; \sqrt{p^2c^2 + (m_0c^2)^2}. \tag{10.26}$$

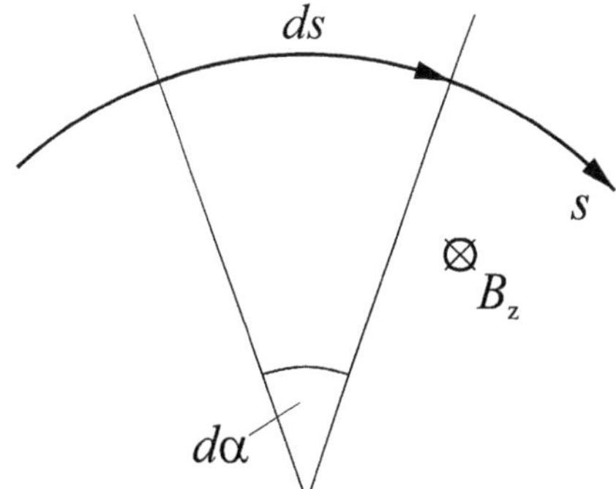

Fig. 10.12 Deflection of a charged particle in a magnetic field.

For extremely relativistic beams with $\gamma \gg 1$, as is almost always the case for electrons, we obtain the simplified expression

$$E = pc = \frac{ec}{\alpha_{\mathrm{tot}}} \int_{\mathrm{path}} B_z \, ds. \tag{10.27}$$

Figure 10.13 shows the design of a simple spectrometer, such as might be installed after a linac to determine the final beam energy.

The angle of the incoming beam must be precisely defined. This can be achieved by using precisely aligned screens to fix the beam position accurately. After deflection the position of the beam and hence the bending angle can be measured using a fluorescent screen, for example. To determine the energy the field integral $\int B_z \, ds$ is again required, which is obtained by a careful measurement of the field as a function of coil current. It is also important to remember that iron magnets undergo hysteresis, which can distort the measurement. In order to avoid this, it is important to travel several times around the hysteresis curve of the magnet, in a careful and well-defined way. This is necessary both when measuring the field and during the energy measurement, in order to define a reproducible magnet 'history'. By convention the required field value is then approached from **below**.

In cyclic accelerators such as storage rings the total bending angle of all the dipole magnets must of course be $\alpha_{\mathrm{tot}} = 2\pi$. It then immediately follows from (10.27) that

$$E = \frac{ec}{2\pi} \oint_{\mathrm{dipole}} B_z \, ds. \tag{10.28}$$

The relation between the field integral and the magnet current must again be found by measuring the field. In storage rings it has proven useful to connect an

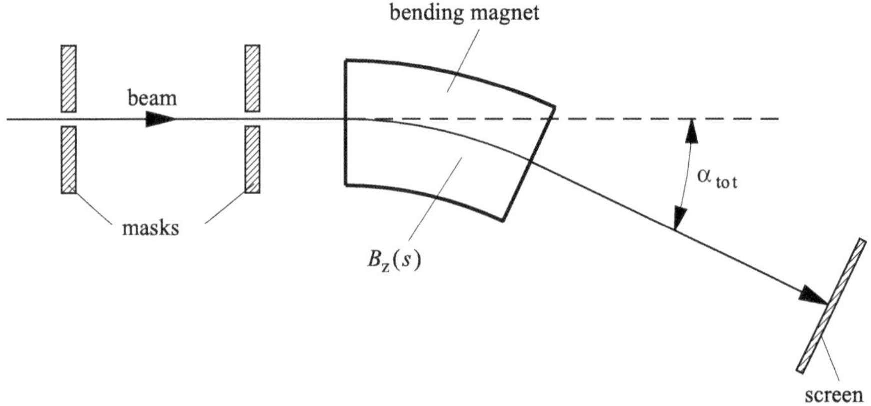

Fig. 10.13 Simplest implementation of a magnetic spectrometer to measure particle energy.

additional identical dipole in series with the accelerator dipoles and to install a precise field gauge within it, such as a nuclear magnetic resonance (NMR) probe. The field and hence also the beam energy can then be continuously monitored, usually to very high accuracy. In this way it is possible to measure the beam energy with a very small uncertainty of $\Delta E/E \approx 2 \times 10^{-4}$.

10.3.2 Energy measurement by spin depolarization

By far the most precise method of measuring the beam energy in an electron storage ring is based upon spin depolarization. A full treatment of this technique is beyond the scope of this chapter, and here we will only outline the most important steps. For an exact treatment and further details of this method, the reader is referred to the literature cited below.

The electrons injected into a storage ring have their spins uniformly distributed in all directions and so are unpolarized. As they circulate they precess in the vertical field of the bending magnets. In 1963 A.A. Sokolov and I.M. Ternov first showed [102] that after a period of time the electron spins align themselves anti-parallel to the magnetic field (parallel in the case of positrons) as a result of emitting synchrotron radiation. Hence they become **polarized**. The polarization P builds up over time according to

$$P(t) = P_\infty \left[1 - \exp\left(-\frac{t}{\tau}\right)\right] \tag{10.29}$$

where

$$\frac{1}{\tau} = \frac{1}{\tau_\mathrm{P}} + \frac{1}{\tau_\mathrm{D}} \qquad \text{and} \qquad P_\infty = \frac{0.924}{1 + \tau_\mathrm{P}/\tau_\mathrm{D}}. \tag{10.30}$$

The polarization time in seconds due to synchrotron radiation is

$$\tau_\mathrm{P} = 98 \frac{R^3}{E^5} \frac{\langle R \rangle}{R}. \tag{10.31}$$

R is the bending radius in metres of the magnets, and $\langle R \rangle$ is the average radius of the machine. The energy E is given in GeV. The polarization time ranges from a few minutes to many hours, depending on the size of the storage ring and the energy of the electrons. The quadrupole and higher multipole magnets and other fields acting on the beam can disturb the spins again, at least partly, leading to depolarization over an approximate depolarization time τ_D [103]. As a result of this depolarization, the maximum achievable polarization is generally limited to $P < 94\%$. In order to maintain sufficient polarization we require $\tau_\mathrm{D} > \tau_\mathrm{P}$.

Once the polarization has built up we may then perturb it, for example using a weak transverse magnetic field which oscillates rapidly at a precise frequency ω_depol[104, 105]. Such a field can be easily produced using a fast kicker magnet. The angle of the precession axis then increases with each revolution in a resonant fashion until it is perpendicular to the direction of the dipole field, at which point the polarization vanishes. The depolarization frequency ω_depol is given by

$$\omega_\mathrm{depol} = |\gamma a \pm n|\, \omega_\mathrm{rev} \qquad \text{with} \qquad n = \text{integer}. \tag{10.32}$$

Here

$$a = \frac{1}{2}(g-2) = 1.159652193\,10^{-3} \tag{10.33}$$

is the anomalous magnetic moment of the electron, which is known to great precision. Note that apart from the particle energy γ, (10.32) contains only the revolution frequency ω_{rev} and the depolarization frequency ω_{depol}, both of which can be measured to very high precision. The details of the magnet structure and the field strength, which is known only to a few parts per thousand, do not matter. Hence it is necessary only to measure the depolarization frequency as accurately as possible in order to determine the beam energy γ from (10.32). This is performed by changing the frequency of the transverse perturbing kicker field in small steps and observing the variation of the polarization level. In this way it is possible to observe the very sharp resonance at which the polarization completely disappears.

To measure the energy a polarimeter is also required in the storage ring, in order to observe the polarization building up and then being deliberately destroyed by the rapidly oscillating transverse magnetic field. There are a variety of spin-dependent effects which may be exploited to make this measurement [106, 107].

As an example we will briefly describe the method of Compton backscattering, which has been successfully used in several experiments [108]. A laser beam is fired at a very small angle to the electron beam, and a few photons are scattered back off the electrons. Due to the highly relativistic velocity of the electrons, the photons reach relatively high energies, in the X-ray region. They are scattered in the direction of the beam, at a very small angle to it. These back-scattered photons may be observed above and below the electron beam by γ detectors such as scintillation counters.

If the electron beam is unpolarized, two detectors arranged symmetrically on either side of the beam axis will record the same rate of backscattered photons. When the beam is polarized, however, one direction will be slightly favoured and the counting rates will become unequal. Comparing the two counting rates yields a measure of the instantaneous polarization of the electron beam as it circulates in the storage ring.

10.4 Measurement and correction of the beam position

For efficient accelerator operation with few particle losses it is necessary for the centre of the beam always to lie as close as possible to the ideal orbit, usually defined by the axes of the quadrupoles. The required tolerances are particularly tight in the case of very low emittance storage rings, such as those used to produce synchrotron radiation. Here the transverse deviation of the circulating beam from the orbit must not exceed 100 to 150 μm. It is therefore essential to measure the transverse position of the beam at as many points around the accelerator as possible and to correct it where necessary.

10.4.1 Transverse beam position measurement

A whole series of different devices exist for measuring the beam position in accelerators and transport lines, collectively known as Beam Position Monitors (BPMs). A good review of these may be found in [109]. In a beam transport line the beam position can simply be measured using a fluorescent screen, as described above, which is inserted into the beam line when necessary. Of course this device is not suitable for cyclic accelerators.

Another very simple way to determine the beam position is to move a blade transversely towards the beam until it begins to scrape against it and remove particles. The beam loss from this **scraper** can then be measured, for example using γ-counters mounted downstream in the vacuum chamber. Since the beam always has a finite thickness which is not always known beforehand, a pair of scrapers must be used, one on either side of the the beam. The central point between the two scraper positions then gives the beam position, assuming a reasonably symmetric density distribution within the beam. This position measurement is relatively accurate, since the scraper positions may be very precisely controlled using micrometer screws. Scrapers may also be used in storage rings, with some restrictions: in particular, the considerable beam losses do not allow this technique to be routinely used to control the beam position.

Magnetic beam position monitor

To continuously monitor the beam position, techniques must be chosen which measure the magnetic field due to the beam and so avoid particle losses. A modified form of the beam transformer described above (Section 10.1.4) is well suited to this task. If the beam does not travel close to the centre of the transformer core, then a stronger field is induced in the region of the iron core near to the beam than on the opposite side. If two separate short coils are wound at opposite points around the core, then the off-centre beam will generate signals of differing strengths in these coils. The beam position can then be obtained simply by measuring this difference. In order to measure the position in both planes simultaneously, four coils are arranged at 90° intervals around the transformer core, as shown in Fig. 10.14.

Close to the beam axis, this monitor gives an output signal for each plane which increases linearly with the beam displacement. For very large displacements there are strong deviations from linearity, but this does not present any problem since the beam should in general lie along the ideal orbit, and exact position measurements are only of interest close to this orbit.

Monitor with four electrodes

The most successful type of position monitor in storage rings has proven over time to be one which couples to the **electric** field. It consists of four electrodes (electrical pickups) arranged symmetrically around the beam axis, as shown in Fig. 10.15.

If this arrangement is compared with the magnetic monitor in Fig. 10.14 it is noticeable that the electrodes are rotated through around 45° with respect

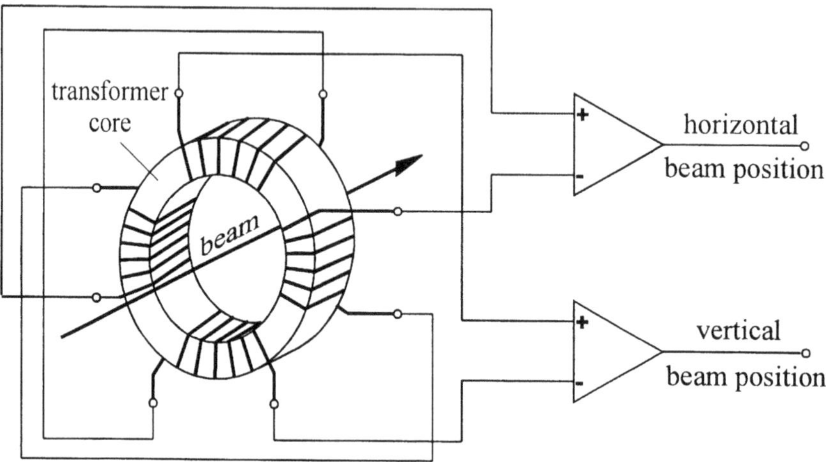

Fig. 10.14 Magnetic beam position monitor. It operates in the same way as a beam transformer. The difference in signals from the two opposite coils within each pair gives a measure of the beam position in that plane.

to the beam axis. This is necessary in electron accelerators to prevent the electrodes being directly struck by synchrotron radiation. The high power of the radiation could destroy the isolated electrodes, and in addition the secondary electrons produced by the photo-electric effect could cause a systematic error in the measurement. A further reason for the tilted arrangement is the geometry of the vacuum chamber. In almost all modern storage rings this is broad and flat, as is clearly seen in Fig. 10.15. The electrodes in the horizontal plane would otherwise be very far from the beam, and hence less sensitive.

The signal at an individual electrode results from the displacement current and so is proportional to the second time derivative of the bunch current. A certain smearing of the signals again occurs due to the finite bandwidth of the electrodes, even though this is relatively high at a frequency of several GHz. The intensity I_{pickup} of the signal depends on the distance from the beam, where this dependence is a relatively complicated function $I_{\text{pickup}}(\boldsymbol{r})$. To a good approximation we may assume $I_{\text{pickup}}(\boldsymbol{r}) \propto 1/r$. In the region near the beam the separation in both planes may be very precisely determined using the following relations

$$\begin{aligned} \Delta x &= a \frac{(I_2 + I_3) - (I_1 + I_4)}{\sum\limits_{j=1}^{4} I_j} \\ \Delta z &= a \frac{(I_1 + I_2) - (I_3 + I_4)}{\sum\limits_{j=1}^{4} I_j}, \end{aligned} \tag{10.34}$$

where I_j $(j = 1, 2, 3, 4)$ denotes the signal intensity in each of the four electrodes. The numbering of the electrodes is as illustrated in Fig. 10.15. The

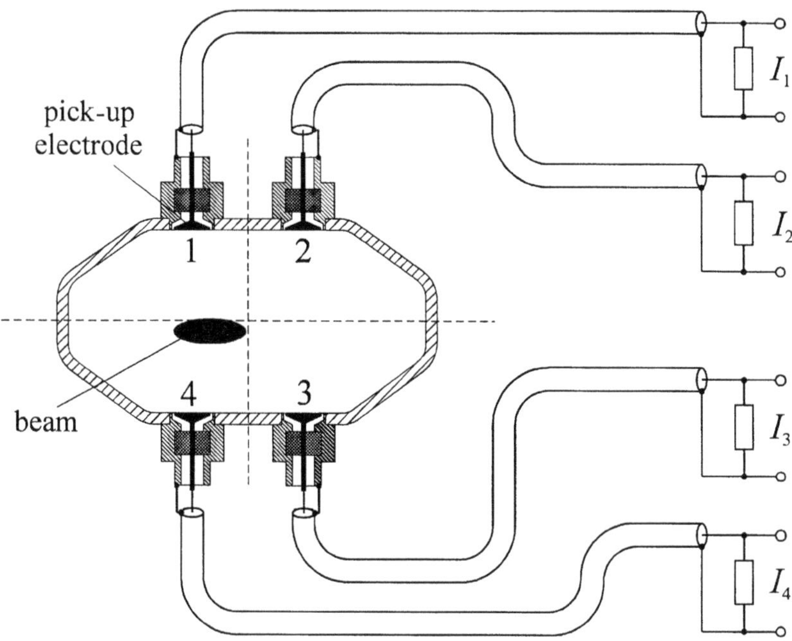

Fig. 10.15 Beam position monitor with four electrodes. The electrodes are tilted away from the beam axis by about 45° in order to reduce the amount of synchrotron radiation hitting them directly, among other reasons.

monitor constant a is determined empirically by calibrating the device in the laboratory.

If the beam lies exactly in the middle of the monitors, then ideally all the signals will have the same intensity, namely $I_j = I_0$. In reality, however, there are variations in signal sizes due to tolerances in the electrodes, the vacuum chamber geometry, the cables, and the electronics which follow. If a signal has an intensity $I_0 + \Delta I$, then according to (10.34) this results in a position error of

$$\Delta x_{\text{error}} = a \frac{\Delta I}{4 I_0}. \tag{10.35}$$

A typical value for the monitor constant is $a = 35$ mm. If the absolute measurement accuracy required is $\Delta x_{\text{error}} < 0.1$ mm then the relative error in an electrode signal may not be larger than

$$\frac{\Delta I}{I_0} = \frac{4 \Delta x_{\text{error}}}{a} < 1.2\%. \tag{10.36}$$

The four electronic channels connected to the electrodes must be very nearly identical, which is rather difficult to achieve because of their broad dynamic range. The orbit measurement must give reliable results for beam currents below 1 mA, but must also operate reliably with currents of several hundred mA.

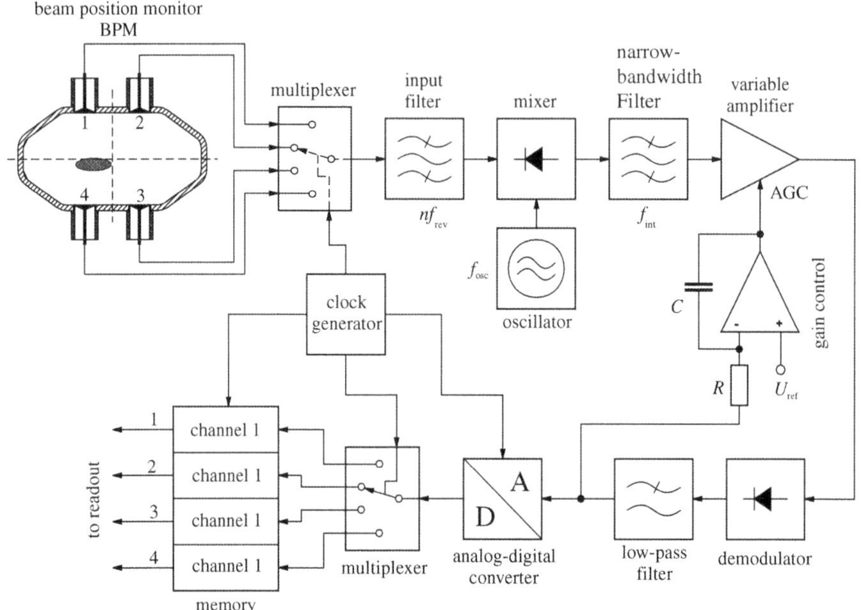

Fig. 10.16 Block diagram of the readout electronics from a beam position monitor. The signals from the four electrodes are switched into the circuit via a multiplexer and after digitization are multiplexed to the four data storage lines.

An expensive regulation system would be necessary to ensure that the channels behave similarly over the whole dynamic range. On the other hand, the measurement does not need to be performed particularly quickly; it is usually sufficient if the measurements are completed within one second. This allows all four signals to be read out one after another using a single channel. Figure 10.16 shows the basic circuit diagram of the readout electronics.

The bunches induce a periodic signal in the electrodes which consists of a combination of many harmonics of the revolution frequency $f_{\rm rev}$. In order to measure this signal with a narrow bandwidth and hence with low noise, a single harmonic is chosen from this spectrum. It does not in principle matter which is chosen, but there are a few important criteria. Since the electrodes yield a differentiated signal, the intensity of the harmonic is proportional to its frequency, and so a relatively high harmonic is desirable. However, the frequency should not be so high that it exceeds the cut-off frequency of the vacuum chamber, which results in propagating waves in the cavity which distort the result. It has proven successful simply to use the frequency of the accelerating cavity $f_{\rm RF} = q\, f_{\rm rev}$ or a value close to it.

First of all a rough selection of the harmonic is performed by a filter at the beginning of the circuit. This harmonic is mixed with a much lower intermediate-frequency signal from a local oscillator, e.g. $f_{\rm int} = 10.7$ MHz. The signal is then passed through a very narrow-bandwidth quartz filter of a type commonly used in communications, tuned very precisely to the chosen harmonic. An adjustable-gain amplifier (AGC) then raises the output signal to a well-defined level, depending on the beam current . After rectification and suppression of the high-frequency components by a low-pass filter, the analogue signal is digitized and stored.

A centrally generated clock drives the multiplexers at the input and output of the circuit, which select the input from the electrodes one after another and write the resulting digital value to one of four data lines. The clock also sends a signal to start the analogue-to-digital converter at the right moment.

The intermediate-frequency amplifier has a very long time constant compared to the period of the central clock. Hence the amplifier gain does not change if the four channels are selected one after another. It is thus possible to arrive at an amplified level corresponding to the average of all four signals. The output signals due to this automatic gain control amplifier are independent of the beam current and effectively correspond to the individual signals divided by the sum. Equation (10.34) then simplifies to

$$\begin{aligned} \Delta x &= a' \; [(I_2 + I_3) - (I_1 + I_4)] \\ \Delta z &= a' \; [(I_1 + I_2) - (I_3 + I_4)] \,. \end{aligned} \tag{10.37}$$

The intensities I_j are now the digitized values stored in the four memories and a' is a monitor constant associated with the electronics.

Position monitors based on this principle have successfully achieved the required precision of $\Delta x_{\mathrm{error}} < 150\,\mu\mathrm{m}$, and relative beam shifts down to below $10\ \mu\mathrm{m}$ can even be measured. There is, however, an intrinsic problem in measuring the absolute beam position, which we must not ignore. Fundamentally, it is not possible to define with arbitrary precision the point **relative to which** the beam position is being measured. The monitor electrodes are connected to the vacuum chamber and this is generally fixed to the magnets. However, the magnets can only be positioned with a tolerance of about $\pm$ 0.2 mm. Moreover, the alignment errors of the quadrupoles also create orbit distortions, as described in Section 3.15. Even if the beam position is adjusted so that it has no offset in any of the monitors, this will not necessarily correspond to the real ideal orbit. Empirical corrections are sometimes necessary to optimize the important parameters of the beam, such as the emittance.

Position measurement with a resonant cavity

The beam position may also be measured using a resonant cavity, in a similar way to the current measurement described in Section 10.1.6. Here a mode must be chosen for which the electric field component along the beam direction is zero. If the beam lies exactly along the orbit then its field cannot induce any oscillation in the resonator, and so there is no output signal. The TM_{210} mode satisfies this condition in a rectangular cavity. An outline of the device is shown in Fig. 10.17.

From the derivation in Chapter 5 or in the literature (e.g. [59], [110]) it can be seen that the electric field component in a rectangular resonant cavity has

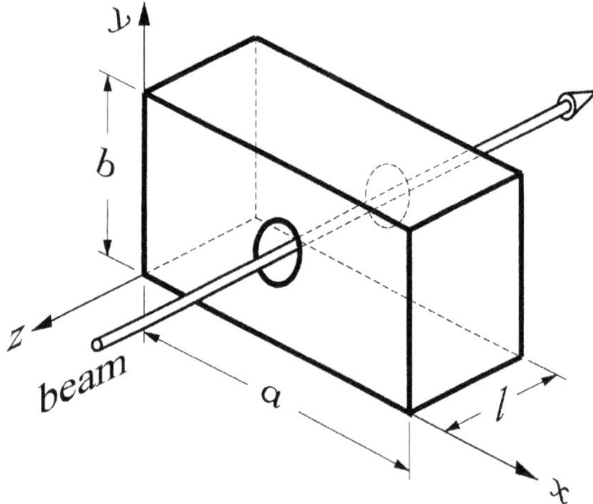

Fig. 10.17 Position measurement with a rectangular cavity in the TM_{210} mode. The z-axis of the resonant cavity runs parallel to the beam direction. In the centre of the beam ($x = a/2$) the electric field component E_z vanishes.

the form

$$E_z = C_\mathrm{e}(k_x^2 + k_y^2) \sin k_x\, x \;\; \sin k_y\, y \;\; \cos k_z\, z \tag{10.38}$$

with

$$\begin{aligned} k_x &= \frac{m\,\pi}{a} \\ k_y &= \frac{n\,\pi}{b} \\ k_z &= \frac{p\,\pi}{l}. \end{aligned} \tag{10.39}$$

The TM_{210} mode is obtained by setting $m = 2$, $n = 1$ and $q = 0$. The form of the electric field is illustrated in Fig. 10.18.

Away from the beam axis the field increases, with opposing sign on either side. To a good approximation $E_z(x)$ may be assumed to be linear near the point at which it passes through zero. A horizontally displaced beam can therefore couple to this mode, with the amplitude of the induced oscillation being roughly proportional to the offset in position, for a constant beam current. The direction of the displacement may be determined from the phase of the oscillation, which changes by 180° as the orbit is crossed. The oscillating signal is coupled out using a short capacitive coupling antenna, and its amplitude and phase measured using a phase detector. The reference phase for this measurement may be obtained, for example, from the accelerator master clock. The basic arrangement is shown in Fig. 10.19.

From (5.37), the resonant frequency of a rectangular cavity is given by

$$f_\mathrm{r} = \frac{c}{\lambda_\mathrm{r}} = \frac{c}{2}\sqrt{\left(\frac{m}{a}\right)^2 + \left(\frac{n}{b}\right)^2 + \left(\frac{q}{l}\right)^2}. \tag{10.40}$$

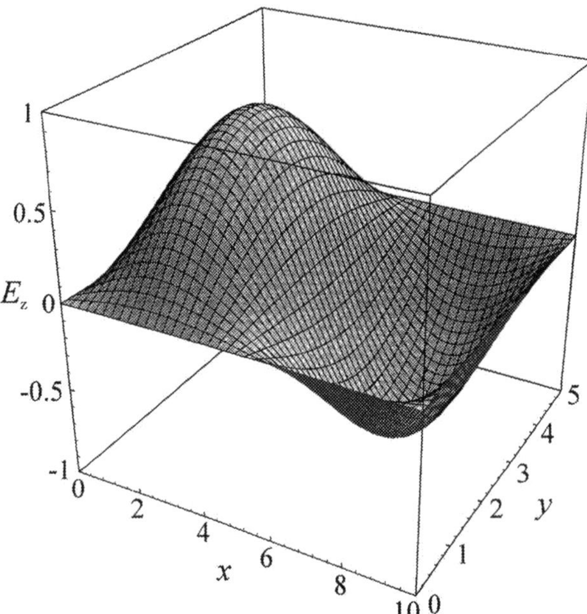

Fig. 10.18 Form of the electric field E_z parallel to the beam axis for the TM_{210} mode in a rectangular resonant cavity.

If we take a resonant cavity with $a = 141$ mm, $b = 71$ mm, and $l = 50$ mm and set $m = 2$, $n = 1$, and $q = 0$ to give the TM_{210} mode, we obtain a resonant frequency of $f_r = 2.998$ GHz. This lies exactly in the range used for S-band linacs. Due to the convenient size of the cavity for this frequency range, this type of monitor is preferred for use in linear accelerators.

In order to measure both the horizontal and vertical beam position, two separate cavities are actually required, with one rotated through 90° relative to the other. However, it is also possible to use a single resonator chosen such that $a = b$. The TM_{210} mode is then degenerate and two waves are excited at the same frequency, each with the field distribution shown in Fig. 10.18 but rotated through an angle 90° relative to one another. Hence one wave crosses through zero in the horizontal direction, the other in the vertical direction. This means that one wave is excited by a horizontal beam displacement, the other by a vertical displacement. To selectively read out the two waves, the coupling antennae must also be arranged at 90° to one another.

10.4.2 Correction of the transverse field position

Due to field errors and misalignments of the steering magnets, the actual beam position is offset from the ideal orbit, as described in Section 3.15. A new stationary beam trajectory results which includes partial betatron oscillations about the ideal orbit in the regions between the individual transverse field disturbances. In order to determine the complete path of the beam in the accelerator, these dis-

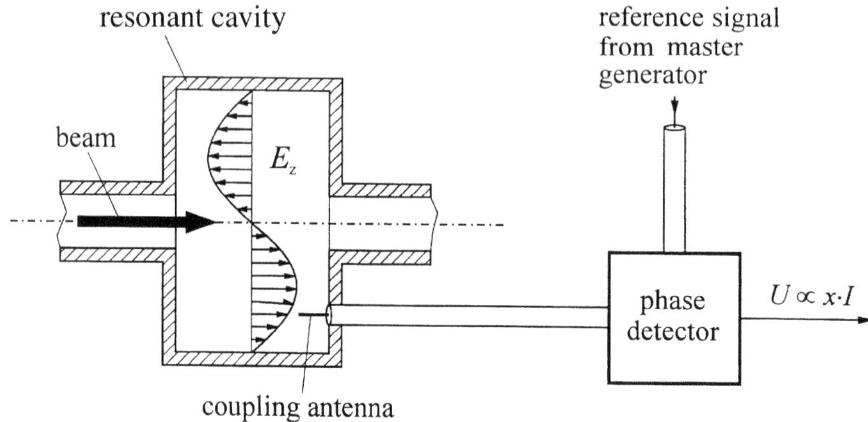

Fig. 10.19 Layout of a position monitor using a resonant cavity in the TM_{210} mode. The oscillating signal induced by the beam is measured by a phase detector.

placements must be measured using beam position monitors (BPMs) described above, placed at sufficiently many points along the beam to give the required precision. According to a general theorem, at least four measurements are required per period, at approximately equal phase separations, in order to unambiguously reconstruct the first harmonic of an oscillation. This means that in order to measure the path of the beam, about four BPMs are required per betatron oscillation. In circular accelerators at least $N_{\mathrm{BPM}} \geq 4\,Q$ monitors must be arranged around the circumference, with a separation chosen according to the betatron phase Ψ rather than the geometric position s around the ring. A compromise is always necessary, since the phase generally evolves differently in the horizontal and vertical planes.

The main purpose of measuring the beam position is to correct the orbit, since the best beam conditions are achieved when the particles travel as closely as possible along the ideal orbit. Large orbit deviations lead in extreme cases to particle losses or to problems such as increased emittance. This is particularly important in electron storage rings used to produce synchrotron radiation, which have extremely strong focusing and require sextupole magnets for chromaticity compensation.

In linear accelerators and beam transport lines, orbit correction is in principle easy due to simple causality. Starting at the point of entry of the beam, the displacement at the first monitor is corrected using a steering coil installed **before** this monitor, at a phase separation of about $\Delta\Psi \approx \pi/2$. The process is then repeated at the next monitor, and so on. The correction procedure is more complicated in cyclic accelerators, since each disturbance always affects the entire machine (Section 3.15). In the following sections we will outline the principal techniques used in this case.

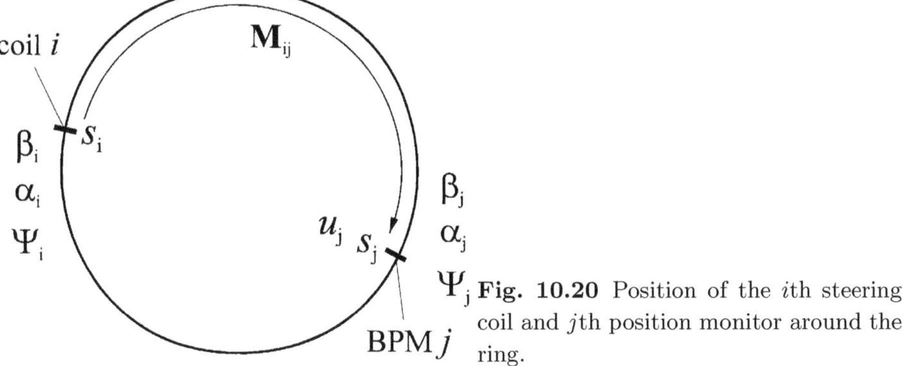

Fig. 10.20 Position of the ith steering coil and jth position monitor around the ring.

Most effective corrector method

An orbit shift is often caused by a particularly strong disturbance at a single point. In such a case it is useful to correct this distortion using a steering coil at or close to this point. The problem, however, is that the disturbance and its position are not known. Hence one attempts to apply a correction with each of the steering coils installed around the ring and measures the remaining error using the individual position monitors. The steering coil with the smallest residual error is chosen. Thus the **most effective corrector** is found.

To describe this method more closely, let us consider a steering coil at a position s_i with the optical parameters β_i, α_i, and Ψ_i and some kind of position monitor at the point s_j with parameters β_j, α_j, and Ψ_j (Fig. 10.20). In total there are n steering coils and m position monitors installed around the ring.

At the various monitors the separations

$$\boldsymbol{u} = (u_1, u_2, \ldots u_m) \tag{10.41}$$

are measured. The steering coil of strength $\kappa_i = \Delta x_i'(s_i)$ results in a beam displacement of x_{ij} at the jth monitor. The coil strength is adjusted to minimize the measured beam offset. To calculate this offset, we first use (3.259) to calculate the beam displacement at the coil. We obtain

$$\begin{pmatrix} x_i \\ x_i' \end{pmatrix} = \frac{\kappa_i}{2} \begin{pmatrix} \dfrac{\beta_i}{\tan \pi Q} \\ 1 - \dfrac{\alpha_i}{\tan \pi Q} \end{pmatrix}. \tag{10.42}$$

The resulting displacement in the monitor installed at the point s_j is then

$$\begin{pmatrix} x_{ij} \\ x_{ij}' \end{pmatrix} = \mathbf{M}_{ij} \begin{pmatrix} x_i \\ x_i' \end{pmatrix} = \frac{\kappa_i}{2} \mathbf{M}_{ij} \begin{pmatrix} \dfrac{\beta_i}{\tan \pi Q} \\ 1 - \dfrac{\alpha_i}{\tan \pi Q} \end{pmatrix}. \tag{10.43}$$

Using the transfer matrix (3.164) and making the substitutions $\beta_0 = \beta_i$, $\beta = \beta_j$ and $\Delta\Psi_{ij} = \Psi_j - \Psi_i$ this becomes

$$x_{ij}(\kappa_i) = \kappa_i \frac{\sqrt{\beta_i\beta_j}}{2} \left[\frac{\cos\Delta\Psi_{ij} - \alpha_i \sin\Delta\Psi_{ij}}{\tan\pi Q} + \sin\Delta\Psi_{ij} \left(1 - \frac{\alpha_i}{\tan\pi Q}\right)\right], \tag{10.44}$$

or simplified

$$x_{ij}(\kappa_i) = \kappa_i \, h_{ij} \tag{10.45}$$

with

$$h_{ij} = \frac{\sqrt{\beta_i\beta_j}}{2} \left[\frac{\cos\Delta\Psi_{ij} - 2\alpha_i \sin\Delta\Psi_{ij}}{\tan\pi Q} + \sin\Delta\Psi_{ij} \right]. \tag{10.46}$$

The best correction is obtained for a strength κ_i, for which the error function

$$f_i(\kappa_i) = \sum_{j=1}^{m} (u_j - x_{ij}(\kappa_i))^2 = \sum_{j=1}^{m} (u_j - \kappa_i h_{ij})^2 \tag{10.47}$$

is a minimum. This means that the sum of the quadratic errors between the measured displacements u_j and the corrections x_{ij} is minimized. This is achieved when

$$\frac{df_i(\kappa_i)}{d\kappa_i} = -2\sum_{j=1}^{m} (u_j - \kappa_i h_{ij})\, h_{ij} = 0. \tag{10.48}$$

The required optimal corrector strength is then fixed, and has the value

$$\kappa_i = \frac{\sum\limits_{j=1}^{m} u_j h_{ij}}{\sum\limits_{j=1}^{m} h_{ij}^2}. \tag{10.49}$$

The value of the residual quadratic errors may be immediately obtained by inserting (10.49) into (10.47). The measured orbit displacement has thus been approximated by the effect of one single steering coil. This process is repeated for all the steering coils, and the one which gives the smallest quadratic error is chosen. This is the **most effective corrector** and the orbit displacement is compensated for by changing the strength of this steering coil by $-\kappa_i$.

By its very nature, this procedure is typically not able to completely correct the beam trajectory. The only rare exceptions are those cases where the orbit displacement really is due only to a single localized disturbance with a correcting coil situated nearby. Very often, however, this process takes a big first step towards the ideal orbit and so offers a good start in correcting the orbit. It can also happen, however, that there are many small disturbances that distort the orbit, and no steering coil is found to give a significantly better improvement than the others. In this case the most effective corrector method cannot be used.

Matrix inversion method

If the most effective corrector method cannot be used, or if it has already been used and further fine-tuning of the orbit is required, then one must try to correct the deviations from the ideal orbit by simultaneously tuning all n steering coils around the ring. The error function in (10.47) is then extended and becomes

$$f(\kappa_1, \kappa_2, \ldots, \kappa_n) = \sum_{j=1}^{m} \left(u_j - \sum_{i=1}^{n} x_{ij}(\kappa_i) \right)^2 = \sum_{j=1}^{m} \left(u_j - \sum_{i=1}^{n} \kappa_i h_{ij} \right)^2 . \tag{10.50}$$

The task now consists of finding the set of corrector strengths $\kappa_1, \kappa_2, \ldots, \kappa_n$ which again minimize the error function. A necessary, and in this case also sufficient, condition for this to be true is that all partial derivatives

$$\frac{\partial}{\partial \kappa_p} f(\kappa_1, \kappa_2, \ldots, \kappa_n) = 0 \qquad \text{with} \qquad p = 1, 2, \ldots n \tag{10.51}$$

should vanish. This condition immediately leads to

$$\frac{\partial f}{\partial \kappa_p} = -2 \sum_{j=1}^{m} \left(u_j - \sum_{i=1}^{n} \kappa_i h_{ij} \right) h_{ij} = 0 \tag{10.52}$$

and hence

$$\sum_{j=1}^{m} \left(u_j h_{pj} - \sum_{i=1}^{n} \kappa_i h_{ij} h_{pj} \right) = 0. \tag{10.53}$$

Writing this expression in the form

$$\sum_{j=1}^{m} u_j h_{pj} = \sum_{i=1}^{n} \left(\sum_{j=1}^{m} h_{pj} h_{ij} \right) \kappa_i \tag{10.54}$$

and setting

$$\begin{aligned} U_p &= \sum_{j=1}^{m} u_j h_{pj} \\ H_{pi} &= \sum_{j=1}^{m} h_{pj} h_{ij} \end{aligned} \tag{10.55}$$

leads to the following matrix equation

$$U_p = \sum_{i=1}^{n} H_{pi}\, \kappa_i . \tag{10.56}$$

This may be written in the alternative form

$$\begin{pmatrix} U_1 \\ U_2 \\ \vdots \\ U_n \end{pmatrix} = \mathbf{H} \begin{pmatrix} \kappa_1 \\ \kappa_2 \\ \vdots \\ \kappa_n \end{pmatrix} \qquad \text{with} \qquad \mathbf{H} = \begin{pmatrix} h_{11} & h_{12} & \cdots & h_{1n} \\ h_{21} & \ddots & & \vdots \\ \vdots & & \ddots & \vdots \\ h_{n1} & \cdots & \cdots & h_{nn} \end{pmatrix} . \tag{10.57}$$

The required set of corrector strengths is immediately obtained by matrix inversion

$$\begin{pmatrix} \kappa_1 \\ \kappa_2 \\ \vdots \\ \kappa_n \end{pmatrix} = \mathbf{H}^{-1} \begin{pmatrix} U_1 \\ U_2 \\ \vdots \\ U_n \end{pmatrix}. \tag{10.58}$$

This is the basic principle of the technique. In practice, however, there are often further conditions to be taken into account. Often the various steering coils vary in their effectiveness because their performance depends on the value of the beta function, and this can vary greatly. In addition the mathematical procedure yields arbitrary corrector strengths without any regard for their technical feasibility. The maximum possible value of κ_i should thus be viewed as a limit.

The corrector strengths calculated from (10.58) are applied again with the opposite sign, which leads to the minimum possible level of orbit errors in a system of purely linear optics, as we have assumed here. In fact this assumption is not completely valid, since the non-linear field components, in particular the sextupoles installed for chromaticity correction, lead to errors. This problem is easily solved, however, by repeating the orbit correction several times until it converges to a minimum error. As a rule this is achieved after three to five iterations.

Correction with local orbit bumps

In very large accelerators with a correspondingly large number of position monitors and corrector coils the limitations of the matrix inversion method are sometimes reached. These are not due to the demand for computing power, which today's workstations can easily provide, but instead to the finite resolution of the monitors and to the finite precision of the corrector alignment. When using this procedure these uncertainties can sum to give large errors, until a usable orbit correction is no longer possible. In addition, it also happens that beam shifts are sometimes deliberately introduced at particular points in the accelerator, for example in the injection region and near to the experiments, and these shifts should be preserved. A solution in such cases is to apply local corrections using closed orbit bumps, as described in Section 3.18.

If a beam displacement u_m is required at a particular position monitor, then the three corrector coils nearest to the monitor are chosen, as illustrated in Fig. 10.21. In this example the monitor lies between the first and second coils, but this is not an obligatory condition. In this case a strength κ_1 in the first coil causes a transverse beam displacement at the monitor of

$$x_m = \kappa_1 \sqrt{\beta_1 \beta_m} \sin(\Psi_m - \Psi_1). \tag{10.59}$$

It is now always possible to apply a corrector strength κ_1 so that the beam displacement at the position monitor disappears, namely $x_m = -u_m$. This corrector strength is

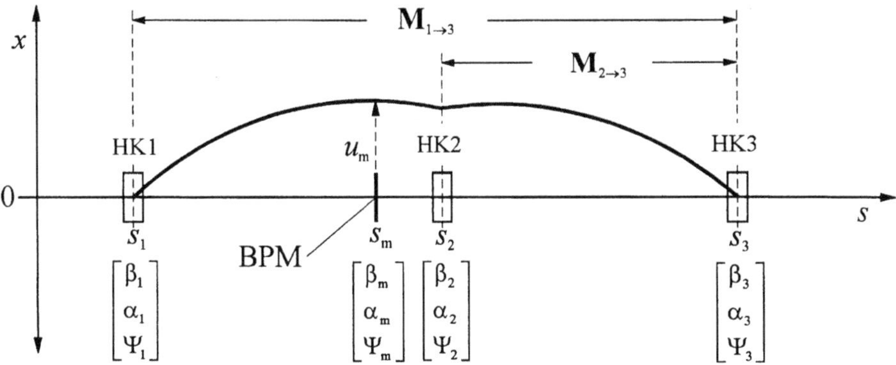

Fig. 10.21 Orbit correction with a beam bump using three corrector coils. Here the position monitor lies between the first and second coils.

$$\kappa_1 = \frac{-u_m}{\sqrt{\beta_1\beta_m}\sin(\Psi_m - \Psi_1)}. \tag{10.60}$$

The strengths κ_2 and κ_3 of the two other corrector coils are obtained from the matching condition for beam bumps given by (3.310) and (3.311). Once the displacement at the monitor m has been compensated for, one proceeds to the next monitor and repeats the process until all the beam shifts have been corrected in turn.

It might seem that that this method allows orbit errors to be almost completely corrected, since the beam position at the monitor can be shifted back onto the ideal orbit to within the measurement accuracy. However, this is only true for the **single** monitor under consideration. The beam bump extends over a betatron phase of approximately π, while the separation of the position monitors is generally not more than $\pi/2$. Hence on average at least two monitors will lie in a bump, but complete trajectory matching can in general only be performed for one monitor. As a result this correction procedure also leads to a non-zero residual error, comparable to that achieved by matrix inversion.

10.5 Measurement of the betatron frequency and the tune Q

As we saw in Section 3.14, the tune Q of a cyclic accelerator is a very important parameter of the beam optics. Once a particular set of beam optics has been installed, the working point, or tune, must be measured in order to check that it lies far enough away from any strong optical resonances. Conversely, measuring the tune allows the detection of any changes in focusing, for example due to a magnetic field imperfection or to the space charge effect in the case of colliding beams. It is thus possible to use the tune to monitor very precisely the stability of the beam focusing during machine operation. Since the tune is defined as the number of betatron oscillations per revolution, measuring it ultimately amounts to measuring the frequency of the transverse beam oscillations. We will now describe this technique.

Consider a particle of nominal momentum, i.e. $\Delta p/p = 0$. Using $K(s) = 1/R^2(s) - k(s)$, the optical equations (3.21) yield the equation of motion in the form

$$x''(s) + K(s)x(s) = 0. \tag{10.61}$$

If we transform this position-dependent equation into a time-dependent one and also take into account a weak damping effect from synchrotron radiation we obtain

$$\ddot{x}(t) + \frac{2}{\tau}\dot{x}(t) + c^2 K(t)\, x(t) = 0. \tag{10.62}$$

Let the damping time τ be very long compared to the duration of a betatron oscillation. The solution of this equation is again found to be

$$x(t) = \exp\left(-\frac{t}{\tau}\right) \sqrt{\varepsilon}\sqrt{\beta(t)} \cos\left[\Psi(t) + \Theta\right]. \tag{10.63}$$

As the damping is very slight, all terms with derivatives of $\exp(-t/\tau)$ may be neglected. The first time derivative of $x(t)$ then becomes

$$\dot{x}(t) = -\frac{c\sqrt{\varepsilon}}{\sqrt{\beta(t)}} \exp\left(-\frac{t}{\tau}\right) \left\{\alpha(t) \cos\left[\Psi(t) + \Theta\right] + \sin\left[\Psi(t) + \Theta\right]\right\}. \tag{10.64}$$

The constants of integration ε and Θ are fixed by the initial conditions for $t = 0$. Using

$$\begin{aligned} \beta(0) &= \beta_0 \\ \alpha(0) &= \alpha_0 \\ \Psi(0) &= 0 \end{aligned} \qquad \text{and} \qquad \begin{aligned} x(0) &= x_0 \\ \dot{x}(0) &= \dot{x}_0 \end{aligned} \tag{10.65}$$

substitution into (10.63) and (10.64) yields

$$\begin{aligned} \cos\Theta &= \frac{x_0}{\sqrt{\varepsilon\beta_0}} \\ \sin\Theta &= -\frac{1}{\sqrt{\varepsilon}}\left(\frac{\sqrt{\beta_0}\,\dot{x}_0}{c} + \frac{\alpha_0\, x_0}{\sqrt{\beta_0}}\right). \end{aligned} \tag{10.66}$$

The transverse motion of the centre of mass of the beam is, as a function of time,

$$x(t) = \exp\left(-\frac{t}{\tau}\right) \frac{\sqrt{\beta(t)}}{\sqrt{\beta_0}} \left\{x_0 \cos\Psi(t) + \left(\frac{\beta_0\,\dot{x}_0}{c} + \alpha_0\, x_0\right) \sin\Psi(t)\right\}. \tag{10.67}$$

This equation describes the motion of the beam along its trajectory. For practical reasons the instantaneous offset in position can only be measured with a fast position monitor at a few points along the orbit, and usually only at one point s_0. If the optical parameters at this point are again β_0 and α_0, and we start from the initial conditions x_0 and $\dot{x}_0$, then by the nth revolution the measured displacement is

$$x_n = \exp\left(-\frac{n\,T_u}{\tau}\right)\left\{x_0 \cos(2\pi Q n) + \left(\frac{\beta_0\,\dot{x}_0}{c} + \alpha_0\,x_0\right)\sin(2\pi Q n)\right\}. \quad (10.68)$$

with $n = 0, 1, 2, \ldots$. If we now write the tune in the form

$$Q = q + a \qquad \text{with} \qquad \begin{array}{rcl} q & = & \text{integer} \\ 0 & \leq & a \leq 1 \end{array} \quad (10.69)$$

then it immediately follows that

$$\begin{aligned} \cos(2\pi Q n) &= \cos(2\pi a n) \\ \sin(2\pi Q n) &= \sin(2\pi a n). \end{aligned} \quad (10.70)$$

This means that because of the ambiguity of the trigonometric functions, a measurement at one fixed point around the orbit is only able to determine the fractional part a of the tune. This uncertainty is termed the stroboscope effect. This is understandable, since the monitor only measures the phase shift from one revolution to the next and is unable to to determine how many oscillations take place during the course of a revolution. What is more, there is another ambiguity in the measurement. Using the elementary trigonometric relations one can easily show that

$$\begin{aligned} \cos[2\pi(1-a)n] &= \cos(2\pi a n) \\ \sin[2\pi(1-a)n] &= \sin(2\pi a n). \end{aligned} \quad (10.71)$$

If we assume that $0 \leq a \leq 0.5$, it is not possible to determine from the frequency measurement in the position monitor whether the tune lies **above** or **below** an integer value. Hence the tune values $Q_1 = 3.18$ and $Q_2 = 3.82$ will give the same frequency at the monitor.

The monitor only ever measures a pulse-like position signal for a brief period of time as the bunch flies past. This type of signal is hard to measure and so the bandwidth of the readout electronics is chosen such that a low-pass filter may be used to smooth out the signal first. This results in a roughly constant signal, with an approximately linear phase dependence. The discrete time points and phases in (10.68) correspond to

$$n\,T_u \Rightarrow t \qquad\qquad 2\pi Q n \Rightarrow 2\pi Q \frac{t}{T_u}. \quad (10.72)$$

The time dependence of the signal from the monitor thus takes the form

$$x(t) = \exp\left(-\frac{t}{\tau}\right)\left\{x_0 \cos\left(2\pi a \frac{t}{T_u}\right) + \left[\frac{\beta_0\,\dot{x}_0}{c} + \alpha_0\,x_0\right]\sin\left(2\pi a \frac{t}{T_u}\right)\right\}. \quad (10.73)$$

This is once again the solution of the oscillation equation, assuming very weak damping,

$$\ddot{x}(t) + \frac{2}{\tau}\dot{x}(t) + \Omega^2 x(t) = 0 \tag{10.74}$$

with

$$\Omega = \frac{2\pi}{T_u} a = \omega_u a \qquad 0 \le a \le 0.5. \tag{10.75}$$

If the beam undergoes coherent betatron oscillations, then the frequency Ω is measured at the fast position monitor. Since the circumference of the machine and hence the revolution frequency ω_u are fixed, (10.75) may be used to determine the fractional part a of the tune. The integer part q is best determined from a position measurement. First of all the beam position is measured using all the position monitors as a reference. The strength of a steering coil is then altered, which causes a standing betatron oscillation about the reference orbit. A second set of position measurements are made and the difference from the reference orbit then gives the betatron oscillations, which are simply counted to obtain the integer q. The only remaining uncertainty is whether the tune lies above or below this integer. However, this may be determined simply by increasing the strength of a **focusing** quadrupole slightly, which always causes the value of Q to increase. If a also increases, then Q lies **above** the integer value; if a decreases, then Q is **below** the integer value.

To measure the tune it is thus necessary to excite the beam into coherent transverse oscillations. This may best be achieved using a fast bending magnet, which produces a periodic field $B(t) = B_0 \sin \omega_{\text{gen}} t$. The magnet is driven by a power amplifier controlled by a sine-wave generator. The Lorentz force

$$F_x(t) = e\,c\,B_z(t) = \dot{p}_x(t) = \ddot{x}_{\text{gen}} m_{\text{e}} \tag{10.76}$$

acts upon the beam, giving an acceleration

$$\ddot{x}_{\text{gen}} = \frac{ec}{m_{\text{e}}} B_0 \sin \omega_{\text{gen}} t. \tag{10.77}$$

Usually the driving magnet is not situated at the same point as the fast position monitor. Furthermore it is very short compared to the circumference of the machine, and so from revolution to revolution it has a pulse-like effect on the beam. This can be described using an effective strength κ_{eff}. The resulting perturbation of the beam is again described by the equation

$$\ddot{x}(t) + \frac{2}{\tau}\dot{x}(t) + \Omega^2 x(t) = \kappa_{\text{eff}} \sin \omega_{\text{gen}} t. \tag{10.78}$$

This is the equation of a forced oscillation. As the damping is very weak, resonance occurs if the generator frequency $\omega_{\text{gen}} = \Omega$. An arrangement used to measure the tune in this way is shown in Fig. 10.22.

The fast bending magnet used to drive the betatron oscillations does not need to be very strong, since the low damping and high quality-factor of the

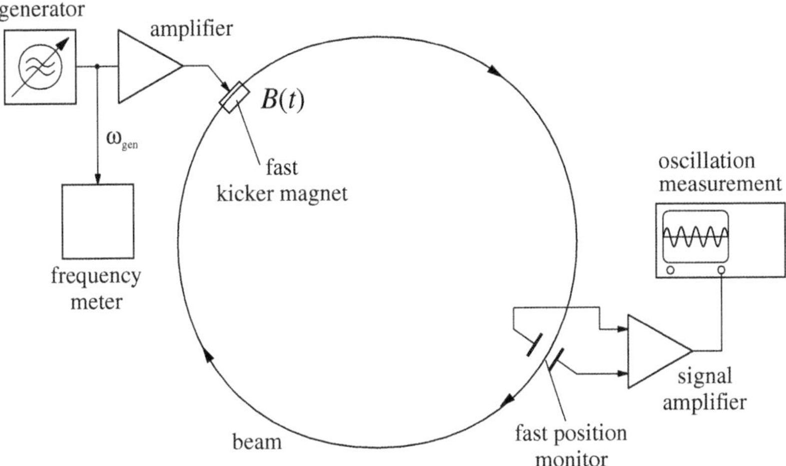

Fig. 10.22 Arrangement used to measure the tune of a cyclic accelerator. A fast kicker magnet stimulates the beam at a frequency ω_{gen}, which is varied until the resonance is found. The amplitude of the induced betatron oscillation is measured using a fast position monitor.

system mean that the amplitude grows very rapidly at resonance. As a rule, an integrated field strength of around $\int B\,ds \approx 10^{-4}$ Tm is sufficient. The magnet is supplied with current by a power amplifier, fed by a sine-wave signal generator with a frequency which may be varied in the range $0 \le \omega_{\text{gen}} \le 0.5\omega_{\text{rev}}$. A special position monitor with fast electronics records the variation in the position, which in the simplest case may be viewed with an oscilloscope. The resonance can also be viewed by observing the synchrotron radiation with a camera. More elegant techniques combine the stimulation and measurement of the oscillations in a spectrum analyser which automatically reads out and displays the resonances.

10.6 Measurement of the synchrotron frequency

As equation (5.81) in Section 5.6 shows, particles in the accelerating field of a cyclic machine perform longitudinal oscillations at the synchrotron frequency Ω. For small amplitudes these oscillations are harmonic, to a good approximation. Hence the frequency may again be measured by resonant excitation of the particle beam. This is done by modulating the phase of the RF generator with a periodic time-dependent function of the form

$$\Psi(t) = \Psi_0 + \Delta\Psi_{\text{mod}}(t) \qquad \text{with} \qquad \Delta\Psi_{\text{mod}}(t) = \hat{\Psi}\,\sin\omega_{\text{mod}}\,t. \tag{10.79}$$

If we again assume small phase oscillations and weak excitation, i.e. $\Delta\Psi \ll \Psi_0$ and $\Delta\Psi_{\text{mod}} \ll \Psi_0$, then (5.64) may be extended to give

$$\Delta\dot{E} = \frac{eU_0}{T_0}\Big[\sin\left[\Psi_0 + \Delta\Psi_{\text{mod}}(t) + \Delta\Psi\right] - \sin\Psi_0\Big] - \frac{2}{\tau_0}\Delta E. \tag{10.80}$$

From this we obtain, in the same way as (5.79)

$$\Delta \dot{E} = \frac{eU_0 \cos \Psi_0}{T_0} \Big[\Delta\Psi_{\mathrm{mod}}(t) + \Delta\Psi \Big] - \frac{2}{\tau_0} \Delta E. \tag{10.81}$$

For $\gamma \gg 1$ it follows from (5.77) that

$$\Delta E = \frac{ET_0}{2\pi q \alpha} \Delta \dot{\Psi}. \tag{10.82}$$

Using this relation we can substitute for ΔE and $\Delta \dot{E}$ in (10.81) and obtain the inhomogeneous phase oscillation equation in the form

$$\Delta \ddot{\Psi} + \frac{2}{\tau_0} \Delta \dot{\Psi} + \Omega^2 \Delta \Psi = -\Omega^2 \Delta \Psi_{\mathrm{mod}}(t) = -\Omega^2 \hat{\Psi} \sin \omega_{\mathrm{mod}}\, t \tag{10.83}$$

with

$$\Omega^2 = -\frac{2\pi q \alpha e U_0 \cos \Psi_0}{E\, T_0^2}. \tag{10.84}$$

The resonant excitation of the beam is again achieved with $\omega_{\mathrm{mod}} = \Omega$. We see that coherent synchrotron oscillations can be excited by modulation of the phase of the generator. A method must now be found to observe these oscillations. The electromagnetic field associated with the beam induces electrical signals in electrodes or measurement cavities, which describe the intensity of the beam and the phase oscillations. These signals consist of a large number of harmonics of the revolution frequency ω_{rev}. If we choose a particular harmonic $\omega_n = n\omega_{\mathrm{rev}}$, the signal in the electrodes can be written in the form

$$U(t) = \hat{U} \sin \Big[\omega_n t + \hat{\Psi} \sin(\omega_{\mathrm{mod}} t) \Big]. \tag{10.85}$$

Using the relations (see, for example, Abramowitz-Stegun [96])

$$\begin{aligned} \cos \Big[\hat{\Psi} \sin(\omega_{\mathrm{mod}} t) \Big] &= \sum_{n=-\infty}^{+\infty} J_n(\hat{\Psi}) \cos(n\omega_{\mathrm{mod}} t) \\ \sin \Big[\hat{\Psi} \sin(\omega_{\mathrm{mod}} t) \Big] &= \sum_{n=-\infty}^{+\infty} J_n(\hat{\Psi}) \sin(n\omega_{\mathrm{mod}} t) \end{aligned} \tag{10.86}$$

it follows that

$$U(t) = \hat{U} \sum_{n=-\infty}^{+\infty} J_n(\hat{\Psi}) \sin(\omega_n t + n\omega_{\mathrm{mod}} t). \tag{10.87}$$

We thus have a typical phase modulated signal with the Bessel sidebands $n\omega_{\mathrm{mod}}$, which can be demodulated and analysed using standard techniques from communications engineering.

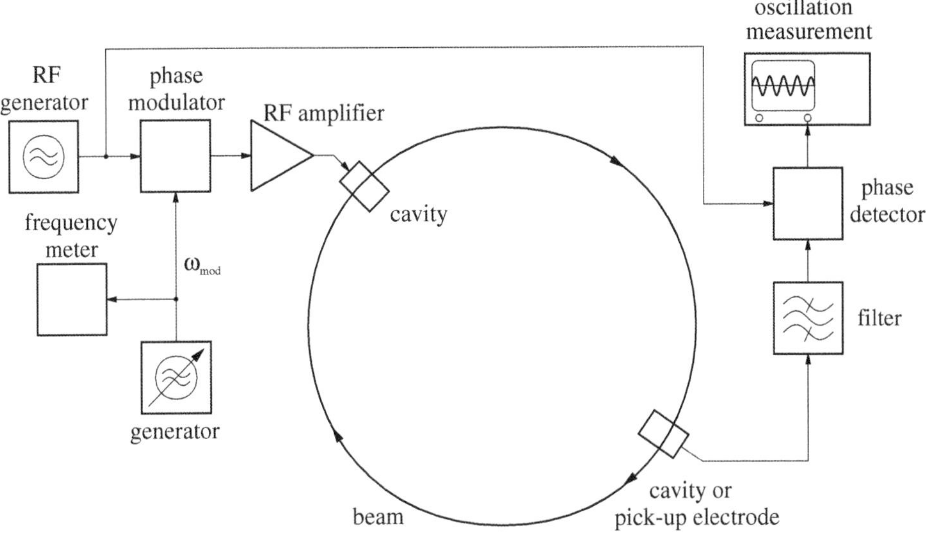

Fig. 10.23 Arrangement used to measure the synchrotron frequency. The oscillations are excited by phase modulation of the generator and measured using a phase detector, which obtains a signal from from an electrode or a measuring cavity.

The basic arrangement used to measure the synchrotron frequency is illustrated in Fig. 10.23. The high-frequency signal supplied by the master generator is modulated with the tunable frequency ω_{mod} using an electronic phase shifter. The periodic modulation signal comes from a signal generator whose frequency is measured by a meter.

As shown above, this phase-modulated RF voltage excites longitudinal oscillations of the beam only if the modulating frequency ω_{mod} lies sufficiently close to the synchrotron frequency. A phase modulated signal, described by (10.85), can be read out using an electrode or a measuring cavity. Demodulation is performed by a phase detector whose output signal may then be displayed on an oscilloscope, for example. The frequency of the signal generator is varied until the resonance of the phase oscillation is found, and this frequency ω_{mod} can then be read out using a meter. This yields the required synchrotron frequency directly. Since this is much smaller than the revolution frequency, there is no ambiguity in this case such as we saw in the measurement of the tune, above.

The phase demodulation shown in Fig. 10.23 always requires a reference signal from a master generator, which must be matched to the phase of the measured signal. In very large accelerators in particular, this requires long cables which can cause the phase to vary with temperature. It it therefore better to measure the phase modulation in some other way without relying on a reference phase. According to (10.85) the time dependence of the phase of the measured signal is

$$\Psi_{\mathrm{meas}}(t) = \omega_n t + \hat{\Psi} \sin(\omega_{\mathrm{mod}} t). \tag{10.88}$$

The instantaneous frequency of the signal is simply the time derivative of the phase, namely

$$\omega(t) = \frac{d\Psi_{\text{meas}}(t)}{dt} = \omega_n + \omega_{\text{mod}}\hat{\Psi}\cos(\omega_{\text{mod}}t). \tag{10.89}$$

Hence we have a **frequency modulated** signal, with a carrier frequency ω_n and modulation $\omega_{\text{mod}}\hat{\Psi}\cos(\omega_{\text{mod}}t)$. It is thus possible to use the technique of **frequency demodulation**, which does not require a reference signal, instead of phase demodulation. One must simply bear in mind that in (10.89) the frequency shift increases in proportion to the modulation frequency. In order to obtain an output signal which is independent of this frequency a suitable low-pass filter or integrator must be employed, with a frequency response proportional to $1/\omega_{\text{mod}}$, as is commonly used in communications systems. The synchrotron oscillations may then be observed merely using an FM radio tuned to the right frequency.

10.7 Measurement of the optical parameters of the beam

The monitors described above also allow other optical parameters of the beam to be indirectly measured. We conclude this chapter by describing the most important examples.

10.7.1 Measurement of the dispersion

The dispersion can be determined from position measurements at several points around the orbit. Let us assume that position monitors are installed at the points s_i with $i = 1, 2, \ldots m$, and these yield m corresponding values of the beam displacement $u^0(s_i)$. The momentum of the circulating particles is then varied by a small amount Δp, while the strengths of the machine magnets remain constant. The beam position thus shifts a distance $\Delta x(s) = D(s)\,\Delta p/p$ onto a dispersive trajectory. The position measurement is now repeated and gives the new displacements $u(s_i)$ at the m monitors. The dispersion at the monitors is then simply calculated from

$$D(s_i) = \frac{\Delta x(s_i)}{\Delta p/p} \qquad \text{with} \qquad \Delta x(s_i) = u(s_i) - u^0(s_i). \tag{10.90}$$

By comparing these local dispersion values at the discrete points s_i with the calculated dispersion function it is possible to check how closely the actual beam optics agree with the calculated design.

The difficulty with this essentially very simple measurement lies in varying the beam momentum p without adjusting the magnets. This is done by altering the **frequency** ν_{RF} of the accelerating voltage by a small amount $\Delta\nu$, which causes a corresponding change in the wavelength λ_{RF}. Since the phase focusing means that the harmonic number q remains constant, the circumference of the particle trajectory changes and hence no longer matches the orbit. The stable particle path thus shifts onto a dispersive trajectory, which entails a corresponding change in momentum Δp.

From the definition of the momentum compaction factor α in equation (3.110), the relation between the relative change in momentum and relative change in path length is given by

$$\frac{\Delta p}{p} = \frac{1}{\alpha}\frac{\Delta L}{L}. \tag{10.91}$$

The path length is an integer multiple of the wavelength of the RF generator, i.e. $L = q\lambda_{\rm RF}$. It follows immediately that $\Delta L/L = \Delta\lambda_{\rm RF}/\lambda_{\rm RF}$. Working in terms of the frequency, which is easer to measure,

$$\lambda_{\rm RF} = \frac{1}{\nu_{\rm RF}} \quad \Rightarrow \quad \frac{d\lambda_{\rm RF}}{d\nu_{\rm RF}} = -\frac{1}{\nu_{\rm RF}^2} \quad \Rightarrow \quad \frac{\Delta L}{L} = -\frac{\Delta\nu_{\rm RF}}{\nu_{\rm RF}} \tag{10.92}$$

we obtain the required relation between the change in radiofrequency and the change in momentum

$$\frac{\Delta p}{p} = -\frac{1}{\alpha}\frac{\Delta\nu_{\rm RF}}{\nu_{\rm RF}}. \tag{10.93}$$

The momentum compaction factor α is well known from optical calculations. In practice α has proven to be very insensitive to variations in the magnet structure, with field errors largely cancelling out. The calculated value is thus very reliable. The frequency of the accelerating voltage $\nu_{\rm RF}$ can be measured very precisely using a frequency monitor, as of course can the change in frequency $\Delta\nu_{\rm RF}$. Thus the relative change in momentum produced by varying the frequency may be calculated using (10.93) and inserted into (10.90) to quantify the dispersion.

10.7.2 Measurement of the beta function

As we saw in Section 3.15, if the strength of a quadrupole changes by an amount Δk then according to (3.274) the tune of a cyclic machine shifts by

$$\Delta Q = \frac{1}{4\pi}\int_{s_0}^{s_0+l} \Delta k\ \beta(s)\ ds.$$

The size of the shift is proportional to the value of the beta function in the quadrupole. Since as a rule the variation of the beta function in the quadrupole is small and the quadrupole strength k is constant along the magnet axis, we may to a good approximation use the average value of the beta function in the quadrupole, namely

$$\Delta Q = \frac{\Delta k}{4\pi}\int_{s_0}^{s_0+l} \beta(s)\, ds \approx \frac{\Delta k}{4\pi}\langle\beta\rangle\, l. \tag{10.94}$$

If we start from a particular set-up of the beam optics and impose a well-defined change in quadrupole strength Δk, then by measuring the tune Q before and after the change the average beta function in the quadrupole may be immediately calculated using

$$\langle \beta \rangle = \frac{4\pi}{l}\frac{\Delta Q}{\Delta k}. \tag{10.95}$$

It is essential to ensure that the quadrupole in question is only adjusted very slightly: as a rule the change should not exceed $\Delta k/k < 1\%$. If the change is any larger, then the beta function itself can sometimes vary considerably, and a misleading measurement could be obtained. The best method is to vary the quadrupole strength in small steps around the nominal value. By making a large number of measurements the function $\Delta Q(\Delta k)$ can be traced out. The gradient of the curve at the nominal value gives the most reliable value for the required quantity $\Delta Q/\Delta k$.

In general k is varied by changing the current from the power supply connected to the quadrupole. Often several quadrupoles are connected in series, reflecting the symmetry of the machine, which means that the shift in Q is proportional to the number of quadrupoles. The value calculated from (10.95) must thus be divided by the number of quadrupoles. In this case the quantity obtained is merely the average value of the beta function in **all** the quadrupoles involved. Hence effects such as asymmetries in the beam optics cannot be identified by this method. For a careful measurement of the complete beam optics of a cyclic accelerator additional, separate 'piggy-back' power supplies are required. Alternatively a variable 'shunt' resistor connected in parallel can be used.

10.7.3 Measurement of the chromaticity

The chromaticity, described in Section 3.16, is defined in (3.291) as

$$\xi \equiv \frac{\Delta Q}{\Delta p/p},$$

which immediately suggests how it should be measured. One must simply vary the momentum of the circulating particles and measure the value of Q in the usual way before and after the change. The momentum is varied according to (10.93) by changing the RF frequency ν_{RF}, in exactly the same way as in the dispersion measurement. Since the relationship between the change in momentum and the tune is far from linear, especially in very strongly focusing accelerators, it is again desirable to measure a function $\Delta Q(\Delta p/p)$, whose value in the region around the nominal value directly yields the chromaticity.

A chromaticity measurement is essential for correct tuning of the sextupoles. The calculation of the chromaticity compensation in Section 3.16 is actually only partially correct, since it does not take into account the sextupole components of the higher multipoles. Furthermore, any variation in a quadrupole leads to a corresponding change in the beta function, which in turn also affects the sextupole compensation. Hence the chromaticity should also be regularly checked during operation, especially in storage rings with strong focusing.

A
Maxwell's equations

Charged particles are accelerated by electric fields $\boldsymbol{E}$ and steered and focused by magnetic fields $\boldsymbol{B}$. These fields may also vary with time. A particle beam consists of a spatial distribution of charge, with a charge density $\rho(\boldsymbol{r})$. The relationships between charge and electromagnetic fields are described by **Maxwell's equations**, which are the fundamental theoretical basis of accelerator physics. Maxwell's equations may be written both in integral and differential form, and both are used where appropriate in the text. For convenience we give both forms here, in SI units. In integral form Maxwell's equations are

$$
\begin{aligned}
\oint_A \boldsymbol{E} \cdot d\boldsymbol{A} &= \int_V \frac{\rho}{\varepsilon_\mathrm{r}\,\varepsilon_0}\, dV \\
\oint_A \boldsymbol{B} \cdot d\boldsymbol{A} &= 0 \\
\oint \boldsymbol{E} \cdot d\boldsymbol{s} &= -\int_A \dot{\boldsymbol{B}} \cdot d\boldsymbol{A} \\
\oint \boldsymbol{B} \cdot d\boldsymbol{s} &= \int_A \mu_\mathrm{r}\mu_0(\boldsymbol{i} + \varepsilon_\mathrm{r}\,\varepsilon_0 \dot{\boldsymbol{E}}) \cdot d\boldsymbol{A}.
\end{aligned}
\tag{A.1}
$$

The integrals on the left-hand side run over a closed surface or closed path, while those on the right-hand side run over the enclosed volume V or the enclosed area A respectively. $\boldsymbol{i}$ denotes the current density. In differential form, Maxwell's equations are written as

$$
\begin{aligned}
\nabla \cdot \boldsymbol{E} &= \frac{\rho}{\varepsilon_\mathrm{r}\,\varepsilon_0} & \mathrm{div}\boldsymbol{E} &= \frac{\rho}{\varepsilon_\mathrm{r}\,\varepsilon_0} \\
\nabla \cdot \boldsymbol{B} &= 0 & \mathrm{div}\boldsymbol{B} &= 0 \\
\nabla \times \boldsymbol{E} &= -\dot{\boldsymbol{B}} & \mathrm{curl}\boldsymbol{E} &= -\dot{\boldsymbol{B}} \\
\nabla \times \boldsymbol{B} &= \mu_\mathrm{r}\mu_0(\boldsymbol{i} + \varepsilon_\mathrm{r}\,\varepsilon_0 \dot{\boldsymbol{E}}) & \mathrm{curl}\boldsymbol{B} &= \mu_\mathrm{r}\mu_0(\boldsymbol{i} + \varepsilon_\mathrm{r}\,\varepsilon_0 \dot{\boldsymbol{E}}).
\end{aligned}
\tag{A.2}
$$

If we consider the time-dependent fields in free space, i.e. $\rho = 0$ and $\boldsymbol{i} = 0$, the last two equations of (A.2) take the form

$$\begin{aligned} \nabla \times \boldsymbol{E} &= -\dot{\boldsymbol{B}} \\ \nabla \times \boldsymbol{B} &= \mu_{\mathrm{r}}\mu_0\varepsilon_{\mathrm{r}}\varepsilon_0\dot{\boldsymbol{E}}. \end{aligned} \tag{A.3}$$

We again apply the operator $\nabla\times$ to the first of these equations and differentiate the second with respect to time. These relations then combine to give

$$\nabla \times (\nabla \times \boldsymbol{E}) + \mu_{\mathrm{r}}\mu_0\varepsilon_{\mathrm{r}}\varepsilon_0\ddot{\boldsymbol{E}} = 0. \tag{A.4}$$

Since $\rho = 0$, the divergence of $\boldsymbol{E}$ is zero according to the first equation of (A.2). Using the vector relation

$$\nabla \times (\nabla \times \boldsymbol{E}) = \nabla(\nabla \boldsymbol{E}) - \nabla^2\boldsymbol{E}$$

(A.4) then yields the **wave equation**

$$\nabla^2\boldsymbol{E} - \frac{1}{v^2}\ddot{\boldsymbol{E}} = 0 \tag{A.5}$$

with phase velocity

$$v = \frac{1}{\sqrt{\mu_{\mathrm{r}}\mu_0\varepsilon_{\mathrm{r}}\varepsilon_0}}. \tag{A.6}$$

A charge Q moving in an electric field $\boldsymbol{E}$ and magnetic field $\boldsymbol{B}$ experiences the **Lorentz force**

$$\boldsymbol{F} = Q(\boldsymbol{E} + \boldsymbol{v} \times \boldsymbol{B}). \tag{A.7}$$

For a detailed treatment of electrodynamics the reader is referred, for example, to the book by J.D. Jackson [29].

B

Some important relations in special relativity

An **inertial frame** is a frame of reference in which a body which is not acted on by a force will move at constant velocity. If a particular reference frame is moving at a constant velocity relative to an inertial frame, then it too is an inertial frame. In any such coordinate system four quantities are needed to specify the position at which a physical event occurs, namely the three spatial coordinates s, x and z plus the time t. If we consider a reference frame K which is at rest and an inertial frame K' moving with a velocity v relative to it, these four quantities transform between the two frames according to the **Lorentz transformation**

$$\begin{aligned} s' &= \frac{s - vt}{\sqrt{1-\beta^2}} \\ x' &= x \\ z' &= z \\ t' &= \frac{t - \frac{v}{c^2}x}{\sqrt{1-\beta^2}} \\ \\ s &= \frac{s' + vt'}{\sqrt{1-\beta^2}} \\ x &= x' \\ z &= z' \\ t &= \frac{t' + \frac{v}{c^2}x'}{\sqrt{1-\beta^2}} \end{aligned} \tag{B.1}$$

where $\beta = v/c$ and c is the speed of light. Here it is assumed, without any loss of generality, that the corresponding axes of the coordinate systems are parallel to one another and that the motion is along the s-axis.

If a particle in the rest frame K has the **four-momentum**

$$P_\mu = \begin{pmatrix} p_t \\ p_s \\ p_x \\ p_z \end{pmatrix}, \tag{B.2}$$

then it transforms into the moving frame K' according to the relation

$$P'_\mu = \begin{pmatrix} \gamma & 0 & 0 & -\beta\gamma \\ 0 & 1 & 0 & 0 \\ 0 & 0 & 1 & 0 \\ -\beta\gamma & 0 & 0 & \gamma \end{pmatrix} P_\mu \tag{B.3}$$

with

$$\gamma = \frac{1}{\sqrt{1-\beta^2}}. \tag{B.4}$$

Due to the Lorentz invariance of electrodynamics, a static magnet at rest in a frame K will produce an electric field as well as a magnetic field in a moving frame K'. If we label the individual field components in these two systems as E_s, E_x, E_z and B_s, B_x, B_z, or E'_s, E'_x, E'_z and B'_s, B'_x, B'_z respectively, the transformation equations for the field components are

$$\begin{aligned} E'_s &= E_s \\ E'_x &= \frac{E_x - vB_z}{\sqrt{1-\beta^2}} \\ E'_z &= \frac{E_z + vB_x}{\sqrt{1-\beta^2}} \\ B'_s &= B_s \\ B'_x &= \frac{B_x + \frac{v}{c^2}E_z}{\sqrt{1-\beta^2}} \\ B'_z &= \frac{B_z - \frac{v}{c^2}E_x}{\sqrt{1-\beta^2}} \end{aligned} \tag{B.5}$$

A particle of rest mass m_0 and velocity v has the relativistic energy

$$\gamma = \frac{E}{m_0 c^2}. \tag{B.6}$$

We may therefore express the momentum and energy of a particle in the form

$$\begin{aligned} p &= \gamma m_0 v \\ E &= \gamma m_0 c^2. \end{aligned} \tag{B.7}$$

The relation between energy and momentum is then

$$E = \frac{c}{\beta} p. \tag{B.8}$$

The relationship between a relative change in momentum and the corresponding relative change in velocity my be obtained by differentiating the momentum p in (B.7):

$$\frac{dp}{dv} = m_0 \frac{d}{dv}(\gamma v) = \gamma m_0 (1 + \beta^2 \gamma^2). \tag{B.9}$$

According to (B.4) we have $\gamma^2 = 1 + \beta^2\gamma^2$. Substituting this into (B.9) yields

$$dp = \gamma^3 m_0 dv. \tag{B.10}$$

Dividing by $p = \gamma m_0 v$, we finally obtain

$$\frac{dp}{p} = \gamma^2 \frac{dv}{v}. \tag{B.11}$$

We also require the relation between the relative change in energy and the corresponding relative change in momentum. We obtain this by differentiating the energy with respect to the momentum in (B.7), namely

$$\frac{dE}{dp} = m_0 c^2 \frac{d\gamma}{dp} \tag{B.12}$$

and replacing γ by $p/m_0 v$

$$\frac{dE}{dp} = c^2 \frac{d}{dp}\left(\frac{p}{v}\right). \tag{B.13}$$

Using (B.10) this expression gives

$$\frac{d}{dp}\left(\frac{p}{v}\right) = \frac{1}{v}\left(1 - \frac{1}{\gamma^2}\right) = \frac{1}{v}\beta^2. \tag{B.14}$$

Substituting this into (B.13) yields

$$dE = \frac{c^2}{v}\beta^2 dp \tag{B.15}$$

which when divided by $E = cp/\beta$ finally gives

$$\frac{dE}{E} = \beta^2 \frac{dp}{p}. \tag{B.16}$$

C

General equation of an ellipse in phase space

To describe an ellipse in the transverse plane x-x' of phase space we need the general equation of an arbirtrarily rotated ellipse, as shown in Fig. C.1. We therefore move to a rotated coordinate system $\tilde{x}$-$\tilde{x}'$ in which the principal axes a and b of the ellipse correspond to the axes of the coordinate system. In this new coordinate system $\tilde{x}$-$\tilde{x}'$ the ellipse is described by

$$\begin{aligned}\tilde{x} &= a\ \cos\varphi \\ \tilde{x}' &= b\ \sin\varphi,\end{aligned} \tag{C.1}$$

where the parameter φ lies within the range $0 \leq \varphi \leq 2\pi$. The transformation into the x-x' coordinate system, which is rotated by an angle Ψ, proceeds using the relations

$$\begin{aligned}x &= a\ \cos\varphi\ \cos\Psi + b\ \sin\varphi\ \sin\Psi \\ x' &= -\,a\ \cos\varphi\ \sin\Psi + b\ \sin\varphi\ \cos\Psi.\end{aligned} \tag{C.2}$$

We now eliminate the free parameter φ by solving these equations with respect to $\sin\varphi$ or $\cos\varphi$ and using the general relation $\sin^2\varphi + \cos^2\varphi = 1$. We immediately obtain

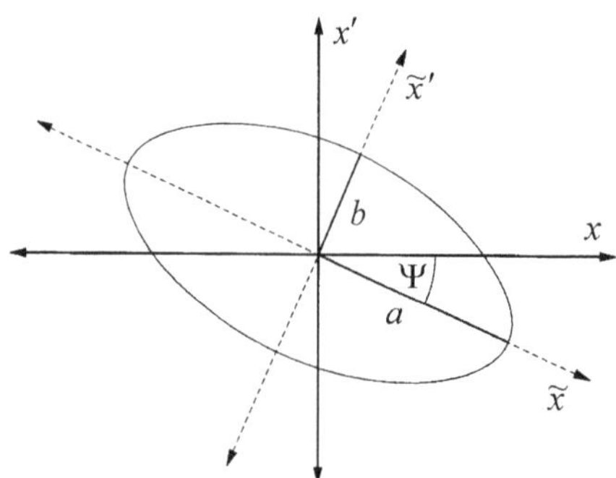

Fig. C.1 Generalized ellipse rotated by an arbitrary angle Ψ in the coordinate system x-x'. The reference coordinate system $\tilde{x}$-$\tilde{x}'$ is rotated by the same angle Ψ as the ellipse.

$$\begin{aligned} a^2b^2 &= b^2(x\cos\Psi - x'\sin\Psi)^2 + a^2(x\sin\Psi + x'\cos\Psi)^2 \\ &= (b^2\sin^2\Psi + a^2\cos^2\Psi)\,x'^2 + 2(a^2-b^2)\cos\Psi\ \sin\Psi\ x\ x' \\ &\quad +(b^2\cos^2\Psi + a^2\sin^2\Psi)\,x^2. \end{aligned} \tag{C.3}$$

We divide these equations by $a\ b$ and define

$$\begin{aligned} \beta &\equiv \frac{b}{a}\sin^2\Psi + \frac{a}{b}\cos^2\Psi \\ \alpha &\equiv \left(\frac{a}{b} - \frac{b}{a}\right)\cos\Psi\ \sin\Psi \\ \gamma &\equiv \frac{b}{a}\cos^2\Psi + \frac{a}{b}\sin^2\Psi. \end{aligned} \tag{C.4}$$

Using the expression for the ellipse area $A = \pi\ a\ b$, the required general equation of an ellipse is then given by

$$\boxed{\frac{A}{\pi} = \beta\ x'^2 + 2\ \alpha\ x\ x' + \gamma\ x^2 = \varepsilon.} \tag{C.5}$$

It is easy to show by substitution of the definitions (C.4) that

$$\gamma = \frac{1+\alpha^2}{\beta}. \tag{C.6}$$

The ellipse is thus uniquely defined by the parameters α, β, and ε. Let us now determine a few important locations on the ellipse. The points of intersection with the coordinate axes are immediately obtained by setting $x = 0$ or $x' = 0$, namely

$$\begin{aligned} x &= \sqrt{\frac{\varepsilon}{\gamma}} \qquad \text{for} \qquad x' = 0 \\ x' &= \sqrt{\frac{\varepsilon}{\beta}} \qquad \text{for} \qquad x = 0. \end{aligned} \tag{C.7}$$

To find the extreme edges of the ellipse we now solve (C.5) for x or x':

$$\begin{aligned} x &= \frac{-\alpha x' \pm \sqrt{\gamma\varepsilon - x'^2}}{\gamma} \\ x' &= \frac{-\alpha x \pm \sqrt{\beta\varepsilon - x^2}}{\beta}. \end{aligned} \tag{C.8}$$

Differentiating these expressions immediately yields the extrema

$$\begin{aligned} x_{\text{extr}} &= \pm\sqrt{\varepsilon\beta} \qquad \text{for} \qquad \frac{dx}{dx'} = 0 \\ x'_{\text{extr}} &= \pm\sqrt{\varepsilon\gamma} \qquad \text{for} \qquad \frac{dx'}{dx} = 0. \end{aligned} \tag{C.9}$$

Bibliography

[1] M.S. Livingston, J.P. Blewett, *Particle Accelerators* (McGraw-Hill, New York, 1962)

[2] J.D. Cockcroft and E.T.S. Walton, *Proc. Roy. Soc.* (London), **A136**, 619 (1932), **A137**, 229 (1932), and **A144**, 704 (1934)

[3] Schenkel, *Elektrotech. Z.*, **40**, 333 (1919)

[4] H. Greinacher, *Z. Physik*, **4**, 195 (1921)

[5] Bellaschi, *Trans. AIEE*, **51**, 936 (1932)

[6] R.J. Van de Graaff, *Phys. Rev.*, **38**, 1919A (1931)

[7] G. Ising, *Ark. Math. Astron. Phys.*, **18**, No. 30, Part 4, 45 (1925)

[8] R. Wideröe, *Arch. Elektrotech.*, **21**, 387 (1928)

[9] J.W. Beams, L.B. Snoddy, *Phys. Rev.*, **44**, 784 (1933);
J.W. Beams, H. Trotter, *Phys. Rev.*, **45**, 849 (1933)

[10] E.L. Ginzton, W.W. Hansen, W.R. Kennedy, *Rev. Sci. Instrum.*, **19**, 89 (1948);
M. Chodorow, E.L. Ginzton, W.W. Hansen, R.L. Kyhl, R.B. Neal, W.K.H. Panofsky, *Rev. Sci. Instrum.*, **26**, 134 (1955)

[11] R.H. Helm, G.A. Loew, W.K.H. Panofsky, *The Stanford Two-Mile Accelerator* (W.A. Benjamin, Inc., New York, 1968), chapter 7

[12] L.W. Alvarez, H. Bradner,J.V. Franck, H. Gordon, J.D. Gow, L.C. Marshall *et al.*, *Rev. Sci. Instrum.*, **26**, 111 (1955)

[13] E.O. Lawrence, N.E. Edlefsen, *Science*, **72**, 376 (1930)

[14] S. Rosander, *The Development of the Microtron*, *Nucl. Instrum. & Methods*, **177**, 411–416 (1980)

[15] B. Schoch, The MAMI Project, *Proc. Electron and Photon Interactions at Intermediate Energies*, Bad Honnef 1984, 440–446 (1986);
T. Walcher, The Mainz Microtron Facility MAMI, *Proc. The Nature of Hadrons and Nuclei by Electron Scattering*, Erice 1989, 189–203 (1989)

[16] D.W. Kerst, *Phys. Rev.*, **58**, 841 (1940), and **60**, 47 (1941)

[17] R. Wideröe, *Arch. Elektrotech.*, **21**, 400 (1928)

[18] E.M. McMillan, *Phys. Rev.*, **68**, 143 (1945)

[19] V. Veksler, *J. Phys. (U.S.S.R.)*, **9**, 153 (1945)

[20] F.G. Gouard and D.E. Barnes, *Nature*, **158**, 413 (1946)

[21] M.L. Oliphant, J.S. Gooden, G.S. Hide, *Proc. Phys. Soc.* (London), **59**, 666 (1947)

[22] M.H. Blewett (ed.), Cosmotron Staff, *Rev. Sci. Instrum.*, **24**, 723–870 (1953)

[23] G.K. O'Neill, Component Design and Testing for the Princeton-Stanford Colliding-Beam Experiment, *Proc. Int. Conf. on High Energy Accelerators*, Brookhaven (1961)

[24] C. Bernadini, U. Bizzarri, G.F. Corrazza, G. Ghigo, R. Querzoli, and B. Touschek, A 250 MeV Electron-Positron Storage Ring: The "A a A", *Proc. Int. Conf. on High Energy Accelerators*, Brookhaven (1961)

[25] G.I. Budger *et al.*, Status Report on Electron Storage Ring VEPP I, *Proc. 5th Int. Conf. on High Energy Accelerators*, Frascati (1965)

[26] B.H. Wiik *et al.*, *Study for the Project of the Proton electron Storage Ring HERA*, DESY HERA 80-01, (1980);
V. Soergel, The HERA Project, *Proc. High Energy Spin Physics*, 575–581, (1986)
G.A. Voss, *Status of the HERA Project, Proc. Lepton and Photon Interactions at High Energies*, 525–552 (1987)

[27] E. Keil, The Large European e^+ e^- Collider Project LEP, *Proc. Particle Accelerator Conference*, Washington, March 11–13, (1981);
H. Schopper, The LEP Project, *Proc. Conf. on High Energy Accelerators*, Novosibirsk 1986, Vol. 1, 39–43 (1986)

[28] H. Wiedemann, The SLAC Linac Collider (SLC) Project, *Proc. Particle Accelerator Conference*, Washington, March 11–13 (1981)

[29] J.D. Jackson, *Classical Electrodynamics*, (Wiley, New York, 1975)

[30] A. Hofmann, *Theory of Synchrotron Radiation*, **38**, SSRL ACD-Note (1986)

[31] A. Liénard, *L'Eclairage Elect.*, **16**, 5 (1898)

[32] F.R. Elder, A.M. Gurewitsch, R.V. Langmuir, and H.D. Pollack, *Phys. Rev.* **71**, 829–830 (1947) *J. Appl. Phys.*, **18**, 810 (1947); F.R. Elder, R.V. Langmuir, and H.D. Pollack, *Phys. Rev.* **74**, 52 (1948)

[33] J. Schwinger, *Phys. Rev.* **70**, 798 (1946), *Phys. Rev.*, **75**, 1012–1025 (1949), and *Proc. Natl. Acad. Sci. USA*, **40**, 132 (1954)

[34] E.M. Rowe and F.E. Mills, *Part. Accel.* **4**, 211–227 (1973)

[35] H. Winick, *IEEE Trans. Nucl. Sci.*, **20**, 984–988 (1973) and *Proceedings of the 9th International Conference on High Energy Accelerators*, Stanford, California, 685–688 (1974)

[36] E.E. Koch, C. Kunz, and E.W. Weiner, *Optik (Stuttgart)*, **45**, 395–410 (1976)

[37] E.M. Rowe *et al.*, Status of the ALADDIN Project, *IEEE Trans. Nucl. Sci.*, **28**, 3145–3146 (1981)

[38] M. Barthes *et al.*, Magnet System for Super ACO, the new Orsay Synchrotron Radiation Source, *Proc. Magnet Technology*, Zürich, 114–117, (1985)

[39] G. Mülhaupt *et al.*, Status of BESSY, an 800-MeV Storage Ring Dedicated to Synchrotron Radiation, *IEEE Trans. Nucl. Sci.* **30**, 3094–3096 (1983)
[40] M.R. Howells, Progress and Prospects at the National Synchrotron Light Source (NSLS), *Nucl. Instrum. & Methods* **195**, 17–27 (1982)
[41] J. Tanaka *et al.*, Design and Status of Photon Factory, *Proc. High Energy Accelerators*, Geneva, 242–246 (1980)
[42] A.L. Robinson, A.S. Schlachter, The ALS: A High Brightness Synchrotron Radiation Source, *Proc. Particle Accelerator Conference*, San Francisco, May 5–6 (1991)
[43] L. Fonda, M. Puglisi, R. Rosei, A. Wrulich, The ELETTRA Project, *Helv. Phys. Acta*, **62**, 633–644, (1989)
[44] DELTA Collaboration, *Status Report of the Dortmund Storage Ring Project DELTA*, University of Dortmund (1990)
[45] B. Buras and S. Tazzari, *European Synchrotron Radiation Facility, Report of the ESRP*, CERN, Geneva (1984)
[46] Y. Cho *et al.*, Conceptual Design of the Argonne 6-GeV Synchrotron Light Source, *Proc. Particle Accelerator Conference IEEE*, 3383 (1985)
[47] SPring-8 Project Team, Spring-8 Project, Jaeri-Riken, Japan (1991)
[48] E.B. Courant, H.S. Snyder, Theory of the Alternating Gradient Synchrotron, *Ann. Phys.* **3**, 1–48 (1958)
[49] E. Persico, E. Ferrari, S.E. Segre, *Principles of Particle Accelerators*, (W.A. Benjamin, Inc., New York, 1968)
[50] K.G. Steffen, *High Energy Beam Optics* (Interscience, New York, 1965)
[51] M. Sands, The Physics of Electron Storage Rings. An Introduction. *Proc. International School of Physics Enrico Fermi*, (ed. B. Touschek) (1971)
[52] A.A. Kolomenski, A.N. Lebedev, *Theory of Cyclic Accelerators* (North Holland, Amsterdam, 1966)
[53] E.J.N. Wilson, *Proton Synchrotron Accelerator Theory*, CERN 77-07 (1977)
[54] H. Bruck, *Circular Particle Accelerators* (Translated from the French), LA-TR-72-10 Rev (Los Alamos National Laboratory, 1966)
[55] S. Turner (ed.), *CAS CERN Accelerator School*, Fifth General Accelerator Physics Course, CERN 94-01, Vol. I and II (26 January 1994)
[56] W. Buckel, *Supraleitung - Grundlagen und Anwendungen* (Physik Verlag, Weinheim, 1977)
[57] Pierce, *J. App. Phys.*, **11**, 548 (1940)
[58] F.M. Penning, *Physica*, **4**, 71 (1937)
[59] O. Zinke and A. Vlcek, *Lehrbuch der Hochfrequenztechnik*, Vol. 1, (Springer-Verlag, Heidelberg, New York, Tokyo, 1986)
[60] H. Gerke, H. Musfeld, *The Radiofrequency System for the PETRA Storage Ring*, DESY M-79/33 (1979)
[61] P.M. Lapostolle, A.L. Septier (ed.), *Linear Accelerators*, (North Holland, Amsterdam, 1970)

[62] M. Chodorow and C. Susskind, *Fundamentals of Microwave Electronics* (McGraw-Hill, New York, 1964)

[63] K.W. Robinson, *Phys. Rev.*, **111**, 373 (1958)

[64] R. Chasman and K. Green, *BNL Report*, BNL 50505 (1980)

[65] H. Winick, Present and Future Synchrotron Raidation Facilities in Japan, *Nucl. Instr. & Meth.* A152: 177–196 (1998)

[66] F. Amman and D. Ritson, *Proc. Int. Conf. on High-Energy Accelerators*, Brookhaven, 262 (1961)

[67] H. Winick and T. Knight (ed.), *Wiggler Magnets, Wiggler Workshop, SLAC*, SSRP Report No. 77/05 (1977)

[68] J. Spencer and H. Winick, *Synchrotron Radiation Research*, ed. H. Winick and S. Doniach (Plenum Press, New York, 1980) Chapter 21, Wiggler systems as sources of electromagnetic radiation.

[69] H. Winick, G. Brown, K. Halbach, and J. Harris, *Physics Today*, **34**, 50–63 (1981)

[70] G. Brown, K. Halbach, J. Harris, and H. Winick, Wiggler and Undulator Magnets - a Review, *Nucl. Instrum. & Methods*, **208**, 65–77 (1983)

[71] D.E. Baynham and B.E. Wyborn, A 5 Tesla Superconductive Wiggler Magnet, *IEEE Trans. Magn.*, **MAG-17**, 1595 (1981)

[72] M.W. Poole, V.P. Suller, and S.L. Thomson, *A Second Superconducting Wiggler Magnet for the Daresbury SRS*, Preprint, Daresbury Lab., DL / SCI / P630A (1989)

[73] K. Halbach, J. Chin, E. Hoyer, H. Winick, R. Cronin, and J. Yang, *IEEE Trans. Nucl. Sci.*, **NS-28**, 3136–3138 (1981)

[74] G. Dattoli and A. Renieri, *Nuovo Cimento*, **B59**, 1 (1980)

[75] G. Dattoli and A. Renieri, *Nuovo Cimento*, **B61**, 153 (1981)

[76] G. Dattoli and A. Renieri, *The Laser Handbook*, Vol. IV (North Holland, Amsterdam, 1985) 1

[77] F. Ciocci *et al.*, *Phys. Reps.*, **141**, 1 (1986)

[78] W.B. Colson, *Phys. Lett.*, **A64**, 190 (1977)

[79] W.B. Colson, One-body analysis of free electron lasers, in *Novel Sources of Coherent Radiation*, ed. S.F. Jacobs, M. Sargent, M.O. Scully (Addison-Wesley, Reading, Mass. 1978)

[80] W.B. Colson, *Nucl. Instr. & Meth.*, **A237**, 1 (1985); W.B. Colson and A. Sessler, *Ann. Rev. Nucl. Part. Sci.*, 25 (1985)

[81] C. Pellegrini and J. Murphy, Introduction to the physics of the FEL, *Proc. Conf. South Padre Island*, Springer, 163 (1986)

[82] S. Krinsky, *Introduction to the Theory of Free Electron Lasers*, AIP **153**, 1016 (1987)

[83] R.B. Palmer, *J. Appl. Phys.*, **43**, 3014 (1972)

[84] L.R. Elias, W.M. Fairbank, J.M.J. Madey, H.A. Schwettman, and T.I. Smith, *Phys. Rev. Lett.*, **36**, 710 (1976)

[85] D.A.G. Deacon, L.R. Elias, J.M.J. Madey, G.J. Ramian, H.A. Schwettman, and T.I. Smith, *Phys. Rev. Lett.*, **38**, 892 (1977)

[86] J.M.J. Madey, *Nuovo Cimento*, **B50**, 64 (1979)

[87] J. Bisognano, S. Chattopadhyay, M. Cornacchia, A. Garren, A. Jackson, K. Halbach *et al.*, *Feasibility Study of a Storage Ring for a High Power XUV Free Electron Laser*, Lawrence Berkeley Lab. LBL-19771 (June 1985)

[88] N.A. Vinokurov and A.N. Skrinsky, *Institute of Nuclear Physics*, Novosibirsk, USSR, Preprint (1977)

[89] M. Billardon, P. Elleaume, J.M. Ortega, C. Bazin, M. Bergher, M. Velghe, *et al.*, *IEEE J. Quantum Electron.*, **21**, 805 (1985)

[90] M.E. Couprie, C. Bazin, M. Billardon, and M. Velghe, *Nucl. Instr. & Meth.*, **A285**, 31 (1989)

[91] I.B. Drobyazko, G.N. Kulipanov, V.N. Litvinenko, I.V. Pinayev, V.M. Popik, I.G. Silvestrov *et al.*, *Proc. 11th Int. FEL Conf.*, Naples, U.S.A. (1989)

[92] T.C. Marshall, *Free Electron Lasers* (MacMillan Publishing Company, New York, 1985)

[93] Charles A. Brau, *Free-Electron Lasers* (Academic Press, 1990)

[94] *Free-Electron Generators of Coherent Radiation*, Physics of Quantum Electronics, Vol. 8, ed. S.F. Jacobs, G.T. Moore, H.S. Pilloff, M. Sargent III, M.O. Scully, R. Spitzer (Addison-Wesley, 1982)

[95] W.B. Colson, C. Pellegrini, A. Renieri, *The Laser Handbook*, Vol. VI, (North Holland, Amsterdam, 1990)

[96] M. Abramowitz and I.A. Stegun, *Handbook of Mathematical Functions* (Dover, New York, 1965)

[97] A. Renieri, *Nuovo Cimento*, **53B**, 160 (1979)

[98] R.W. Allison, R.W. Brokloff, R.L. McLaughlin, R.M. Richter, M. Tekawa, J.R. Woodyard, UCLR-19270, Berkeley (1969);
S. Yencho, D.R. Walz, *Proc. Particle Accelerator Conference, IEEE Trans. Nucl. Sci.*, **NS-32**, No. 5, Vancouver (1985);
C.D. Johnson, *CERN/PS/90-42 (AR)* (1990)

[99] R.T. Avery, A. Faltens, E.C. Hartwig, *Part. Acc. Conference, IEEE Trans. Nucl. Sci.*, **NS-18**, No. 3, Chicago (1971);
T. Naito, H. Akiyama, J. Urakawa, T. Shintake, M. Yoshika, *8th Symp. on Acc. Science and Technology*, Wako, Saitama (1991)

[100] F. Loyer, T. André, B. Ducodret, J.P. Rataud, *Part. Acc. Conference, IEEE Trans. Nucl. Sci.*, **NS-32**, No. 5, Vancouver (1985)

[101] K. Unser, *Part. Acc. Conference, IEEE Trans. Nucl. Sci.*, **NS-28**, No. 3, Washington (1981);
G. Burtin, R.J. Colchester, C. Fischer, J.Y. Hemery, R. Jung, M. Vanden Eynden, J.M. Voillot, *2nd Europ. Part. Acc. Conf., (EPAC 90)*, Nice (1990)

[102] A.A. Sokolov and I.M. Ternov, On Polarization and Spin Effects in the Theory of Synchrotron Radiation, *Sov. Phys. Doklady* **8**, 1203 (1964)

[103] V.N. Baier, Radiative Polarisation of Electrons in Storage Rings, *Sov. Phys. Uspekhi*, **14**, 695 (1972)

[104] S.I. Serednyakov, A.N. Skrinskii, G.M. Tumaikin, Yu.M. Shatunov, Study of the Radiative Polarisation of Beams in the VEPP-2M Storage Ring, *JETP*, **44**, 1063 (1976)

[105] R. Neumann and R. Rossmanith, A Fast Depolarizer for Large Electron-Positron Storage Rings *Nucl. Instrum. & Methods*, **204**, 29 (1982)

[106] V.N. Baier, *Proceedings of the Workshop on Advanced Beam Instrumentation*, KEK, Tsukuba, Japan, 365 (1991)

[107] A.N. Skrinskii and Yu.M. Shatunov, Precision Measurements of Masses of Elementary Particles Using Storage Rings with Polarized Beams, *Sov. Phys. Uspekhi*, **32**, 548 (1989)

[108] D.P. Barber *et al.* A Precision Measurement of the Y' Meson Mass, *Physics Letters*, bf 135B, (1984)

[109] D.A. Goldberg, G.R. Lambertson, *US Particle Accelerator School, 1991, AIP Conf. Proc.*, 249 (1992)

[110] G. Lehner, *Elektromagnetische Feldtheorie*, (Springer-Verlag, Berlin, Heidelberg, New York, 1994)

Index

The manufacturer's authorised representative in the EU for product safety is Oxford University Press España S.A. of el Parque Empresarial San Fernando de Henares, Avenida de Castilla, 2 – 28830 Madrid (www.oup.es/en or product.safety@oup.com). OUP España S.A. also acts as importer into Spain of products made by the manufacturer.

www.ingramcontent.com/pod-product-compliance
Ingram Content Group UK Ltd.
Pitfield, Milton Keynes, MK11 3LW, UK
UKHW012156240726
13966UKWH00002B/379